An Introduction to

FARM ORGANISATION and MANAGEMENT

SECOND EDITION

Titles of Related Interest

BUCKETT
Introduction to Livestock Husbandry, 2nd Edition

DILLON
Analysis of Response in Crop and Livestock Production, 2nd Edition

HILL
An Introduction to Economics for Students of Agriculture

KENT
Technology of Cereals, 3rd Edition

LOCKHART & WISEMAN
Introduction to Crop Husbandry, 6th Edition

NASH
Crop Conservation and Storage, 2nd Edition

SHIPPEN, ELLIN & GLOVER
Basic Farm Machinery, 3rd Edition

An Introduction to FARM ORGANISATION and MANAGEMENT

SECOND EDITION

by

MAURICE BUCKETT, B.Sc. (Hons.), C.Biol., M.I.Biol., F.R.Ag.S.

Head of Agriculture and Farm Director
The West of Scotland Agricultural College, Ayr, U.K.

PERGAMON PRESS

OXFORD · NEW YORK · BEIJING · FRANKFURT
SÃO PAULO · SYDNEY · TOKYO · TORONTO

U.K.	Pergamon Press, Headington Hill Hall, Oxford OX3 0BW, England
U.S.A.	Pergamon Press, Maxwell House, Fairview Park, Elmsford, New York 10523, U.S.A.
PEOPLE'S REPUBLIC OF CHINA	Pergamon Press, Room 4037, Qianmen Hotel, Beijing, People's Republic of China
FEDERAL REPUBLIC OF GERMANY	Pergamon Press, Hammerweg 6, D-6242 Kronberg, Federal Republic of Germany
BRAZIL	Pergamon Editora, Rua Eça de Queiros, 346, CEP 04011, Paraiso, São Paulo, Brazil
AUSTRALIA	Pergamon Press Australia, P.O. Box 544, Potts Point, N.S.W. 2011, Australia
JAPAN	Pergamon Press, 8th Floor, Matsuoka Central Building, 1-7-1 Nishishinjuku, Shinjuku-ku, Tokyo 160, Japan
CANADA	Pergamon Press Canada, Suite No. 271, 253 College Street, Toronto, Ontario, Canada M5T 1R5

First edition 1981

Second edition 1988

Library of Congress Cataloging in Publication Data

Buckett, M.
An introduction to farm organisation and management
Bibliography: p.
Includes index.
1. Farm management. I. Title.
S561.B78 1987 630'.68 87–18982

British Library Cataloguing in Publication Data

Buckett, Maurice
An introduction to farm organisation and management.—2nd ed.
1. Farm management
I. Title
630'.68 S561

ISBN 0–08–034203–5 (Hardcover)
ISBN 0–08–034202–7 (Flexicover)

Printed in Great Britain by A. Wheaton & Co. Ltd., Exeter

PREFACE

THE aim of this book is to provide a basic introduction to the study of farm management. It has been primarily designed for students following post-certificate and diploma courses. The text should also be appropriate as background reading for students at undergraduate level, and it is hoped that many farmers will find the work of value.

Management is a subject which has to be practised to be fully appreciated. Its function is primarily decision making, in the short and long terms, both at the general husbandry and the overall business levels. Hypothetical decisions are more easily taken than those in a real life situation which can have an impact on one's future livelihood. Anyone who aspires to management should therefore take every opportunity to practise decision making, and the more real the circumstances the better.

Many decisions have to be taken with incomplete information. The biological nature of farming and the variation in climate between years make it a business which cannot be run entirely by a set of rules. Plans made in the comfort of an office can appear to be tarnished when things are going wrong in the field or, for example, the cows are still housed long after normal turnout dates, winter fodder supplies are exhausted, the grass is not growing, and the bank overdraft is already high. Yet it is on just such occasions that good managers can distinguish themselves by their determination and ability to take sound decisions.

The book is comprehensive, ranging from basic economic principles, through records and accounts, to planning, implementation and control. An example business, "Church Farm", is used to illustrate many of the sections.

Generally, principles relating to farm business management change little, but in a dynamic industry such as agriculture many other aspects, and in particular prices, change rapidly. Thus, many of the data used for illustration in this book will be dated by the time it is available in print. Several annual publications are available which give up-to-date financial information. It is suggested that one of these be used in conjunction with this book, and that the reader uses this text to learn the basic principles.

The book has been designed to be used in association with teaching courses. Many readers may read it in stages as their course progresses. For this reason it has been a deliberate policy to have a minor element of repetition, with further development, of certain topics to provide reinforcement of important information.

Book knowledge is no substitute for experience, and the acquisition of experience requires time, sometimes many years. However, book knowledge can make the learner more observant and thus enable experience to be acquired more quickly.

PREFACE TO SECOND EDITION

THE interval between the production of the first and second editions of this book witnessed a significant change for food producers in the European Economic Community. They moved from a period when there was a stimulus, through the Common Agricultural Policy, for increased production to an era of food surpluses and pressure for a reduction in food supplies. This is shown in one way by the fact that in the chapter on budgeting in the first edition it was possible to illustrate an example by the expansion of a dairy herd whilst in this edition it was necessary to consider a reduction in herd size to comply with milk quotas.

This overall change in the fortunes of the agricultural industry, and the challenges which it produced, increased the need for sound business management. The viability of some farms was put into jeopardy and a number did not survive. The importance of establishing realistic business objectives, of undertaking sound strategic and tactical planning supported by appropriate investment appraisal, and of exercising control at all stages of production was increased. Whilst the "economic" pressure faced by individual businesses differed with such things as the amount of interest they had to pay on borrowed capital, and the type of product being produced, it was clear that there was a more positive need for all farmers to consider how to make optimal use of the resources available to them if they were to maintain or increase profits.

ACKNOWLEDGEMENTS

I WISH to take this opportunity to thank the following people who have been of assistance in the preparation of the second edition of this book:

David Arnot, Stuart Ashworth, Roger Evans, Alan Gill, Joy Gladstone, David Godfrey, Rod Gooding, Chris Groves, D. Howat, Robert Laird, Andrew Leggate, J. Turnbull and Gary Wallace.

In addition I am particularly grateful to Miss Vivien France for her work in typing the manuscript.

CONTENTS

CHAPTER 1

THE FARMER AND HIS MANAGEMENT FUNCTIONS

CONTENTS

1.1. THE PROCESSES AND FUNCTIONS OF MANAGEMENT

1.1.1. Introduction

Those who are new to the study of management frequently venture the opinion that it is concerned with making decisions which will result in more profit. Whilst this is basically true there are several shocks in store. It comes as a great surprise to find that it is possible for a farmer to be making what, in some circumstances, might be described as a satisfactory profit and still be in financial difficulties. Management lecturers frequently tell their students that technical skill in the production of crops and livestock, whilst being very significant, does not necessarily ensure financial success. Disappointment may then result when the students find that the so-called principles and techniques of management will not always produce absolute solutions to problems but simply provide guidelines to assist in the formation of a sound judgement. Occasionally surprise is evoked by the discovery that there is more than one solution to a problem, each of which is suitable in appropriate circumstances.

At the outset in the study of management it has to be recognised that whilst farms must be planned and run on business lines, the biological nature of agriculture, together with an inherent variability and uncertainty, frequently requires decisions to be taken and implemented on the basis of incomplete information. However, it is important to obtain as much information as possible to increase the chances of success.

1.1.2. The Processes of Management

Each farmer has objectives for his business. Management is concerned with ensuring that these objectives are attained. Every business has three primary resources; capital, which is used to obtain other resources; land (including buildings); and labour. Superimposed on these is what might be termed the key factor—management ability.

Assuming that the farmer has completed the essential first step—that of defining his objectives—the rest of the processes of management can be conveniently summarised in diagrammatic form (Fig. 1). Every farmer has to consider the "organisation" of his resources into a suitable plan. At each stage he must take particular notice of the amount of capital required and the possibilities of obtaining it. Next the plan has to be put into operation and "managed". Hence the term "organisation and management" is frequently used for this subject.

Once the plan is in operation the results have to be recorded. These are then analysed and appraised to establish what they indicate. Such results can be used as a basis for control of the plan. They may suggest that modifications are desirable to exploit strengths and remove weaknesses. Any changes have then to be put into operation if deemed appropriate.

Sometimes minor amendments to the plan are insufficient to enable the business to comply with the objectives. It may then be necessary to undertake a complete replanning exercise for the whole farm.

In practice once the plan is operating most of the above processes are going on at the same time and are closely interrelated. These processes are applied to the "subjects" of the management, i.e. the capital, land, labour, crop, stock, and machinery. Environmental factors, which can be physical, economic, political, and sociological, may influence these management subjects and the way in which the management processes are applied to them. To run a business effectively an additional process must be added. This involves forecasting both the performance of the management subjects and the influence of the environment. Thus, there are five main processes: (i) forecasting; (ii) planning; (iii) implementation, which is allied to operation; (iv) recording; (v) controlling.

1.1.3. The Functions of Management

The functions of management can be presented diagramatically (Fig. 2). They are primarily concerned with the decision making involved in the processes given above.

(i) Planning

Long term strategic planning and shorter term, day-to-day, tactical planning are both necessary. Thus, under the heading of production the farmer has to decide which products, together with the quantities, that he will produce. Later, when the plan is operating he has to take tactical decisions involving such things as the timing of operations, e.g. sowing in relation to soil conditions, and time of year; technique of operation, e.g. sowing method and depth of sowing; adjustments within production processes, e.g. fertiliser levels or concentrate usage; and adjustments to production processes, e.g. if the price of bacon pigs falls relative to that of pork, the bacon producer may decide to change to pork production for a limited time.

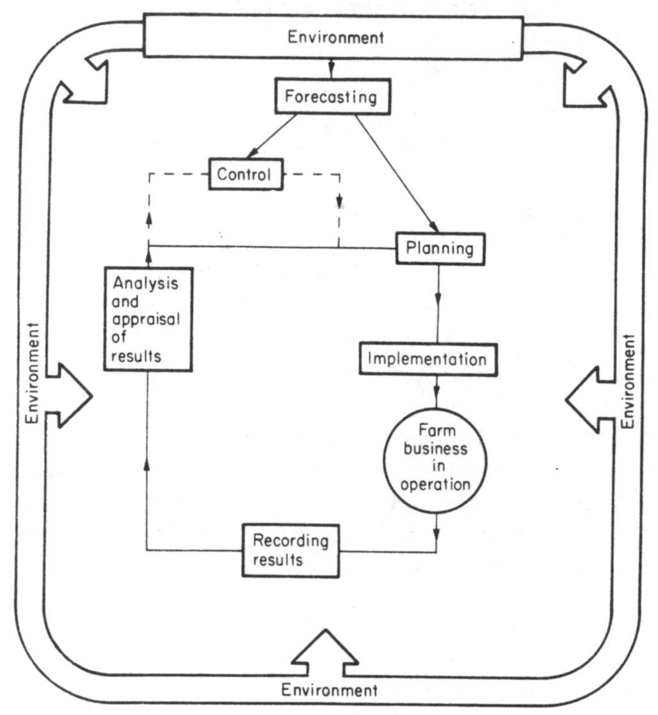

FIG. 1. The processes involved in farm management.

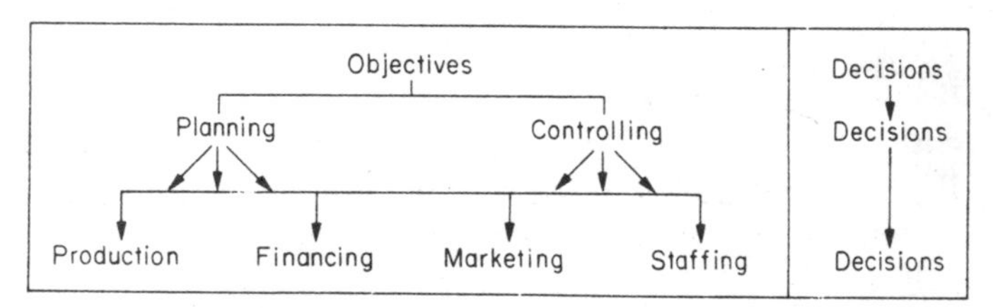

FIG. 2. Functions of management—decision areas.

In essence, the farmer, having formulated his plan, has to put it into action. He has to control his resources and exercise his judgement to ensure that the production processes are carried out with as little friction as possible between the different activities in their demand for resources at the one time, and to ensure that production takes place to comply with the plan and the objectives for the business.

(ii) Financing

With regard to financing, at the planning stage it is necessary to examine all the sources of capital, to decide how much capital is required, when it is wanted and which source should be used, and how and when it has to be repaid.

At the operational stage the capital has to be deployed in accordance with the plan and allocated to those activities which produce greatest benefit to the business. It is necessary to ensure that money is available to meet the demands for continuing production, to see that there is sufficient finance for living expenses, taxation, and interest plus repayment of capital. At the same time the farmer must consider the future needs for capital to develop the business.

The farmer has to keep within his borrowing limits and he has to ensure that the capital is put to work effectively. This requires a great deal of skill in timing production and marketing to ensure that money is available when required.

(iii) Marketing

Successful marketing, both buying and selling, is one of the key functions of management. At the planning stage it is essential to ensure that the markets are, and will remain, available. At this point cognisance must be taken of the type and quality of the product which it is intended to produce. Consideration must also be given to the timing of sales.

At the operational stage it is necessary to make use of every opportunity to market to best advantage. Marketing contracts may be negotiated. Produce has to be selected for sale and presented in the right condition, when it is in demand, at the right market. Equal care has to be taken in the purchase of goods required for production.

(iv) Staffing

When planning labour it is necessary to consider the number of staff required, the skills that will be necessary, the availability of workers, the degree of authority and responsibility which will be given to each person, and the wages and conditions which will be provided.

Once the plan is in operation, quite apart from managing his own time, the farmer has to direct the staff he has selected, supervise their work, keep them motivated, and maintain good labour relations. Many factors are involved in successful man management, and these will be covered in Chapter 8 on labour. It is essential to recognise that farm staff can markedly influence the results obtained from a farm.

The farmer may do all his own recording but many staff are motivated by monitoring the progress of the enterprises that they work with. After recording has been undertaken, the point may arise when it is necessary to consider changes in strategy or tactics so that the business best complies with its objectives.

This book is primarily concerned with the processes and functions of management, but first it is necessary to consider the difference between farmers and the managerial skills which may be available.

1.2. THE FARMER AND MANAGEMENT SKILLS

1.2.1. The Source of Management

On many farms the owner of the business, or the person who carries the ultimate risk, provides most of the management skills, either as tenant of his own land or of someone else's. On other farms management staff are employed, but the amount of authority delegated to them differs enormously from business to business. For convenience, this book will use the term farmer unless there is a point requiring qualification.

The success of management is very much determined by the quality of judgement in relation to the decisions that have to be taken. It is this factor which separates the good farmer from the bad if they are working under similar conditions.

1.2.2. Objectives

Clear objectives must be formulated for a business if it is to have positive direction. Many factors influence their selection, and in practice, most farmers modify what may be their initial objective, that of long term profit maximisation.

Present and future requirements for the business, policies for development, age and the provision for successors, attitudes to risk and uncertainty, personal preferences, knowledge, judgement and ability, tradition, status as owner-occupier or tenant, taxation, and many other factors may be relevant.

The first need of any farmer is to provide enough money to live on after his expenses, including taxes and servicing costs of borrowed capital, have been deducted from his income. Farmers with a limited area of land may be forced to aim for high profits per hectare to justify this need. Those with large farms can place more emphasis on return on capital and can consider less intensive systems. It must be realised that profitability per hectare is not necessarily synonymous with return on capital.

Once the basic needs are satisfied, more consideration can be given to long term growth of the business. Farmers differ in their resolve to do this, and some accept management pressures which others would not consider. Occasionally farmers adopt plans which reduce profits in the short term but in the long run produce greater benefits. A typical example could be the purchase of extra land, partially with borrowed money. The essential point is to avoid insolvency by ensuring that there are sufficient funds to finance current activities until the benefits from increased outputs arise.

Farmers who have high rents or expensive mortgages and those with large amounts of borrowed capital from other sources, at high interest rates, have particular pressures placed upon them. They may have less opportunity to indulge some of their personal preferences than more fortunate colleagues on similar farms.

Visitors to farms must take care in forming judgements about the farmer's policy and objectives unless they know all the circumstances. Whilst it is frequently good advice to put as much money as possible into productive units such as cows, there comes a time, which is not always at the point of absolute necessity, when it is judicious to invest in buildings and machinery according to the particular financial and taxation position. This must not be confused with situations where investment is more a matter of prestige, to keep up with other leading farmers, and where the investment is not always justified.

Care must also be taken not to over-emphasise one's own attitudes when considering another person's objectives. A farmer who is happy with his situation and can remain viable should not necessarily be forced into the management difficulties which might be attendant to profit maximisation.

1.2.3. Differences between Farmers

Caution is necessary when judging the respective merits of farmers. Clearly some have greater ability than others to take decisions, and there are differences in their knowledge, willingness to research techniques, their ability to evaluate and to innovate.

The danger is to neglect careful consideration of the land and other factors prevailing for a given farmer. Each farmer is working under a different set of conditions. Farmers doing moderately well on extremely poor land may be just as good at management as those doing very well on good land.

Attitudes to risk and uncertainty can be an important factor in influencing decisions. When a farmer is planning, his knowledge of the future is imperfect. Although the finite differences between risk and uncertainty can sometimes be a little vague, where enough observations of something happening can be made to establish the chances, in quantitative terms, of it arising again, the situation falls within the realms of risk. There are different degrees of risk. Thus a car insurance company knows from statistics that a student is more likely to have an accident than his more mature lecturer and charges him a higher premium accordingly. Uncertainty refers to situations where the probable outcome cannot be expressed in quantitative terms.

In practice few farmers have enough data to accurately assess the risks involved in all the decisions for their particular farm, but in an empirical way they attempt to do so. Not surprisingly two farmers may attribute totally different degrees of risk to the same situation.

Some farmers are prepared to take greater risks than others. The rewards for taking risks can sometimes, but by no means always, be high. The penalties for failure can be serious. Examples range from basic aspects such as hay making in changeable weather to major capital investments. The reason why high risk activities may be very profitable is that not enough farmers are prepared to operate them, and if their products are in short supply relative to demand, prices can be high.

Many factors influence attitudes to risk. The farmer with a well-established business in which he owns most of the assets, and which is making good profits, might be considered to be in a better position to take chances than his colleague who is new to farming, and who has not achieved the same scale of business and profitability. In reality the latter person has sometimes to take significant chances just to get established. Some established farmers are guilty of following tradition and are reluctant to adopt new techniques until they are confident that they can assess the risks. They may therefore miss the opportunity for increased profits, although in some cases they will be wise to avoid a technique just because it is new.

There are farmers who feel that in a changing world there is little point in planning for the future. They claim that there is too much uncertainty about prices and the availability of resources, quite apart from the difficulty of predicting crop and animal performances, to make it worthwhile. One aim of this book is to present the case that whilst looking into the future can be fraught with difficulties, planning can be of great value.

1.2.4. Diversification and Specialisation

To avoid the risks involved in "putting all their eggs into one basket", many farmers maintain a diversity of enterprises. They hope that if something goes wrong with the performance or product price of one enterprise, the others will be able to cover the losses. Some farmers also assert that they are justified in diversifying because they need to select a range of enterprises to suit the different types of land found on their particular farm. Others point to advantageous interaction effects between enterprises which result in higher production. Examples quoted include improved land fertility, which may be derived from livestock using a grass break in a cropping rotation, the use of arable by-products to feed stock, and in other cases better weed, disease, and pest control. Claims are sometimes made that it is easier to plan labour on a mixed farm because the demands for staff by the different enterprises can be made to dovetail if care is taken at the planning stage. Similar points are made about the requirements for working capital and the spread of income over the year, which might be better with several enterprises rather than just one.

Specialisation and simplification can result in the concentration of management and other resources such as labour on to a limited range of activities, perhaps with greater productivity because of attention to detail. Farmers who specialise do, however, have to accept the risk that if things go wrong heavy losses can result.

Marketing may be easier in specialised situations because not only will the farmer be able to devote more of his time to one product, but the quantities of goods he sells or buys will be larger, with the result that he may be able to obtain better terms.

The question of unit costs is particularly important. Large machines such as combine harvesters are expensive. The depreciation cost and interest charges on the capital invested in them have to be met irrespective of the area of land on which they are used. These costs contribute towards the unit costs, i.e. the costs per hectare, and the larger the area over which the machine is employed the more its costs are spread and the lower the unit costs. Many farmers therefore decide on at least a degree of specialisation to help justify machinery and perhaps buildings and other equipment.

1.2.5. The Real Significance of Management

People who are new to the study of management are frequently confused about its real importance. Many farmers claim that they have managed a farm successfully "without any of that management out of books". Students of the subject may be faced by lecturers with the proposition mentioned on the first page of this chapter, that technical skill in the production of crops and livestock does not necessarily ensure financial success. They interpret this as extolling the virtues of the management principles and techniques to which they are about to be introduced.

The truth is that successful management depends very largely on personal experience for the formation of sound judgements. The part that the so-called academic knowledge can play varies from situation to situation.

Management should be regarded as a skill which utilises a wide range of abilities. Students of agriculture will find that it requires, and indeed is, a logical combination of all facets of their course, whether it be the farm practical, the husbandry or science classes, or the management and economic studies. They will have to realise, however, that successful management principally depends upon the efforts and ability of the individual to utilise his experience to fully analyse and interpret situations. Knowledge of a few facts and techniques alone cannot do this.

Many farmers without formal academic training are exceptionally good managers. Years of experience, a good training from father, innate ability, luck, personal research or self-teaching, use of advisers, and many other factors play their part. Students should not necessarily assume that they can reach high standards overnight but should use their courses to shorten the time taken to become good farmers, and to help make themselves better ones than they might otherwise have been.

They are advised to become highly proficient in some of the main skills of husbandry as an integral part of good management. It is difficult to become proficient at them all, but generally skills are transferable or easily learned by someone with ability and enthusiasm. They might be wise to interpret and apply the council of one hill farmer who received a visit from the author's students. His lambing percentage, with good quality lambs, was at least 40% higher than that of comparable farms. When asked for his secret he replied: "I don't just look at my stock, I look after them." His ability to achieve much more efficient production than others was largely a product of attention to detail, the ability to observe and interpret, and the knowledge and experience to take a high percentage of correct decisions. What had to be explained to the students was that, in addition, the pressures of modern business management had forced him to study every detail of his business to examine which of the traditional practices were essential, which could be dispensed with, and which new techniques were viable in his situation. He was not just achieving good physical results but also financial success.

It is important to realise that all the above factors cannot always be applied directly by the farmer but must be expressed through his staff who receive a large part of their motivation from him.

Successful husbandry can cover up many deficiencies in other aspects of management, but many argue that it is more important to have a well-organised business which is properly structured and efficiently operated. Usually the argument is fatuous since if the best results are to be achieved all aspects should be good.

Many farmers, and especially those with new businesses, face enormous pressure. When farming is experiencing a successful period land prices may rise and present additional problems for new entrants, especially if interest rates are high. If input prices then increase relative to those for outputs the need for skilful management also increases. In fact this generally applies irrespective of the time the farmer has been in business, but it can be particularly true for those with large amounts of borrowed money.

When farming enters a difficult period survival may become a problem for some and such people may have to consider whether they should "sell-up" and take some money from the business whilst they can.

Farmers with a lot of borrowed capital generally must be efficient even in "good times" but they can face special problems in "bad times".

All farmers would be wise to stand back from time to time and look objectively at their businesses to assess not only the reasons for their present performance but also their future viability. Planned changes may not always be successful but forced changes might be more serious. Sometimes planning can make it possible to achieve good results, future viability, and at the same time allow the farmer to have a more satisfying life.

The principles and techniques of management will not take decisions for anyone, but if properly applied should provide sound guidelines which will not only give some idea of the risks involved in particular courses of action, but at the same time promote an increased number of correct decisions.

CHAPTER 2

ENVIRONMENTAL FACTORS AND FORECASTING

CONTENTS

2.1. THE BACKGROUND TO TRADE

2.1.1. Introduction

A farm business has to trade with agencies outside the farm gate and cannot operate in isolation. This means that quite apart from a consideration of the physical environmental factors of land, soil, and climate, the farmer must consider economic, political, and social influences that might affect agriculture and his business in particular. As an individual he may not be able to directly alter their impact nationally but may be able to use his vote, lobby, or act in concert with his fellow farmers in an attempt to bring about changes. Sometimes there are steps which he might take to minimise any adverse influences or exploit beneficial possibilities for his own farm. This is particularly significant in relation to forecasting the future for the various aspects of his business.

It is therefore very important that farmers are fully aware of the factors affecting the agricultural industry.

2.1.2. Supply and Demand

In order to understand the full significance of the environmental factors it is necessary to be aware of the interrelationship between supply, demand, and the price of products. Here the term "product" is used, but remember that one business's product may become another business's input, e.g. barley sold for stock feed.

It might be considered that the ideal price is that at which demand for a product equals the supply since that is the point at which price would be most stable. In practice the amount demanded rarely equates with the amount supplied although there are many forces which work to bring this about.

In general the demand for products increases as their price falls because people are prepared to buy more at the lower price (Fig. 3a). (By convention the price is expressed on the vertical axis and quantity demanded at any given price along the horizontal axis.)

The supply of a product tends to rise as the price goes up because businesses are willing to produce and sell more (Fig. 3b). If the supply and demand curves are now shown together (Fig. 3c), the point *E* can be obtained. This is the price which equates supply and demand. At prices above *pe*, surpluses develop due to supply exceeding demand, and at prices below *pe* shortages occur due to demand exceeding supply.

At this point it is essential that the reader realises that the graphs depicted in Fig. 3 show the quantity of product that businesses are willing to supply and that purchasers are willing to buy. Supply and demand curves represent the change in amount supplied and demanded *in response to a change in price.* This type of change must be differentiated from a change in supply or demand *caused by a movement* of the demand or supply curve.

The latter point can be illustrated by reference to a year which produces a poor cereal harvest, but is an average year for livestock. Assume that stock farmers are not prepared to alter their attitudes on the price that should be paid for barley. The aggregate demand curve for barley (*DD*) therefore remains the same (Fig. 4a). However, because of the poor harvest total supplies of barley are lower.

The supply curve (*SS*) for a normal barley harvest moves to the left as shown by the diagram and takes up the new position (S_1S_1) shown by the dotted line. At this new position it can be seen that the price which equates the new supply and demand is pe_2. Price pe_2 is higher than pe_1 because cereal growers offer less barley than formerly at all price levels whilst supply is less than demand, thus forcing prices upwards. As the price rises, stock farmers demand less grain until at pe_2 the amount demanded and supplied is in equilibrium.

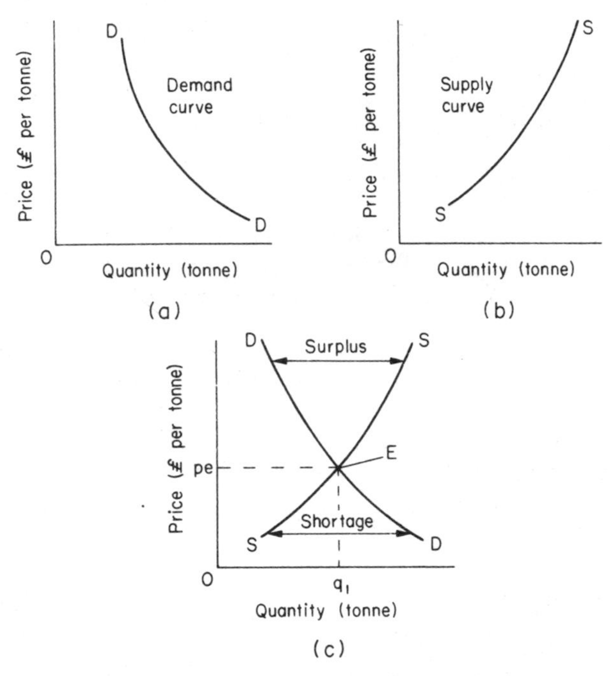

FIG. 3. Supply and demand for a product.

In a good harvest year the opposite effect will occur (Fig. 4b), i.e. the price will be lower at all quantities supplied.

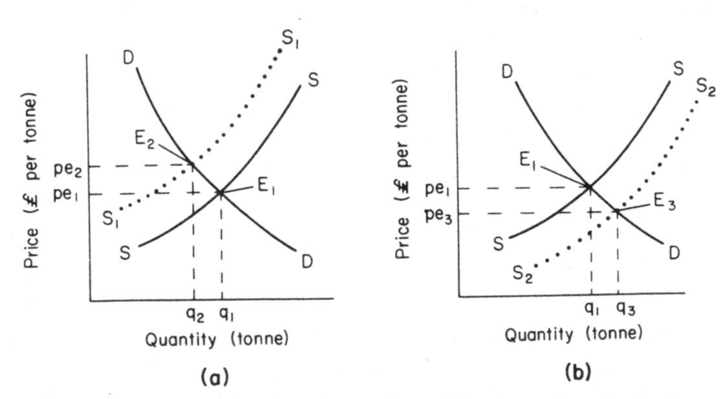

FIG. 4. Changes in supply of a product with demand unchanged.

It might be considered naive to assume that in a time of shortage the stock farmers would not be forced to change their attitude somewhat about the price they would pay for barley because their stock would still require to be fed. Remember that many would look for alternative feeds, some beef producers would adopt less-speedy methods of finishing, and others who forecast the increased price of barley might have reduced the number of cattle they bought, or purchased barley in advance.

In practice, however, the stock farmers would have to pay more for the reduced supply.

The reader will probably be able to visualise graphs comparable to Fig. 4, which would depict movements in the demand curve for barley due to stock farmers who had increased the numbers of animals at a time when the barley supply curve (*SS*) remained static because of an average harvest year. In such a situation the price of the barley would rise because the demand curve would move to the right and a shortage of grain would develop.

A review of the above principles shows that anything which affects supply and demand is worth the attention of farmers as it will affect the price of their products and inputs.

An important point arising from what has been said above is the response in supply or demand to a change in price. This is usually referred to as the price elasticity of supply or demand.

2.1.3. Price Elasticity of Demand

The percentage change in the quantity demanded of a given product relative to the percentage change in price gives a measure of its price elasticity of demand. If the percentage change in demand is less than the percentage change in the price, the demand is relatively inelastic, i.e. Ed=less than 1.0. When the percentage change in demand is bigger than that of the price, the demand is elastic, i.e. Ed=more than 1.0.

What does this mean and how can it affect the farmer? Let Fig. 5a show the demand for milk and Fig. 5b the demand for potatoes. If the price was to rise in each case from p_1 to p_2, the response in demand would be greater in the case of milk than potatoes, i.e. in Fig. 5a demand fell from q_1 to q_2 units and in Fig. 5b from q_3 to q_4 units. The difference in fall in demand between milk and potatoes is due to the difference in the slope of their demand curves.

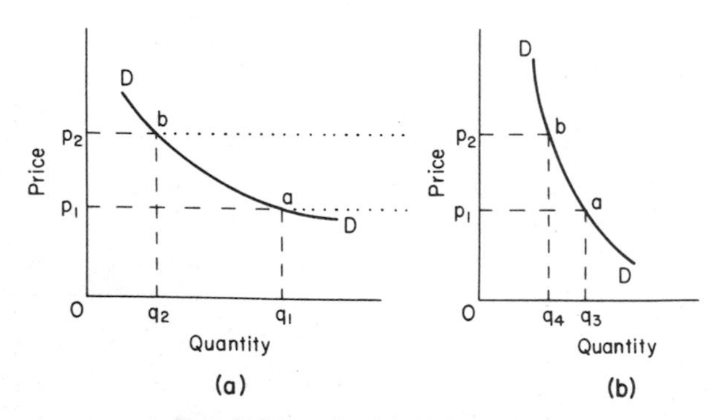

FIG. 5. Price elasticity of demand.

In Fig. 5a the quantity demanded has contracted very quickly like a piece of elastic. Consumers have opposed the rise in price by reducing consumption and it has become more difficult to sell at higher prices.

In Fig. 5b the slope of the demand curve is steeper. Here the price of potatoes could rise to p_2 without such a large reduction in the demand compared to the milk in Fig. 5a. This means that potatoes are more of a necessity, and even when prices rise consumers have little option but to buy. The demand has not contracted so swiftly so that it is considered to be fairly inelastic.

This response in demand to a price change is important to the farmer because in Fig. 5a his total revenue has fallen due to the price rise, but in Fig. 5b it has risen. This can be seen if in the case of milk the size of the rectangles p_1aq_1o and p_2bq_2o are compared and in the case of potatoes the rectangles p_1aq_3o and p_2bq_4o are compared. These rectangles represent the total revenue at each price.

In some poor seasons for potatoes there are many cases of farmers obtaining much greater income than in average seasons because the enormous increase in price per tonne has more than offset the drop in yield, i.e. there has been a move in the supply curve to the left. There have, however, been years when the price has risen so much that consumers have resisted buying potatoes, making the demand more elastic, replacing them by eating pastas and rice.

When potatoes are over-supplied, prices can fall significantly, and the Potato Marketing Board may step in to buy and then sell potatoes for stock feed in order to maintain demand at acceptable levels.

Before progressing further it is pertinent to note the following points with reference to the demand for agricultural products:

(i) The demand for some products can be said to be a derived demand. This means that the requirement for them is derived from the demand for other products, e.g. the demand for feed barley depends upon the demand for beef, lamb, and other livestock products.

(ii) The increase in demand for agricultural products tends to be slow because of the pattern of spending on food by consumers. Significant increases in demand can come with increases in the population, but in Britain the birth rate is slowing down.

(iii) Where there is elastic demand for a product, demand will decrease markedly if the price rises.

(iv) Where there is inelastic demand for a product, total revenue will increase if a supply shortage occurs.

(v) Inflation apart, there is a tendency for the price and demand for many products to remain fairly static. This encourages farmers to introduce new technology to raise yields to increase their incomes. Increased yields tend to produce surpluses, which tend to push down prices, so the farmers have again to increase yields to maintain or increase their incomes. This produces what can be described as the "treadmill syndrome".

2.1.4. Price Elasticity of Supply

The price elasticity of supply can be explained in much the same way as that for demand. Basically it deals with the way in which supply changes in response to an increase or a decrease in price. One factor to note is that in farming there is frequently a time lag between the change in price and the response because of the length of the biological cycle needed to produce the product. Another point is that a farmer has to take care before deciding to reduce his production because many of his overhead costs, such as rent, may still exist irrespective of his level of production.

2.1.5. Substitution

Reference was made earlier to farmers looking for substitutes when the price of barley was high and to housewives substituting pastas and rice for potatoes. Substitution is clearly one of the factors which can affect the slope of the demand curve and therefore warrants examination when forecasting price trends.

2.1.6. Perfect and Imperfect Competition

The farmer should know who his competitors are and be aware of their strength. One of the main conditions for perfect competition is that no one business is big enough to have an effect on the market on its own. When a limited number of businesses can market sufficient produce to influence prices, imperfect competition occurs, and if there is a single seller, a monopoly exists.

There are big dangers for consumers from monopolistic type situations because sellers can dictate prices for products. The severity of the situation depends upon the possibility of using substitutes for the particular product and upon the ease with which new producers may begin to supply it if its price rises to a high level.

Farmers must be aware of the development of monopolistic situations in the feedingstuff, fertiliser, and farm machinery industries. Their ability to prevent them occurring is limited.

Usually individual farmers do not buy sufficient quantities of goods or sell sufficient products to influence national prices. It is conceivable for one or a limited number of farmers to be able to influence prices within an area for a limited time, but such cases are infrequent especially in view of modern transport which could bring in goods in short supply.

Large imports of foreign produce can have a significant effect on prices and cause concern to farmers. This is why many are prepared to make stringent efforts to minimise imports of certain products which they can produce. The monopolistic control, which the Milk Marketing Boards have in connection with milk, is well known, and British farmers are anxious to ensure that imports of raw or UHT milk do not take place.

2.1.7. Comparative Advantage

The law of comparative advantage states that "a product tends to be produced in areas where its ratio of advantage over other areas is higher, or its disadvantages are lower, than any other product".

Physical factors such as soil type, topography, altitude, and climate can be important in this context. Farmers in the west of Britain have a more favourable climate for grass production, but those in the east have the advantage in the case of cereal growing. In the Channel Islands, Cornwall, and parts of western Wales and Scotland, it is possible to produce early potatoes because of freedom from frosts at critical times.

Biological factors can also convey advantage. The production of seed potatoes is facilitated in certain parts of Scotland by the fact that the climate in these areas minimises the existence of aphids and hence the diseases spread by these insects.

Obviously the ability to market potatoes early or to sell high quality seed confers economic advantage. This in turn makes these crops highly competitive, indeed usually superior, to other enterprises in these areas.

The advantage of close proximity to market has been reduced by modern transport developments but it can still exist in certain situations, especially with perishable foods. Farmers near large cities may be able to market stored potatoes and other vegetables when those further away cannot because of adverse weather conditions. Prices are usually high at such times. They will also have lower transport costs than their distant colleagues but have these advantages offset by the fact that, being near to industrialised areas, they have to pay higher wages to attract labour.

The key factor is that each farmer must look to see if he has any advantages which he can exploit. They may enable him to produce a better product, to produce it at less cost, to produce more of it, or to sell it at a higher price than his competitors.

2.2. POLITICAL AND ECONOMIC FACTORS

2.2.1. Introduction

These two factors have so many interrelated facets which affect farming, either directly or indirectly, that they can be considered together. Their social implications must also be reviewed. Indeed, social factors precipitate many political decisions.

It has to be realised that government at the national level represents the interests of all branches of the community and not just farmers. However, the strength of the farmers' political vote, i.e. the number of farmers who can vote at parliamentary elections relative to people in other walks of life, can have a remarkable effect on many politicians' decisions.

In some countries farmers have to produce their goods to sell in a free market situation. This means that the laws of supply and demand are allowed to operate freely to influence the level of production. In other countries politically inspired economic action is taken to modify the normal supply/demand forces. There is, however, little point in confining the study of political and economic factors to the national level, particularly in a group such as the European Economic Community, since decisions taken in one country can affect individual farmers in another.

2.2.2. Gross National Product, Balance of Payments

Governments are interested in the Gross National Product (GNP) of their country because it gives an indication of the performance of the nation's economy. A crude definition of GNP is that it measures the quantity of goods and services produced by a nation in a given period, usually one year. A fall in GNP is undesirable, and serious falls can be associated with a depression, unemployment, and hardship for many in the population.

At the same time, governments are concerned about their country's balance of payments. This gives a measure of the flow of money into and out of the country. In order to establish the flow it is necessary to look at the visible trade in items such as motor-cars, the invisible trade, typified by funds received from foreign organisations for insurance with home firms, and the movement of capital into and out of the country. It may be undesirable for the visible trade balance to be negative for any length of time unless the country is receiving capital from abroad, such as foreign investments, or significant invisible earnings.

There are links between the GNP and the balance of payments, and both of them can be important to farm prices. If the GNP is low there may be unemployment and, depending upon the level of unemployment pay, there can be a reduction in the demand for goods and services. Although reduced demand will initially be for luxury goods such as cars and televisions, food items such as beef steaks, and later even more basic foods, may be affected. On the one hand, the unemployment will help the balance of payments by reducing consumption of imported goods, but, on the other hand, it will do little in a positive way for exports and for the creation of real wealth for the nation.

The contribution which governments feel that agriculture can make to the GNP and to the balance of payments through import saving is significant to the incomes of their nations' farmers. Some countries concentrate very much on the production of industrial goods such as cars, machines, computers, and military weapons with the intention of selling a high proportion to other countries.

They use the money so generated to buy raw materials necessary for the production of such goods and for the purchase of food from nations which rely upon selling foodstuffs for their international trade.

There is a tendency for some industrialised nations to channel labour away from agriculture into industry to foster greater exports. They consider that the extra industrial goods exported will more than offset the cost of any additional food imports that might be necessary. Over a period of years, however, government policies can change, precipitated by changes in world trade.

In Britain the June 1968 report of the National Economic Committee, "Little Neddy", entitled Agriculture's Import Saving Role, re-alerted the Government to the place of British agriculture as an "import saver" in a balance of payments crisis. The white paper Food from our own Resources, published in 1975, indicated the Government's intentions to promote home food production even if they took insufficient positive action in the view of many farmers.

The mid-1980s witnessed both a period of difficult conditions in world trade and the over supply of foodstuffs in the European Common Market. For a time at least Britain had a trade deficit in manufactured goods for the first time since the industrial revolution. Although British farmers could have continued to increase the productivity of their farms the cost of support for agricultural products forced the politicians to introduce milk quotas and to consider ways of reducing supplies of other farm products which were in surplus.

2.2.3. National Wage Levels

Any economy which depends upon the export of industrial products relies very much upon its manufacturers being able to produce goods at prices which are competitive on world markets. Wages are a major factor influencing the cost of production. Each government must therefore be conscious of anything which might influencc the demand for wage rises. Some governments have taken the view that holding down food prices is one way to contain wage demands.

This may not be entirely a bad thing for farmers since lower wages can hold back increases in that portion of their costs, such as machinery, which can be attributed to increases in industrial wages, as well as holding down farm wages. Depressed farm wages might, however, influence productivity if workers feel that they are not being fully rewarded for skill and effort. On the other hand when farm workers obtain wage rises farmers may not always obtain corresponding increases in the prices for their products and some are forced to reduce their work force.

2.2.4. Import Control Measures

A government can protect its farmers from cheap imports in a variety of ways. It may ban them altogether or have an import quota system, but there is a risk of other countries taking reciprocal action. Frequently a price control mechanism is employed. This can involve a levy or tariff on imports which makes the imported goods less competitive with home-produced supplies. Such a system was adopted by the European Economic Community as part of its Common Agricultural Policy (CAP).

The EEC established a mechanism for fixing threshold prices, prices deemed adequate to protect its farmers, for certain items each year. Any of these items subsequently imported would attract a levy determined by the difference between the respective threshold price and the minimum import offer price of the product from the world market. Such a system protects home producers from the artificiality of the world market. In protecting the efficient it also protects the inefficient. Many farmers, who would otherwise have gone out of business, found protection from EEC policies which, whilst putting up the price of food to consumers, kept prices at the farm gate higher by reducing the competitive position of imports. Critics of the policy have asked if it was established because of social concern for continental peasants and part-time farmers or because of the need to attract the farmers' considerable political vote in certain countries.

In addition to the above measure the EEC introduced a common external tariff on agricultural products from non-EEC countries. The rates of duty varied with the particular product and ranged from 0% to 25%.

The food surpluses created by the EEC's import protection and agricultural support systems are well documented. Some members of the public, concerned at the high cost of support, advocated a policy which would allow market forces, i.e. supply/demand/ price, to operate but this was not politically acceptable.

2.2.5. Fiscal Measures

Taxation can affect both investment in agriculture and farmers' attitudes to production on their farms. In Britain favourable rates of Inheritance Tax for farmers have influenced the price of land and caused some landlords to take farms in hand rather than make them available to rent. Taxation allowances for machinery and buildings have often precipitated decisions to invest in these items especially in the past when these were at high rates.

Some farmers do claim that they are not prepared to accept further risks and incur the additional management associated with attempting to increase production because the extra profits would attract high rates of taxation. The reality is that they do not always follow what they claim.

2.3. GOVERNMENT SUPPORT MECHANISMS

2.3.1. Price Guarantees and Production Subsidies

Governments can give positive support to home food production in a variety of ways. One example was provided by the British Agriculture Act of 1947. During the 1939–45 war, and indeed up to 1954, the British Government through the Ministry of Food controlled the purchase and rationing of all essential foodstuffs. There were fixed prices and assured markets for farmers during the war with the emphasis on quantity of production.

The 1947 Act was designed to provide a sound post-war agriculture policy at a time when there was a need for expansion of food production. Its main objectives were to promote and maintain "by the provision of guaranteed prices and assured markets . . . a stable and efficient agricultural industry capable of producing such part of the nation's food and other agricultural produce as in the national interest it is desirable to produce in the United Kingdom, and of producing it at minimum prices consistent with proper remuneration and living conditions for farmers and workers in agriculture and an adequate return on capital invested."

The Government agreed that it would hold an annual review to determine the prices and it assured markets for the commodities listed under the Act. It also shielded home producers to some extent through certain import tariffs and controls.

During the time that the 1947 Act was in operation there were several ways through which the guarantees were implemented. In many cases it was through the price of the product. Other methods included acreage payments to cereal producers, fertiliser subsidies, calf subsidies, beef cow and hill cow subsidies, and hill sheep subsidies.

In the initial period up to 1954 the policy was one of fixed guaranteed prices with the emphasis on quantity. Agricultural output increased to a level 50% above that obtained prior to the war. Farmers were able to finance expansion of their businesses partially through their own profits and partially from banks which considered agriculture to be a sound investment because of the guarantees.

A change of Government in 1951 signalled the beginning of amendments to the way in which the Act was operated although the main changes did not occur until 1954. The Government wanted a less rigid system of fixing prices, some limitation to Exchequer liability, and freer marketing. The deficiency price system was introduced so that cereal and fatstock producers received the difference between average prices and guaranteed prices as a supplement to what they achieved by their own marketing ability. Milk Boards resumed their powers and the guaranteed price for milk was limited to standard quantities fixed at the price reviews. Production above these levels did not receive support.

The 1957 Agriculture Act made amendments to that of 1947 and limited the cuts in guarantees which could occur in any one year. It also introduced grants for farm improvements and amalgamations. Although production continued to rise there was less urgency for it to do so, and more emphasis was put on efficiency of production.

Early in the sixties the cost of Exchequer support had grown considerably, and changes in 1964 saw the introduction of a policy of more selective expansion and greater emphasis on efficiency and market competition. A standard quantity system was introduced for cereals and there were tighter controls on the support for pigmeat.

Efforts were made to protect British farmers from the dumping of cheap imported food. Minimum import prices were introduced for cereals and eggs, and long term supply contracts were fixed for butter from New Zealand and for bacon from Denmark and the Irish Republic.

In general the policy implemented by the 1947 Act kept food prices to British housewives below what they might otherwise have been. It can be argued that in its original form it was more a policy providing consumers' subsidies than producers' subsidies. Remember that taxes were necessary to pay the subsidies.

When Britain had to vote on entry to the EEC in the early seventies the proposition was put to the public that entry to the Community would boost Britain's international trade and in turn this would significantly increase wages. It was claimed that although food prices on the Continent were substantially higher, the full impact on British shop prices would be delayed by gradually phasing Britain into the EEC system. This would give time for wage rises to materialise and so offset food price rises. British farmers were encouraged to think of the very high prices being received in the EEC for farm products, although there was not the same emphasis in some quarters on the increase in costs that would occur.

2.3.2. Market Price Support System

The EEC adopted a market price support system for its CAP. The original aim of the EEC was to establish a market with free trade and competition so that goods and resources could move freely to where they would be best utilised. The aims of the CAP were:

(i) to increase agricultural productivity thereby
(ii) ensuring a fair standard of living for the agricultural community;
(iii) to stabilise agricultural markets;
(iv) to guarantee regular supplies of food;
(v) to ensure reasonable prices to consumers.

Within the EEC British farmers found a significantly different support system to that under the 1947 Act. Instead of a mechanism based largely on production subsidies they encountered the CAP which operated through support of the wholesale market price. The protectionist practice of import levies described above in Section 2.2.4. was in force. In addition they met a policy employing a system of target prices, or for some products guide or basic prices, and intervention buying.

Not all products were covered by the CAP, however, notably mutton and lamb, potatoes, and wool. Apart from France, sheep did not feature significantly in the economy of the original EEC countries.

The way in which the system operates varies slightly from product to product. The target, guide, or basic prices for the respective products are established by the Council and reviewed each year. These are in no way guaranteed prices. The intention is that producers should receive these returns in the wholesale markets.

Intervention prices are fixed for each product at a level below the target, guide, or basic price level. Special intervention authorities in each Member State are given the responsibility of buying up produce when market prices fall to intervention levels, and in some cases even above these levels. The goods they buy may be stored at intervention centres. Essentially they are buying up surpluses which would otherwise tend to depress market prices.

The individual farmer cannot be guaranteed that he will get even the intervention price. Although in general anyone who can satisfy the quantity and quality requirements can sell into intervention it is usually only carried out by large wholesaling and co-operative organisations. Farm gate prices are often below intervention prices principally because costs of transport from the farm to the intervention centre must be deducted from the intervention price to derive in effect a "farm gate intervention price".

2.3.3. Market Price Support for Barley

The way in which the support system operates can be illustrated by reference to barley (Fig. 6). An intervention price, a target price, and a threshold price is set annually by the Commission.

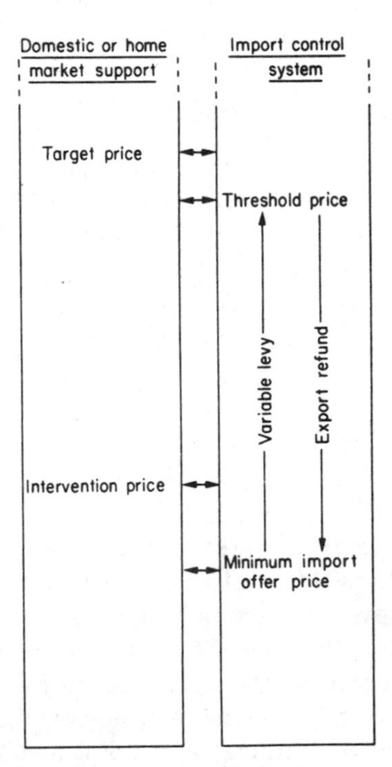

FIG. 6. The market support system for barley operated by the EEC.

A variable levy is imposed on imports to the Community. This is the difference between the minimum import offer price and the threshold price. The levies can vary daily according to the minimum import offer prices. Exports from the EEC to non-EEC countries attract export refunds based on the difference between the Community price levels and world market price levels.

Intervention authorities are obliged to buy all barley offered to them, subject to quantity and quality conditions, at the intervention price. The intervention prices relate to the prices at the Ormes Intervention Centre in France, the centre in the EEC with the greatest surplus of cereals sold wholesale. The target price which is derived reflects what the Community authorities consider to be an appropriate price for barley in the next year. It includes an element for the transport costs from Ormes to Duisburg, the centre with the greatest shortage of cereals.

The cost of transporting barley from Rotterdam to Duisburg is deducted from the target price before arriving at the threshold price. The net effect is that prices at Duisburg should not be below target prices.

In practice different monthly prices are published to encourage some storage and so foster orderly marketing during the whole year.

2.3.4. Market Price Support for Beef and Veal

Beef and veal can be used to illustrate some of the differences in terminology and basic principles employed (Fig. 7). Instead of a target price, a guide price for live adult cattle is established annually. This is the price which it is considered they should make when sold for slaughter.

Reference prices, which are the average wholesale market prices, are calculated each week for the EEC as a whole and for each Member State using prices realised at selected markets.

Typically the invention price for beef is 90% of the guide price. When the reference price in a Member State is equal to or less than the intervention price for 2 weeks, permanent intervention is introduced. If the reference price is above intervention price for three successive weeks, intervention is stopped in that Member State.

Import levies are calculated each week and based on the difference between the guide price and offered import price plus any customs duty. The amount of basic levy is varied, however, according to the level of the EEC reference price in relation to the guide price (Fig. 7). For example, if the reference price as a percentage of the guide price was less than 90%, the percentage of the basic levy payable would be 114%. At 98–100% it falls to 100%. The reference price has to increase to 106% of the guide price before the basic levy payable is zero. The very high rates of levy are designed to particularly penalise imports when market prices are depressed.

Any Member State may opt for either a variable premium system or a calf premium system. The United Kingdom chose the first option, unlike Italy who made the latter choice. The UK target price is set at 85% of the EEC guide price. This is broken down into weekly target prices with the highest prices in months when beef is in shortest supply. The average weekly target prices equal the annual target price. The variable premium is paid to farmers selling cattle in any week in which the average fat cattle price is less than the target price for that week.

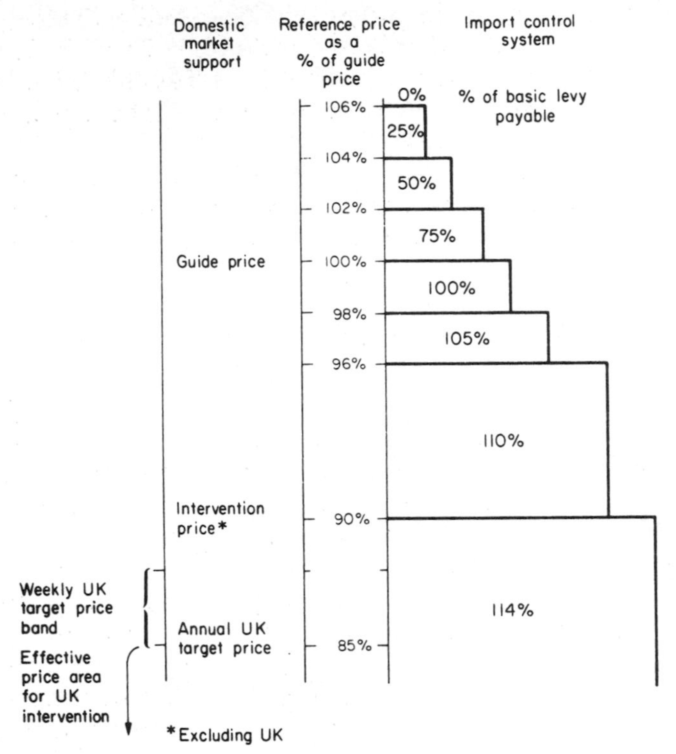

FIG. 7. EEC market support for beef and veal.

Intervention for the UK beef market only operates at a price equal to the target price minus the estimated variable premium.

2.3.5. Surpluses and Support for Exports

Countries that have a surplus of food which they wish to export may subsidise the exports to make them more competitive on world markets. Intervention buying by the EEC created the so-called "mountains" of food in intervention stores. The Community provided "export restitutions" which enabled the high-priced EEC food to be sold to non-EEC countries at low prices. In other cases food suitable for human consumption was converted into animal feed or even destroyed.

The enormous cost of the CAP, particularly for intervention buying, storage and export restitutions resulted in pressure from some member countries for a reform of the system. This was opposed, at least in part, by other countries in the EEC which had much to lose. The costs of the CAP to Britain were being compared in terms of such things as "how many hospitals could be built with the money that could be saved".

The mid-1980s marked the start of a period of concern for many farmers as pressure intensified for changes to the CAP. A significant development was the introduction of milk quotas restricting farmers to a production level based on the quantity of milk they sold in 1983, with the proposal that further cut-backs would follow.

Suggestions were made for both the reduction in intervention prices and an increase in quality standards for a range of farm products. Many other ideas were advanced to reduce production. These included a "set aside" programme for cereal producers. In essence this was similar to the system which had been practiced in America whereby farmers were paid not to grow the same area of corn which they had grown previously.

One proposal which was accepted was the introduction of a co-responsibility levy on the sale of cereals. Such a levy had previously been introduced for milk and sugar. The concept was that funds produced would be used to explore new markets, to help cover the cost of export restitutions, and also to finance research into alternative crops.

Concern was expressed that unless the EEC took an overall view of agriculture farmers would simply transfer the problem from activities in surplus, such as milk and grain, to those which were not currently over supplied, such as lamb production.

Claims were made that up to 1.25 million hectare would have to be taken out of food production in Britain alone. Numerous proposals were made for the alternative uses of the land, in particular forestry. The policy makers faced the problem of how to minimise the social problems, particularly in less favoured and other rural areas, which cut-backs would create. They had to remember that the jobs of many people were associated with agriculture either dealing with its products or supplying it with goods and services.

2.3.6. Capital Grants

There are several reasons why Governments may give capital grants to farmers. Frequently they are used to encourage the development of projects which will foster future production of agricultural goods or to improve marketing arrangements. They can, however, be employed to encourage people to go out of farming, or to stop producing certain items. The milk "outgoers" scheme illustrates the last point whereby farmers were offered compensation to stop producing milk.

For more than a decade after entry to the Common Market Britain's farmers enjoyed what may, in retrospect, be regarded as relatively attractive grant schemes funded in one case by the UK Exchequer and in the other supported by the European Farm Fund (FEOGA). Over the years these schemes were amended both in title and in what they supported. The move away from grant aiding buildings for such things as milk production, to grants under the heading of "Environmentally Positive Investment", reflects the change in policy to agricultural production which had taken place by the mid-1980s.

2.3.7. Cheap Loans

In some countries it is Government policy to provide loans to farmers at very low levels of interest, but this has not been the practice in Britain. In France Credit Agricole, which is state-owned has, over the years, offered loans to farmers on very attractive terms. Belgium has a government institution, the Agricultural Investment Fund, which helps to finance farmers, whilst Germany has a system of Co-operative Banks.

2.3.8. Regional Support

In many countries there are regions where agricultural production is extremely difficult. The reasons include topography, soil type, climate, inaccessibility, costs of transport, and labour shortage.

Special support measures may be employed for these less-favoured areas to prevent depopulation or hardship. Hill and mountain areas are particularly good examples. On entry to the EEC the United Kingdom was able to continue to give support to hill cattle and hill sheep farmers through a payment based on numbers of stock. The payment of increased capital grants compared to other areas was also possible.

2.3.9. Socio-economic Advice

The introduction of socio-economic advice through the EEC was particularly valuable to farmers in less-favoured areas. In part this was designed to encourage farmers and their workers to improve their farming abilities by such means as training courses. It was also concerned with the provision of advice on the acquisition of supplementary income from non-agricultural sources, e.g. tourism; labour problems and relationships; relationships with the Community and future prospects for farmers and farm workers.

2.4. THE VALUE AND COST OF MONEY

2.4.1. Currency Values

The strength of a country's currency on the world market can have an important impact on farming. It may affect prices of farm products directly or have an indirect effect by influencing such things as imported raw materials which in turn precipitate price changes for farm goods. In the past the United Kingdom has all too frequently seen the devaluation of the pound. This means that overnight the pound is worth less on the world money markets. It automatically takes more pounds to buy the same amount of imported feedingstuffs and such items as imported oil cost more. Rises of this kind can produce a chain reaction. In the case of oil any process directly or indirectly using oil may be affected, and in turn the cost of farm machinery, fertilisers, and even the men's wages can be influenced.

The reasons why a country might devalue its currency and risk inflation, are extremely complex and beyond the scope of this book. One benefit the country does gain is that its exports become cheaper relative to other nations' exports because, for example, a nation buying British goods after the devaluation of the pound finds that its own currency is stronger and so buys more pounds or their equivalent in goods. This might result in increased exports which could help the employment situation in the exporting country. With the extra money supply in the public's pocket prices may rise, including the price of food.

Some countries, including the United Kingdom in the 1970s, decide from time to time to let their currency float, which means that it can fluctuate to find its own level on the world market without the rigidity of control through government revaluations. After a period of weakness, somewhat equivalent to a period of devaluation, the pound became relatively strong in the late 1970s. It was also strong in the mid-1980s. This reduced the competitiveness of British exports.

The main concern of the farmer is that the prices of his products should rise by an amount which will at least compensate for any increases in his costs. In times of inflation replacement tractors and other factors of production cost more. One benefit he may obtain at such a time is that if he borrows £10,000 when he repays it, say 5 years later, the £10,000 will be worth less in real terms.

2.4.2. Representative or Green Rates of Exchange

On entry to the EEC British farmers found that they not only had to concern themselves with the value of the real pound but also the green pound. At its inception the EEC decided that if free trade was to occur in the Community the relative values of the currencies of the Member States should be established and stabilised as far as possible. They invented the unit of account (UA), which was purely a theoretical concept. The currency of each Member State was related to it so that 1 UA was worth so many French francs, so many West German marks, and so on. Prices in the Community were then declared in UAs.

While the Member States stuck to a fixed rate of exchange for their real money, there was parity of prices between the different countries. However, when some countries altered their exchange rates this meant that the prices received by their farmers would change relative to those in other Member States.

To avoid this it was agreed that a representative rate should be established for each of the "real" currencies of the Member States. These rates were to be used only for agriculture, and the term "green" was employed to differentiate them from the real rates. Britain found that it had a green pound just as West Germany had a green Deutsche mark and France had a green franc.

In 1979 the European Monetary System (EMS) was established, although Britain refused to join. The European Currency Unit (ECU) replaced the Unit of Account. A system was adopted involving the calculation of the value of the ECU daily with reference to the money market exchange value of a "basket" of specified amounts of the currencies of EEC countries. The green "currencies" were retained.

The value of the green pound is of importance to both the farmers and the housewives of Britain. It can be revalued independently of the real pound. Assume that the intervention price of butter is 3500 ECU/t and the green pound equals 1.6 ECU. The intervention price in Britain is therefore 2187 green pounds per tonne. If the green pound was devalued to equal 1.5 ECU this would increase the intervention price to 2333 green pounds per tonne. This means that devaluations of the green pound increase support prices for farmers. The British Government has to be careful before making application for the value of the green pound to be reduced, however, because this results in an increase in prices in the shops which in turn can stimulate demands for wage rises.

2.4.3. Monetary Compensatory Amounts (MCAs)

When agricultural products are sold from one EEC country to another they are traded in real and not green money. Assume for example that Germany sells butter to

Britain. The price received is influenced by the support price for butter in Britain which reflects the value of the green pound. The Germans, however, receive real pounds and the value in Deutschmarks (DM) is obtained from the exchange rate between the real and not the green "currencies".

If the intervention price of butter in Britain is 2187 green pounds/t based on 3500 ECU/t and the real pound sterling equals 1.7 ECU then there is a strong influence for butter to sell at £2059/t real money. If the rate of exchange between the real currencies was £1 = 3 DM then the price the Germans are likely to receive from Britain for butter is 6177 DM/t.

This must be compared with the price butter is probably going to sell at in Germany. The intervention price in Germany, as in Britain, is 3500 ECU/t. If the real DM equals 0.46 ECU then there is an influence for butter to be sold in Germany at 7609 DM/t, or 1432 DM more than the price obtained in Britain. On this basis it would pay the British to sell butter to Germany and not the reverse way round.

To prevent this situation from distorting trade butter being exported to Germany would face a tax or MCA by the Germans whilst butter coming into Britain from Germany would receive MCA as a subsidy.

The Agriculture Council, as part of its rationalisation programme announced in the mid-1980s that its aim was "the restoration of a single market by dismantling of the monetary compensatory amounts". It was envisaged that by 1992 there would be no MCAs.

2.4.4. Interest Rates

The availability of loan capital and attendant interest rates are of fundamental importance to most farmers. Many factors control the cost and supply of capital. International money markets play a significant part but individual government reaction to current economic situations quickly affect farmers. There are indirect influences even on farmers without loans.

Governments may, to use British terminology, employ the "Base Rate" or "Minimum Lending Rate", (MLR), to control the demand for money. When the minimum rate of interest at which money can be borrowed is forced up by a government it should, in theory at least, reduce the demand for money and so control spending. Sometimes it has to be pushed very high to achieve its effect. Such a move may reduce investment in industry with possible losses in employment and precipitate changes in supply and prices of goods and services to farming. Also at times when interest rates are high farmers must be additionally careful to examine their own borrowing to ensure that it is justified.

It is very much easier to contemplate borrowing money to invest in such things as buildings and machinery when interest rates are low, but unless the money is borrowed at a fixed rate of interest the interest payable may increase substantially later. The farmer must test the sensitivity of his plans to such possible rises.

In Britain and some other countries the effective rate of interest is reduced because it can be entered into accounts as a cost and in turn attract tax relief. In other countries cheap or interest-free loans are provided to some farmers.

2.5. TECHNOLOGICAL AND BIOLOGICAL FACTORS

2.5.1. Technological Improvements

The significant development in agricultural technology during the last 30 to 40 years has fostered an enormous increase in the output from farming. There is still plenty of potential for further improvement, but the food surpluses within the EEC (which can be attributed to a combination of improved technology and attractive prices) and much of the rest of western world have put into question the cost/benefit of further increases in production. Individual farmers must still examine new methods in relation to their own businesses to see if they can be profitably applied at the prevailing supply/demand/price situation. It could be that new techniques could result in greater efficiency at production levels which are forced on farmers by Government inspired control mechanisms.

The progress within the agricultural industry has not only contributed to surpluses but has brought other problems. Increased intensity of production has required higher management standards. It has placed greater dependence on the pharmaceutical, plant protection, and fertiliser industries. Evaluation of capital investment, and finding and organising a better trained work force have become more important in influencing success.

Destruction of soil structure by use of heavy machines on land at times when it should not be worked, and the task of disposing of slurry and farm effluents have added to management worries. The growth in demand from agriculture for energy is also a subject which has attracted debate, with suggestions that recycling of scarce resources and alternative energy sources warrant further investigation. Experiments are continuing on such things as the production of methane from manure to heat buildings and on the growth of algae on liquid wastes to feed livestock. These are integrated into projects designed to reduce pollution.

2.5.2. The Farmer and The Natural Environment

Farmers are facing increasing pressure from the public to pay more attention to animal welfare, social, environmental and ecological aspects. Authors like Rachel Carson (*Silent Spring*), Ruth Harrison (*Animal Machines*) and Marion Shoard (*Theft of the Countryside*) did much to attract the attention of the public to the effects of agrochemicals, to animal welfare, and to the loss of wildlife habitats.

Although the "Green Party" did not have the political success in Britain which it achieved in Germany there was a growing awareness, even by some farmers, that there was a responsibility to preserve the flora and fauna of the countryside for future generations. Concern was expressed at the destruction of wildlife habitats by land reclamation, the filling in of ponds and removal of hedgerows, leaching of nitrogen into water courses, pollution of rivers by farm wastes, and the build up of residues used to control pests and diseases.

As far back as the 1940s the Government created the Countryside Commission (CC) and the Nature Conservancy Council (NCC) to protect the British landscape and wildlife respectively. It also created the first of the National Parks in England and Wales. The NCC was responsible for identifying wildlife areas, the creation of Nature Reserves with the aid of local Wildlife Trusts, and the designation of sites of Special

Scientific Interest (SSSI's). These still only account for a small percentage of the countryside of the United Kingdom.

The Wildlife and Countryside Act (1981) confirmed Government support for conservation. It gave statutory protection to a wide range of wild birds, other animals and plants. The second part of the act related to SSSI's, heightening the awareness of these sites and of potentially damaging farming operations. It also allowed for management agreements to be reached with the possibility of compensation for farmers.

Since the mid-1960s farmers, landowners, and conservationalists have been co-operating in an organisation called the Farming and Wildlife Advisory Group (FWAG) (Forestry FWAG in Scotland). Its main aim is to encourage conservation on farms and estates whilst also sponsoring research work and supporting county advisers.

The whole subject of conservation of the natural environment is one which should be viewed objectively both by those who make a living from the land and those who wish to enjoy and preserve the amenity and other values that the countryside can offer.

For some farmers struggling to maintain financial viability, wildlife conservation and preservation are understandably low in priority. Equally those farmers with more successful businesses have much to offer in terms of developing animal and crop production systems which are more environmentally sympathetic. They may also be able to foster educational projects such as family days and nature trails.

2.5.3. Biological and Natural Factors and Supply/Demand

Many factors which could be included under this heading have to be neglected in a book of this nature. The significance of soil type, fertility, climate, normal disease patterns, regular shortages of water, and a great many related aspects can only be given minimal attention. What has to be mentioned is the importance of changes to normal situations, whether these be temporary or the start of long term trends. It is these changes which can significantly alter supply/demand patterns, sometimes providing the farmer with immense problems but at others providing opportunities which he can exploit to advantage.

In comparison to many countries Britain has few climatic disasters although those who experienced the winters of 1947, 1963 and 1979 may not agree. They would be joined by those who farmed through the drought of 1976 and the wet summer of 1985. In a bad season, however, not all farmers suffer because as sometimes happens with potatoes, when supply is limited, prices increase.

Farmers cannot afford to be insular in their observations of changes in production levels because events across the world may ultimately affect their businesses. World supplies of feedingstuffs, cereals, protein foods such as soya and fish meal and even finished animal products such as beef, can all be important. In some cases export markets will be reduced by oversupply and competition, and in other instances imports of certain items, such as soya, may become more costly because of crop failure.

2.5.4. Relationships between Agriculture and the Food Industry

In the marketing section of this book emphasis will be placed on the need for the farmer to produce goods of a type and quality in demand by the consumer. For many products he will be under the dictates of the food-manufacturing industry. He must be aware of their changes in demand and amend his production and technology to best advantage.

Each stock farmer must examine the argument that using high-priced grain to feed livestock and so produce secondary protein is inefficient. Clearly he must evaluate both the economic and moral issues of this in relation to his own situation.

Hopefully the farmer does not have too much to fear from food substitutes although they are bound to find their place. Adaptability and flexibility might have to be the important words in the agriculture of the future. New products for feeding either to stock or directly to humans are bound to be developed. Not too long ago few would have thought foodstuffs would be derived from the oil industry or that artificial steaks would be made from soya beans. There is likely to be a place for the farmer in the human food chain for a very long time and not just in the luxury category of food produced by "natural methods".

There may even be new outlets for some farm products. Already in parts of the world sugar is being used to replace products from the petro-chemical industry. Items produced from sugar include ethanol to fuel cars, bio-degradable low-foaming detergents which will not pollute rivers, a barrier cream which does not produce allergies, and chemicals used in making paints and plastics.

2.6. FORECASTING

2.6.1. Advice and Forecasting

Forecasting is concerned with decisions which may have short, medium, or long term implications. Typical examples might be: What price will beef be at next week's market? Should the wheat be sold at harvest or will it be at a significantly higher price next spring? If a new building is erected for pigs are future profits from bacon or pork going to justify the money spent?

In a dynamic world it is very difficult to predict the outcome of events some years or even months ahead. The individual farmer usually does not have sufficient time to devote to the sophisticated methods which are available for forecasting. He should therefore make use of the various agencies which will provide advice and predictions, taking care to realise that they also have difficulty in predicting the future and are not infallible.

The main advisory services in England and Wales are provided by the Agricultural Development and Advisory Service (ADAS). In Scotland the work is undertaken by the Scottish Agricultural Colleges. Each of these organisations have specialists who, in addition to keeping the general advisers informed about new technical developments and the outlook for the future, produce reports which are available to the farmer. Some of these reports can be employed in forecasting as can those of the various universities. Reports from the EEC, from the Food and Agricultural Organisation, and other international sources can also be added to these.

Advice is available from many other bodies, including commercial organisations, and most issue their prophecies for the future from time to time. The Milk Marketing Boards, Meat and Livestock Commission, Potato Marketing Board, Home Grown Cereals Authority, and many others fall into this category. The National Farmers Union also issue statements and reports in language which is easy to understand. Add to this the various reports issued by the Government or government-sponsored bodies and one might be tempted to think that the farmer has no need to do any forecasting for himself.

2.6.2. The Farmer and Forecasting

In reality he has to analyse advice and forecasts which are issued in general terms to see how they apply to his business. All farmers attempt, to varying degrees, to do some forecasting themselves. There can be great merit in this but some danger, especially when a farmer has imperfect knowledge of a situation. Every effort must therefore be made to obtain as much information as possible.

Short term decisions necessitate regular research in order to keep up to date with the current situation. The use of the Teletex system on television should prove of great value, supplementing information which can be obtained from the marketing sections of farming papers, radio farming programmes, and the type of report which is available by telephone from the Meat and Livestock Commission and some auctioneers.

As with any type of forecasting, the farmer has to estimate the supply and demand situation and interpret how this will affect the price both for his products and for the inputs he wishes to buy. He can add his own knowledge of local conditions to any information he obtains. For example, he may decide not to market cull dairy cows when local factories and schools are on holiday because of the reduced demand for this type of beast at that time. The factors he considers may not all be local or even within his own country. In the 1970s the price of British fat lamb was very much influenced by the French who would from time to time stop imports from Britain.

When making medium and longer term forecasts there are various factors which the farmer can take into account. Certainly a knowledge and careful interpretation of the political, economic, sociological, and other factors outlined earlier in this chapter can help. Statistics indicating future supplies should be studied. In Britain the June returns made by farmers list the area of crops being grown and the stock numbers. Analysis of these can give some indication of future supplies provided cognisance is taken of natural biological cycles, e.g. young animals recorded have to mature before they can produce or be sold for slaughter. If EEC and perhaps world statistics are added to these, a picture of the future supply of goods can be built up. This data can be supplemented by predictions of crop yields and information about commodities in storage.

Knowledge of supply must be related to demand. Reference to financial newspapers and the radio financial report can prove of value because between them they not only comment on trends in the country's economy, employment, statistics of quantities traded, money supply, and other useful information, but they report on the future's markets. The latter can be valuable in the medium term. The subject of futures will be covered later under cereals but basically it refers to prices being offered for commodities, e.g. grain to be purchased some time ahead.

Studies of production for some agricultural products reveal a cyclical trend if quantity is plotted against time. This is frequently accompanied by price fluctuation, price being inverse to supply. The so-called "pig cycle" is the classic example. When prices for pig products were at their highest more farmers went into pigs, supply increased, prices fell, farmers went out of pigs, supply became short of demand, prices increased, and the cycle was repeated. In Britain this cycle is not now as true as it was. The reason is that there are no longer as many producers of small numbers of pigs as there were and the large producers are so geared up that they cannot easily expand or contract. This last factor must be borne in mind when looking for opportunities based on cycles of production for any product.

Econometricians use trend curves, including those with non-cyclic patterns, to predict future production employing mathematical techniques. Some of these techniques have been adapted for use by farmers.

Clearly there is no way of predicting the future with absolute certainty unless there are clear government or other guarantees in respect of such things as quantities and prices. In spite of this some people achieve remarkable success by interpreting the points discussed above. Occasionally, although not always, they may do the opposite to the trend of what others are doing but judiciously so. Many times they foresee the application of new technologies before their colleagues and are more efficient at an earlier stage than others. Sometimes it takes courage as well as vision if investment opportunities are not to be missed.

The prediction of yields, inputs, and all factors relating to production within the farm gate is yet another aspect of forecasting, but this is best left to the relevant sections of the book which follow.

CHAPTER 3

BASIC ECONOMIC PRINCIPLES OF PRODUCTION

CONTENTS

3.1. THE APPLICATION OF ECONOMIC PRINCIPLES TO AGRICULTURE

3.1.1. Introduction

Some people are initially confused by economic principles and therefore have little time for them. However, if carefully interpreted in relation to a given farm situation they can provide guidelines to help the farmer take decisions necessary for successful, economic production. The reader must be prepared on some occasions to find that literal application of economic concepts is made difficult by imperfect information.

An attempt is made in this chapter to give a simplified presentation. It is necessary to introduce some economic terms, but the potential farmer is advised to concentrate on those sections covering practical application, and from these the merits of economic principles can be appreciated.

When formulating a production programme a farmer has to make many fundamental decisions. He has to decide which products to produce, the quantity of each product to produce and the method of production, including the amount of each input factor (e.g. seed, fertiliser) to use. Economic principles can assist in each case.

Many practical points arise, but one of the most significant which can serve as an example is the difference between technical and economic efficiency.

3.1.2. Technical Efficiency

Some farmers are technically efficient at production without necessarily being economically efficient. The measurement of technical efficiency is undertaken by calculating the ratio of output to input in physical terms. Thus if one bacon producer obtains 1 kg of liveweight gain from his pigs for every 3 kg of feed, he is technically more efficient than another farmer who feeds 3.2 kg to obtain 1 kg of gain.

3.1.3. Economic Efficiency

When measuring economic efficiency it is common practice to use profit as the objective, relating inputs to outputs in financial terms. It is possible, however, to simply use the cost of one input factor in relation to the value of the output. For example, instead of using all the costs of pig production the cost of the main input factor—feed—can be related to the output. It could be that the second farmer quoted in Section 3.1.2. was paying less for his feed than the first and economically he might be just as efficient, or more so, than the first.

Marginal analysis can, amongst other things, be used to differentiate between technical and economic efficiency.

3.2. PRODUCTION DECISIONS IN RELATION TO ECONOMIC PRINCIPLES

3.2.1. Marginal Analysis

The data in Table 1 represents the results of a hypothetical survey of several farms and shows the response of winter wheat to spring applications of nitrogen. Fields which received no application on average produced 1.5 t/ha because of soil nitrogen, but Table 1 only shows yields above this in response to top-dressing.

A cropping farmer is always interested to find out if the cost of adding that "last extra bit" of fertiliser is justified by its effect on that "last extra bit" of income it produces, if any, from the crop. The term last extra bit is clumsy, so the term "marginal" is used instead. Thus it can be seen from Table 1 that the fertiliser has been considered in 20 kg increments up to the 180 kg level, and the cost of each increment is £6.40. Above 180 kg the increments are of 5 kg. The total product is the total yield at each particular level of fertiliser, and the marginal revenue is the increased value of grain for each fertiliser increment.

Table 1. *Returns of winter wheat to spring applications of nitrogen (hypothetical case)*

Nitrogen used (kg/ha)	Total product (yield) due to nitrogen application (t/ha)	Marginal cost of nitrogen (£)	Marginal product change in yield (t/ha)	Marginal revenue increased value of grain (£)
20	0.60	6.40	0.60	60
40	1.50	6.40	0.90	90
60	2.90	6.40	1.40	140
80	3.80	6.40	0.90	90
100	4.40	6.40	0.60	60
120	4.85	6.40	0.45	45
140	5.20	6.40	0.35	35
160	5.46	6.40	0.26	26
180	5.65	6.40	0.19	19
185	5.69	1.60	0.04	4
190	5.70	1.60	0.01	1
195	5.69	1.60	−0.01	−1
200	5.05	1.60	−0.64	−64

Note: nitrogen valued at 32p per kg; wheat sold at £100 per tonne.

3.2.2. Law of Diminishing Returns

A study of Table 1 shows that there is an increase in the marginal product with each increment of fertiliser up to the 60 kg level. After this the marginal product progressively diminishes with additional nitrogen as does the marginal revenue. Meanwhile total product goes on increasing up to about the 190 kg application. This is illustrated graphically in Fig. 8 and shows that it complies with the law of diminishing returns. The latter states that *if the quantity of one variable input* (or more than one variable input) *is increased by equal increments whilst other inputs are held constant, the increments in total product may increase at first, but a point is reached where the marginal product from each successive increment of input will decline.* Observe the first point of inflection in Fig. 8 where the change occurs.

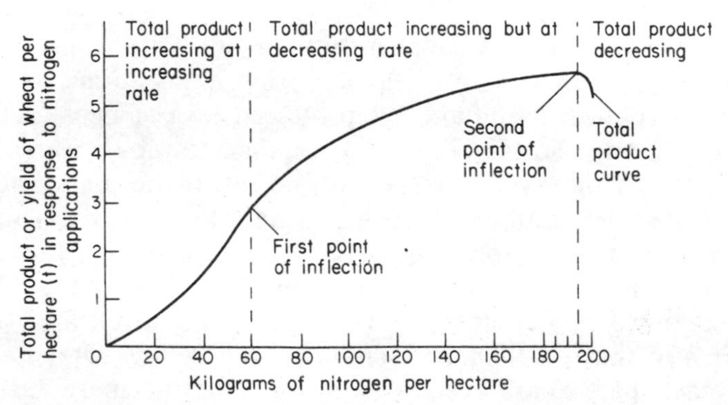

FIG. 8. Illustration of the law of diminishing returns based on data in Table 1.

Each extra tonne of wheat is valued the same as the previous tonne. It is therefore possible to illustrate the response to nitrogen by yield. The curve produced represents what is called a "production function", i.e. it shows the relationship which exists between various levels of an input (known as a factor of production) and a product. This is an example of a factor/product production function, but factor/factor, and product/product production functions also exist.

3.2.3. The Average Product

The data in Table 1 can be examined to calculate the average product which is the ratio of the total output to total input at a given point. In Table 2 this has been undertaken in financial terms, e.g. at 20 kg per hectare of nitrogen, divide £60 by £6.4 to obtain £9.37.

Table 2. *Calculation of the average product*

Nitrogen used (kg/ha)	Total value of grain (£)	Cost of nitrogen (£)	Average product (revenue) (£ of grain per £N)
20	60	6.4	9.37
40	150	12.8	11.72
60	290	19.2	15.10
80	380	25.6	14.84
100	440	32.0	13.75
120	485	38.4	12.63
140	520	44.8	11.61
160	546	51.2	10.66
180	565	57.6	9.81
185	569	59.2	9.61
190	570	60.8	9.38
195	569	62.4	9.11
200	505	64.0	7.89

It can be seen that the highest average product is obtained at comparatively low levels of nitrogen. In the unlikely event that nitrogen was scarce, or if the farmer could only afford a limited quantity, it would obviously pay him to try to obtain the highest possible average product by spreading the fertiliser over a large area rather than use high amounts on only part of the crop. (If he was growing two crops he would split the available fertiliser between the two in a way which would give him the maximum return.) Clearly if he can obtain sufficient nitrogen for higher rates of application to the wheat, more should be applied; although each additional increment after the first point of inflection would, on this evidence, result in a progressively lower average product.

Care is necessary when drawing analogies. For example, if lime was in short supply for applying to very acid land where barley was to be grown, it would be unwise to spread it too thinly because the pH of the whole area might still remain low, and total crop failure could result. This is a different case to the wheat example because the point of maximum average product would not have been reached with the barley. It would be best to achieve this on a limited area rather than none at all. Equal care would have to be taken in a case where winter stock fodder was limited. It might be necessary to spread the fodder over all stock to keep them alive rather than finish a few for slaughter.

Figure 9 shows the classical presentation of the relationship between the total product, the average product, and the marginal product.

It can be seen from this that the average product is at its maximum when its curve is cut by the marginal product curve. This can be likened to a cricket batsman. Whilst he continues to increase his scores in each innings, i.e. his marginal additions to his total score, his average score in all innings increases. However, if each of his scores progressively decline, his overall average falls. Meanwhile his total score for a lifetime keeps on rising. From Fig. 9 it can be seen that the highest total product is at the point where marginal product is zero and cuts the axis.

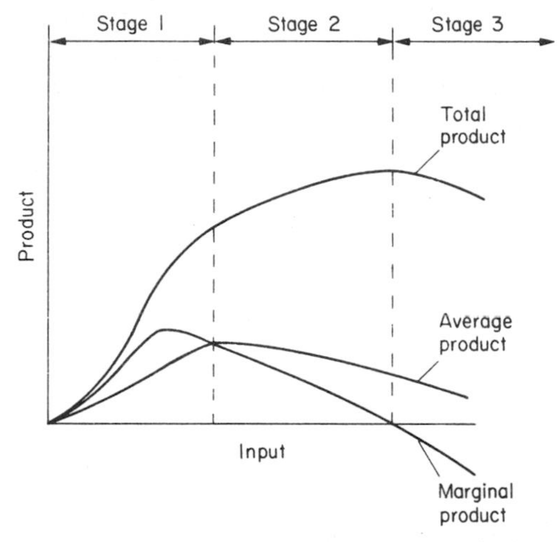

FIG. 9. Total product, average product, and marginal product.

The three stages of production are also shown in Fig. 9. Stage 3 is said to be irrational because extra input results in a reduction in total product. Stage 1 is also irrational because extra units of input will give a more than proportionate increase in output. If the production of this product is profitable at all, it will certainly pay to at least go as far as the point where the average product is at the highest point. Stage 2 shows the rational area for production by a farmer, but not all points in it are necessarily profitable.

3.2.4. Inputs in Relation to Maximum Profit

The farmer has to establish how much of an input factor such as fertiliser to use to produce maximum profit. In other words he requires to know the point in Stage 2 of Fig. 9 at which he should produce.

Examination of Table 1 together with Fig. 8 reveals that the marginal increase in the value of grain produced exceeds the incremental cost of nitrogen associated with it up to an application of somewhere between 185 and 190 kg/ha. The highest total yield, or product, occurs at around the 190 kg level, but at this point the value of the marginal amounts of grain produced does not offset the cost of the marginal amount of fertiliser required to produce it.

This means that maximum profit has occurred before maximum yield has been attained. In this case maximum economic efficiency has been obtained just before maximum technical efficiency.

Examples can be found in agriculture where the point of maximum profit occurs long before maximum yield because marginal cost has equalled marginal revenue considerably before yield has been maximised. In other words the highest yield possible does not always result in the highest profits. The optimum point can be obtained graphically.

Maximum profit occurs where the difference between total revenue and total cost is greatest. In Fig. 10a the total revenue and total cost curves have been drawn on the same graph. A vertical line drawn at the point of maximum distance between the two lines indicates the input which will produce maximum profit. A tangent to the total revenue curve gives the same point (Fig. 10b). In the case of fertiliser application to grain, the slope of the tangent is given by the price ratio of the fertiliser to the grain.

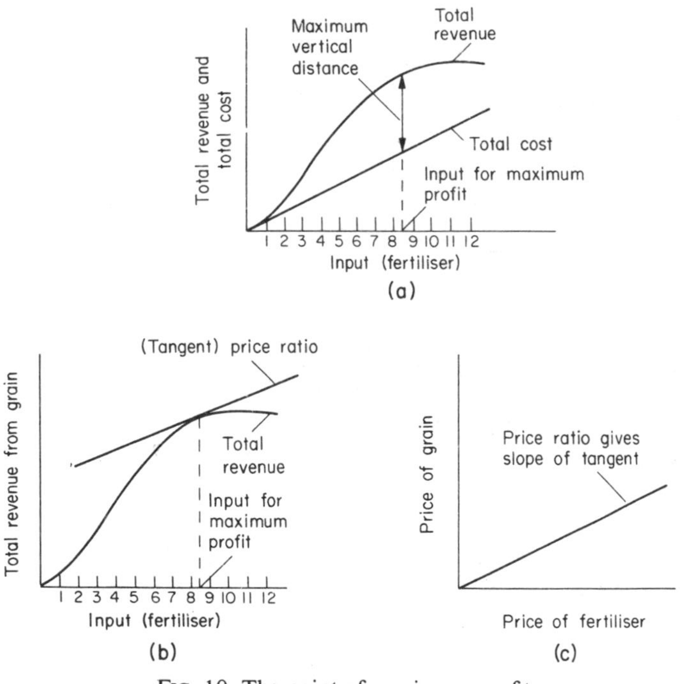

FIG. 10. The point of maximum profit.

Figure 11 emphasises the fact that whilst the point of maximum profit per unit of input is greatest when the average revenue (synonomous with average product if each unit of product is valued the same) is greatest, it pays to continue adding inputs up to the point where the marginal revenue curve cuts the marginal cost line, i.e. where marginal cost equals marginal revenue. Thus the average revenue is greatest at point

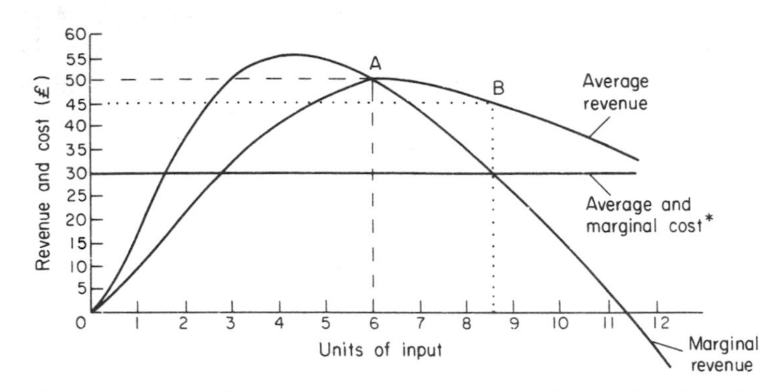

FIG. 11. Profit at point of maximum average return *A* and profit at point of maximum total profit *B*. (*N.B.—Assumes that each extra unit of input costs the same as the previous unit. Discounts for buying large quantities would alter this line.)

A. Here the input is 6 units and the margin between the average revenue and the cost, given on the left hand scale, is £50−£30=£20, i.e. each of the 6 units of input gives a profit of £20, producing a total profit of £120.

The marginal revenue curve cuts the marginal cost line when average revenue is at point *B* with 8.5 units of input. On the left hand scale the average revenue per unit of input is shown to be £45, and each unit of input costs £30. The profit per unit of input is £15, and the total profit 8.5 units×£15=£127.50. If the cost per unit of input increases, say by £2, and the price of the product remains the same, it can be seen from Fig. 11 that the horizontal marginal cost line rises and the point at which the marginal revenue curve cuts this line moves to the left. In other words it no longer pays to use as many units of input as before.

This same point can be illustrated from Fig. 10. If the price ratio of the fertiliser to the grain changes it will alter the slope of the tangent so altering the point where it touches the curve in Fig. 10b. The farmer is entitled to ask if this has definite practical application, and he might be slightly disappointed with the answer. The reason for this is that because of imperfect knowledge he usually cannot predict exactly what the most profitable level of input will be. Frequently many variable factors control production and not just one. In the case of nitrogen application to wheat it is impossible to predict the weather ahead which will influence the crop's response. This does not mean that the relevant economic principles should be totally discounted since they have indicated what to do when a resource is in short supply, and that maximum profits can be obtained nearer to the point where marginal cost equals marginal revenue than when average product is highest.

3.2.5. Constant or Linear Returns

Constant or straight line returns exist when each extra unit of an input factor produces the same increase in product as the previous units of that factor. G. W. Cooke in his book *Fertilising for Maximum Yield*, published by Crosby Lockwood Staples, suggests that responses of grass, beet, cereals, and potatoes to fertiliser nitrogen for individual fields (as opposed to results from surveys of several farms), tend to be of a linear nature up to the transition point (Fig. 12).

The suggestion is that up to this stage nitrogen is the main limiting factor to yield. Additional applications after this point (diagrammatic licence in Fig. 12 might suggest that it is at an identical rate per hectare in all four crops, but this is not so), result in a reduction in response. What happens is that with an increase in nitrogen other factors are altered so that it is not the only limiting factor. For example, competition for light, and possibly carbon dioxide and water, play their part.

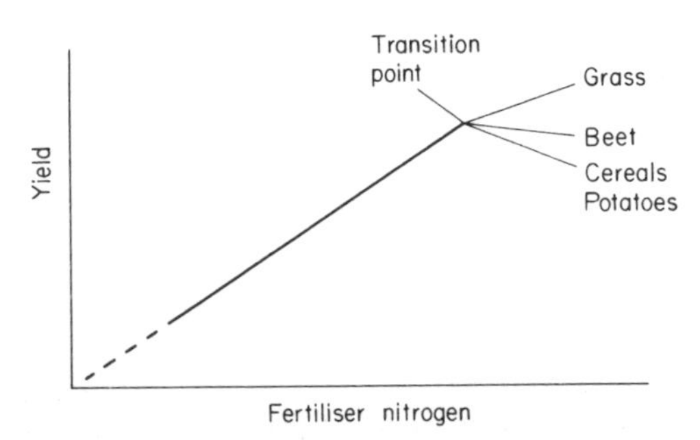

FIG. 12. Response of crops to fertiliser nitrogen (after Cooke).

Taken literally, Cooke's suggestion is that only minimal change takes place in these other factors until the transition point. His suggestion requires explanation. When plotting results from an experiment on a graph the points tend to have some scatter. This is shown in Fig. 13, produced by D. Reid from his work at the Hannah Institute, which depicts the response of grass, in terms of dry matter, to fertiliser nitrogen.

Sometimes it is difficult to know whether to draw a straight line or a curve through the points. Examination of diminishing return curves shows that many have large sections which are almost straight. What Cooke is suggesting is that for all practical purposes the application of nitrogen at rates employed by most farmers is in the region where the curve is virtually a straight line. This indicates very little diminishing response until the point where other factors come into significant play around the transition point. Much work has been done with crops in relation to seed rates and plant populations. This aspect will be left until the specific section on crops.

Stock farmers also wish to know if responses to their input factors are linear or if they conform to the law of diminishing returns. It would ease decisions in dairy cow management if a cow could be expected to produce milk in a straight line response to the feed she eats up to the point where she could eat no more. Unfortunately, it is not as simple as this. Several factors come into play. They include the fact that as feed intake increases body liveweight also tends to rise. Diets of high yielders include more concentrates and less roughage. This can influence faecal energy losses and rumen fermentation.

The net result is that although 1 litre of milk of a given composition requires the same amount of energy to produce it as any subsequent litre of the same composition, in practice, because of the various factors that come into play, progressive increases in units of energy do not result in the same increases in milk yield.

With growing and fattening stock the composition of the gain being made, the liveweight of the animal and the rate of gain all have to be considered. In general, however, at any given liveweight, except possibly for the initial period after birth, there tends to be a diminishing return of liveweight gain to feed intake.

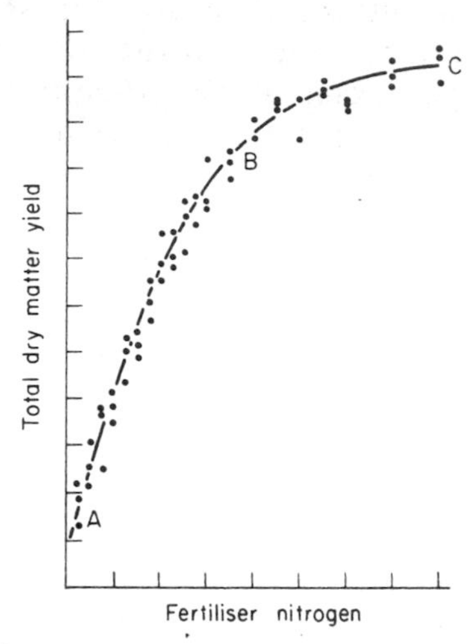

FIG. 13. Scatter of results from trial work. (A straight line joining points adjacent to *A* and *B*, and a second straight line joining points adjacent to *B* and *C* would present the case shown in Fig. 12.)

3.2.6. Substitution of One Input Factor for Another

The above paragraphs raise the whole question of substitution. They have mentioned factor, factor substitution, i.e. the substitution of one input for another. Clearly farmers are always interested in finding the "least cost" or cheapest way to produce a given level of product. Sometimes alternative inputs can be used to produce a given product. The cheapest method of production might be the use of just one input factor and none of another, but in other cases a combination of the two factors may be cheapest.

If the same amount of product can be obtained whilst one input factor X_1 is gradually replaced by another X_2 in such a way that each time the amount of X_2 which replaces X_1 is the same, then the marginal rate of substitution between the factors is constant or linear.

Consider the case of a farmer who mixes his own feed compounds. He decides that the main source of energy will be either barley or maize, and establishes that 0.93 kg of maize supplies the same energy as 1 kg of barley. Call the maize factor X_1 and the barley factor X_2 in Fig. 14a. Assume that the energy required is represented by the product line, and that this energy can be supplied by 100 kg of barley or 93 kg of maize. (The term iso is used to indicate that all points on the product line represent equal amounts of energy.)

From time to time the price of barley relative to that of maize will change. If the price of energy from maize is cheaper per unit than from barley then it pays to use maize.

When the price of 0.93 kg maize equals that of 1 kg barley and a price ratio tangent constructed from this relationship is applied to Fig. 14a it will be seen that it touches the iso-product line at all points. This means that it does not matter if the barley or the maize is used, i.e. any combination of maize or barley would result in the same cost.

In many cases in agriculture the marginal rate of substitution between input factors takes place at a diminishing rather than a constant rate. This is illustrated in Fig. 14b.

If a tangent based on the price ratio between these two inputs (Fig. 14c) was applied to the curve it would touch it at a point which would represent the optimum, or least cost, combination between the two.

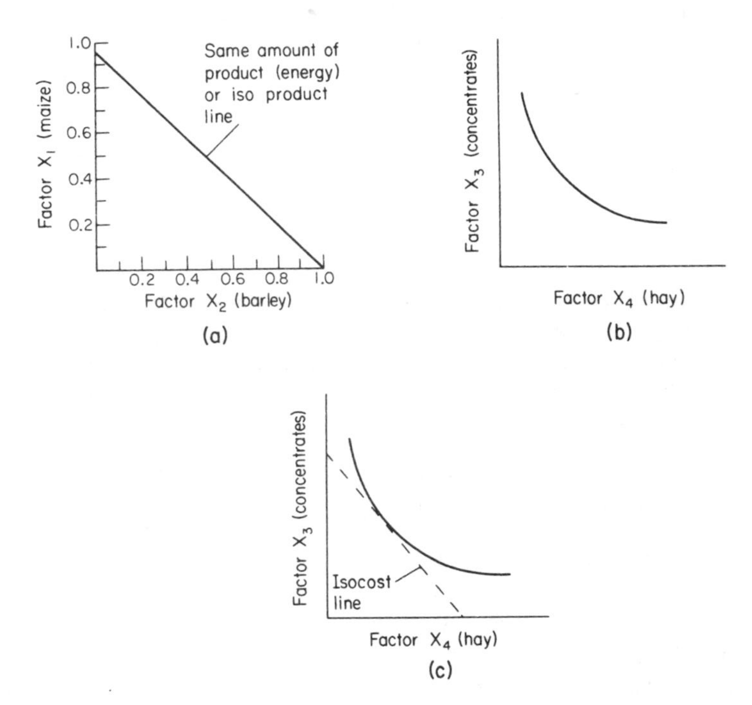

FIG. 14. Factor/factor substitution.

3.2.7. Product Selection

Farmers do not have unlimited supplies of most of the resources necessary for production. They therefore have to take great care in selecting the most appropriate enterprises and also in allocating the resources that are available to them.

In economic circles it is usual to talk about the product rather than the enterprise. The selection of the products which a given business should produce is very much influenced by the relationship which exists between the particular enterprises under consideration. Enterprises, or products as we will now refer to them, can be competitive, supplementary or complementary to each other. In addition they can be antagonistic, or in other cases joint product situations occur.

Competitive circumstances arise where two products compete at the same time for the same input factors, and an increase in one product results in a decrease in the other. Assume that autumn labour is critical on a farm growing potatoes and sugar-beet. If an increase in potatoes means that less sugar-beet can be grown the two products are in competition with each other.

Supplementary situations occur where two products can be produced in such a way that their impact on the requirement of the resources of the farm is independent. Thus a barley beef enterprise on an arable farm might use surplus buildings and labour and not interfere with the arable crops. If the enterprise became too big it might become competitive with the other enterprises.

Complementary situations appear when the production of one product helps the production of another. Well known examples include the use of sugar-beet pulp and tops by stock and the introduction of a break crop on an all cereal farm. If the break crop became very profitable it might be expanded and become competitive with the other enterprises.

Joint product situations occur when one product is dependent upon the production of another, e.g. milk cannot be produced without the birth of a calf, wool cannot be produced without sheep production. In other words they form part of one production process.

Antagonistic problems between products occur when they not only compete for resources, but such things as diseases are common to both. The production of one can add to the problems of another, e.g. *Salmonella* from bought-in calves which are competing for resources with home-reared dairy heifers might add to the troubles of the latter.

3.2.8. Competitive Product Situations

Examination of competitive situations in greater detail reveals that the nature of competition is determined by the respective factor/product relationship for each of the enterprises concerned.

Consider the case of a farmer who has sufficient feed to winter either 100 light suckler beef and sell them as stores in the spring, or to finish 64 heavier stores for slaughter. Buildings are not a constraint. Alternatively he can have a mixture of both light and heavy beasts. For every 10 light animals not kept out of the possible 100, 6.4 heavy stores could be kept as far as is practicable, e.g. he could have 50 light stores and 32 heavy beasts. In other words the two enterprises are competitive with a constant rate of substitution.

Examination of the factor (feed)/product relationship (Fig. 15a) shows that as more feed is devoted to each type of animal the number of beasts which can be kept increases constantly. If the product/product relationship is drawn (Fig. 15b), assuming a given amount of feed, call this x, it shows that the "production possibility line" is linear. The dotted lines in Fig. 15b indicate one of the combinations of heavy stores which can be kept together with a specific number of light sucklers if the total amount of feed available is fixed at x units. If the amount of feed changed the production possibility line would change and so would the number of stock which could be kept.

Assume that the gross margin (i.e. profit before deduction of overhead costs), is £24 per head from the light animals and £40 each from the heavier ones. The highest total gross margin would be derived by keeping 64 heavy beasts (64 × £40 = £2560) and only £2400 would be derived from 100 lighter stores.

This illustrates that in a competitive situation where the products substitute at a constant rate the usual advice is to produce all of the most profitable product and none of the other.

This all seems very obvious. However, frequently the rate at which one product

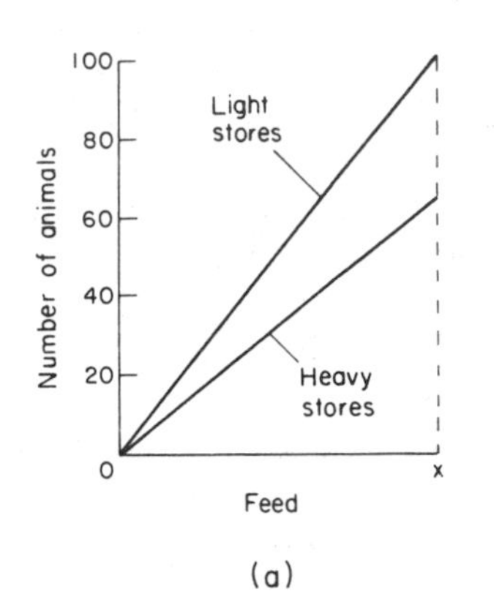

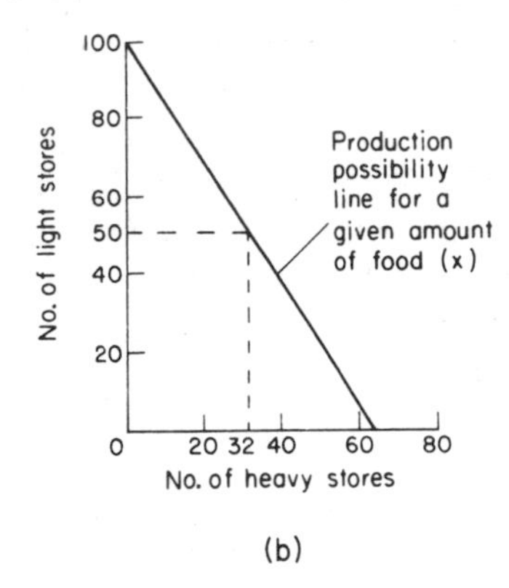

FIG. 15. Factor/product relationship and product/product relationship for a competitive situation with constant rate of substitution.

substitutes for another in a competitive situation is not constant. In some cases the marginal rate of substitution increases, i.e. the more of one product, call this Y_2, that is produced the more of the other, call this Y_1, which has to be given up to produce one unit of Y_2 (Fig. 16a).

The optimum combination of the two products can be obtained by drawing a tangent to the production possibilities curve as shown in Fig. 16c. The slope of this tangent is obtained by establishing the price ratio of product Y_1 to Y_2 and constructing the iso-revenue line shown in Fig. 16b. Thus the optimum combination is 25 units of product Y_1 and 32 units of product Y_2.

In practice it will be found that the substitution between most farm products is more complicated than this. For example, consider a farmer who has barley beef production from bought-in calves and the production of three-month-old calves, also from bought-in stock, as his possible products. It is probable that only part of the curve for the substitution between these two would conform to that shown in Fig. 16c. The construction of the appropriate curve is more the task of the economist, but the lesson to be learnt by the practical farmer is that some effort is necessary to find the optimum combination between products.

3.2.9. Complementary Products

Numerous examples exist of situations where profit is maximised if two enterprises are kept rather than one, because of beneficial interaction between the two. Figure 17 shows a situation for a fixed area of land on which early potatoes are grown each year. Assume that rape is sown after the potatoes are lifted and that it is used to finish store lambs. The fertility created by this activity helps to improve the performance of potatoes in the next year.

This situation remains true in the range *AB* where the relationship between the potatoes and the sheep can be said to be complementary. An increase in sheep numbers beyond point *D* requires special crops of swedes to be grown in order to provide sufficient feed. The sheep therefore become competitive with the potatoes at this point and remain so with increasing numbers of lambs as shown by the curve *BC*.

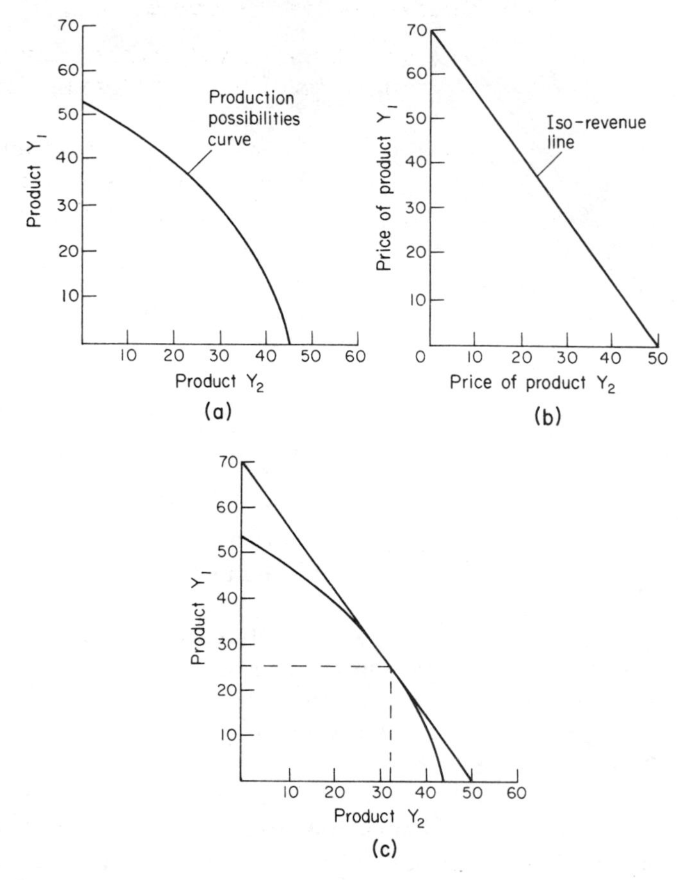

FIG. 16. Optimum combination between two products with an increasing marginal rate of substitution.

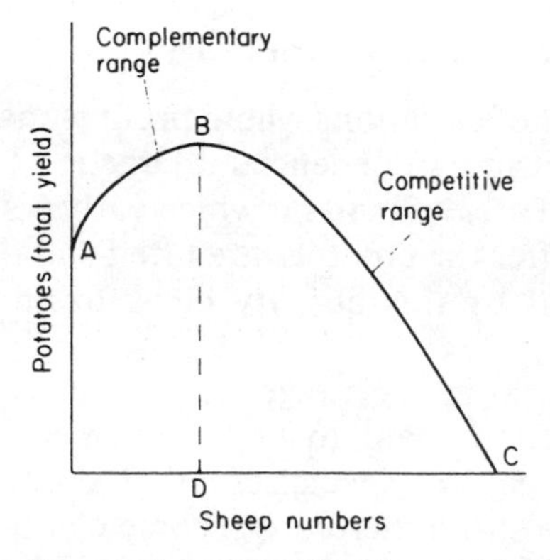

FIG. 17. An example of a complementary situation which becomes competitive.

3.2.10. Supplementary Products

In a situation where the resources available are fixed and one enterprise can be expanded without changing the performance of another, the two products are supplementary.

Consider the case of a farmer who has a fixed amount of spring labour on his farm which produces barley and sheep. Figure 18 shows that the sheep and barley are supplementary to each other in the range *AB* and in the range *CD*, but that they are competitive for the labour between *B* and *C*. This means that if *M* hectares of barley are grown *N* ewes can be kept without any labour problems; of if *P* sheep are kept, *R* hectares of barley can be grown with the available staff. With more ewes in the first case, or more barley in the second case, competition arises.

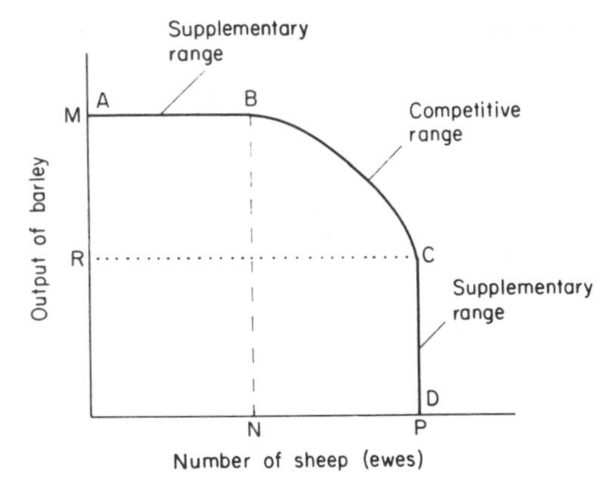

FIG. 18. Example of production possibilities curve for two supplementary products assuming a fixed amount of resource.

3.2.11. Costs of Production

Practical farmers are more familiar with costs of production than the production functions illustrated earlier in this chapter. It is important to understand that by costs of production we mean the cost of producing a certain amount of product in a given time.

(i) Fixed Costs

Costs are of two main forms. There are those which are incurred irrespective of what production, or more specifically what output, is obtained, at least in the short run. For example, rent for land and buildings, regular labour, and depreciation on machinery. These are called fixed costs.

(ii) Variable Costs

Some costs vary with the scale of production, only being incurred if production takes place. They are said to vary directly with the level of output. This point is qualified in Chapter 4. Such costs are known as variable costs and include feed, fertiliser, seed, and sprays.

It is interesting to note that if the resources generated by most fixed-cost items are not used at the time when they are available they are lost for ever, e.g. when labour is under-utilised or buildings are left empty. In the case of most variable costs the resources produced can be stored for future use if not employed at once, e.g. feed. The way in which fixed and variable costs are used in farm management means that there are exceptions so that it is wise not to use this difference in the definition of fixed and variable costs. Fencing costs are treated as fixed, but fencing materials can be stored. The cost of the vets visit is classified as variable, but could not be stored.

(iii) Relating Variable Costs to Output

Three main ways in which variable costs can relate to output are shown in Fig. 19. Figure 19a illustrates a case where each additional unit of output requires the same additional variable cost to produce it. Figure 19b shows another case where each marginal addition to output needs a progressively lower marginal increase in variable cost. Figure 19c depicts the reverse situation where each marginal addition to output requires an increase in the marginal addition to variable costs. Figure 19b is an example of decreasing marginal costs and Fig. 19c an example of increasing marginal costs.

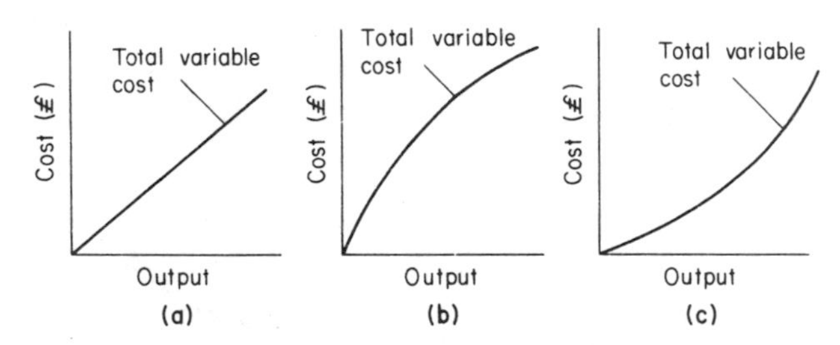

FIG. 19. Relating variable costs to output.

It is possible for both decreasing and increasing marginal costs to apply at different output levels in the same production situation (Fig. 20a). If the fixed costs are added to the variable costs the total costs can be shown in Fig. 20b.

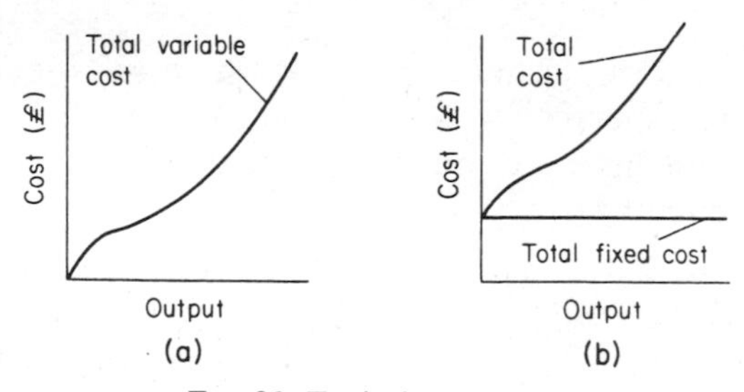

FIG. 20. Typical cost curves.

3.2.12. Equimarginal Returns and Opportunity Cost

Two important principles need highlighting in relation to the points already discussed in this chapter. They are particularly significant in relation to the use of capital.

Most farmers have limited capital and they question where it should best be allocated within a business. The principle of equimarginal returns can provide relevant answers. It says that resources should be employed in such a way that the marginal return to the last unit of the resource invested should be the same from each of the enterprises in which it is employed.

For example, assume that a farmer has money available to invest at the autumn sales in store lambs or store bullocks. He assesses that, up to a certain number, determined by the amount of silage he has available, bullocks would pay better than sheep. Hay would have to be purchased if more than this number of cattle was bought. At this stage store lambs may therefore become more profitable. When the marginal return on the lambs is higher than that on the last bullock bought, lambs should be purchased until the resources run out or the marginal returns on additional bullocks or sheep are equal.

In essence the law of equimarginal returns suggests that funds or resources should be invested where they will make most profit, and that funds should continue to be invested until the marginal returns from each enterprise is equal.

This is very sound in theory in such a simple case, but it is not always practicable. Sometimes the farmer may not want the management problems and the risks attached to investment in the more profitable alternative. Clearly if a farmer has not invested in something, he has foregone the opportunity of obtaining a return from it.

A farmer could put his money into a building society, but if he invests it in his farm instead he has foregone the interest from the society. The interest he would have got from the society is the opportunity cost of investing the money in his farm.

Carrying this argument further, assuming that he invested his money in the bullocks, he has foregone the opportunity of investing in the sheep, and probably a whole range of other things such as fertiliser, machinery, fencing, and so on. He knows that up to a certain number of bullocks the opportunity cost of not investing in the sheep is less than the profit from the bullocks. However, the opportunity cost of not investing the money in more dairy cows, or something else, may be greater.

Opportunity costs can be illustrated diagrammatically. Assume that two products are being compared, Y_1 and Y_2. They could be suckler calves and store cattle for finishing. The total costs of the two could be presented as shown in Fig. 21a. A point where their total costs are equal is found from the left hand scale and the level of output from each product associated with this cost is obtained by drawing vertical lines as shown. This shows a relationship which can be used to construct Fig. 21b, which depicts the iso-cost curve for the two products.

Iso-cost curves can be viewed as opportunity cost curves as shown in Fig. 21c. Essentially these show costs in terms of an alternative product given up. The product "lost" can be viewed as the input and the product "gained" as the output.

Thus in Fig. 21c the opportunity cost of producing *ab* more units of suckler calves is the cost of *cd* units of store cattle foregone.

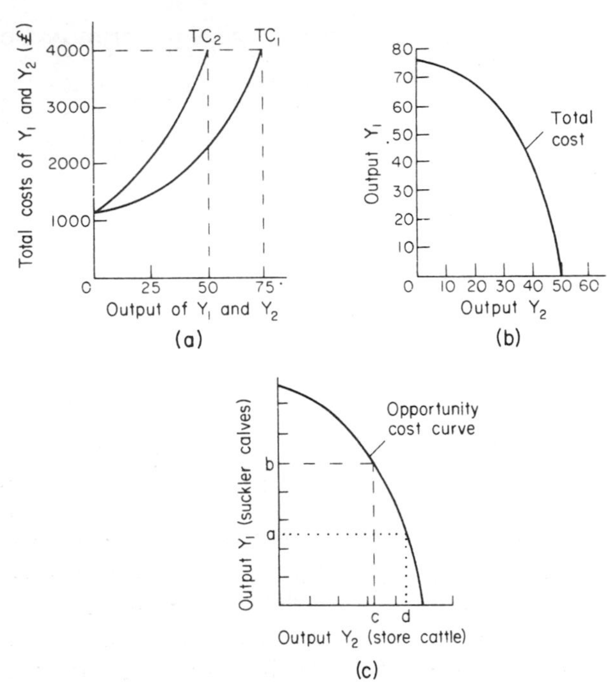

FIG. 21. Opportunity cost.

3.2.13. Profit Maximisation

Profit is maximised when the marginal revenue from an output is equal to its marginal cost. This is illustrated in Fig. 22. Assume that the price, or revenue, received per unit of output, say a pig, is *OP*, and that this price is the same for all pigs sold. Thus *OP* equals the average revenue, and because it is the same for all pigs it equals the marginal revenue. This is represented by the straight line, *P-----AR=MR=P.*

The line cuts the marginal cost curve at A. If the vertical *AY* is drawn it gives the optimum level of output *OY*, which can be translated into numbers of pigs. The average total cost of each of the units of output or pigs, is represented by *OC*. Thus the difference between *OP* and *OC* (i.e. *PC*) times the number of units of output equals the total return at the level of output *OY*.

The average and marginal revenue line would fall with a drop in the price of pigs. If it dropped so far that it cut the points where the marginal cost curve and the average total cost curve cross, the return would just equal the cost of the output.

3.2.14. Loss Minimisation

Although at first sight it may appear to be a negative approach there are times when profits are unattainable and it may be better to minimise losses rather than stop production altogether. Some pig producers, especially those on specialist holdings, will be familiar with the situation. It will not be unknown to certain farmers with broiler chickens or those who rear calves for sale at about 3 months of age.

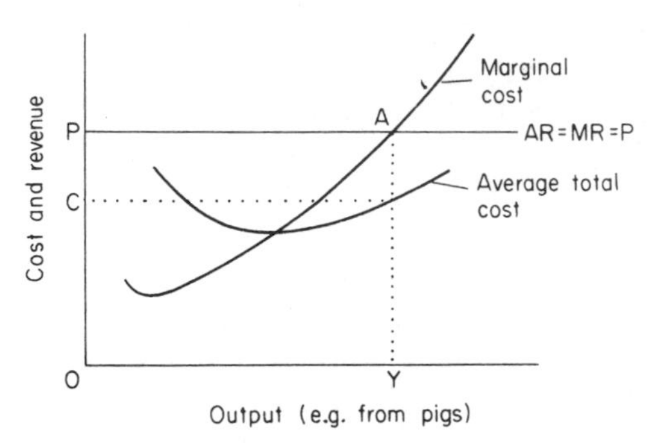

FIG. 22. Profit maximisation.

To understand this idea it is necessary to recall the difference between fixed and variable costs. Short of giving up farming, selling machines, or dismissing men, most of the fixed costs cannot be removed without prejudicing production when enterprises become profitable again. The farmer is thus committed to a certain level of fixed costs.

He therefore has to decide how to employ variable costs such as feed and fertiliser to obtain an output. If the selling price of the particular product is higher than the variable costs required to produce it, then there is some margin to at least cover part of the fixed costs. If, however, the price of the product is less than the variable costs, it is not worth carrying on production.

How long a farmer can withstand this loss depends upon his financial situation, the attitude of his banker, and the prospects of prices increasing. A farmer who has diversified will usually be in a better position than a specialist producer since other enterprises might be able to carry the one currently experiencing difficulty.

The principle of loss minimisation can be explained diagrammatically as shown in Fig. 23. Consider pigs as the example. The average fixed costs can be seen to fall as pig numbers are increased. The average variable costs fall at first but, because of such factors as increased disease, start to increase as more pigs are kept. Together the average variable costs, and the average fixed costs, equal the average total costs.

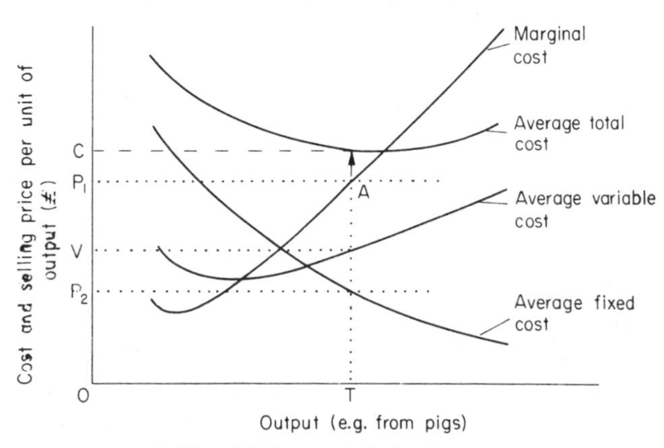

FIG. 23. Loss minimisation.

The marginal cost curve shows the change in cost concerned with an increase in each unit of output.

Assume that the price of each pig, that is the average revenue, is OP_1. A dotted line (which is both the average revenue and the marginal revenue line) has been drawn from P_1 to cut the marginal cost curve at point A. The vertical dotted line was drawn next through the point where these two intersect. This gives the optimum output OT when the price of pigs is OP_1. The vertical line has been projected upwards to establish the average total cost which is OC with OT units of output.

The total revenue for this situation is OP_1 multiplied by OT units of output. This is less than the total cost which is $OC \times OT$. It does however, exceed the total variable costs, which are $OV \times OT$. There is therefore some contribution to cover part of the on-going fixed costs.

Now assume that the price per pig drops to P_2. It can be seen that this price is below the variable costs per unit of output. In this case the farmer should cease production.

CHAPTER 4

FARM BUSINESS RECORDS

CONTENTS

4.1. OBJECTIVES OF RECORDING

4.1.1. Introduction

In practice recording follows after the implementation of a plan, but it is more convenient to cover it at this point since it can serve as a medium by which much of the basic terminology of farm management can be introduced. Recording can involve either financial (monetary) factors or physical factors. The latter include such things as seed, fertiliser and feed use, yields, areas, numbers of pigs reared per sow, or movement of stock to and from the farm.

4.1.2. Legal Requirements

In Britain, as in many other countries, accounts have to be compiled annually for each farm business as a basis for assessment for taxation. Many farmers engage an accountant to prepare the actual accounts. They either record the information required by the accountant themselves or employ some form of secretarial service.

Records have also to be kept for Value Added Tax (VAT) and, if labour is employed, Pay As You Earn (PAYE), tax data, National Insurance statements, and contracts of employment must be maintained. Other information which has to be kept includes livestock movement records and sufficient facts to enable each farmer to complete the forms involved in the compulsory Agricultural and Horticultural Returns.

4.1.3. Value of Records to Management

The correct use of accurate records is an essential part of good management. They can be employed to monitor the performance of a plan once it is put into operation and act as a diagnostic tool to highlight strengths and weaknesses which should be exploited or removed as appropriate. Management can learn from experience assisted by records.

Benefits can be both short and long term. Good records assist cash control, and in the long term can help in obtaining credit or loans. Inputs can be monitored in relation to outputs and, especially where the inputs are frequent, amendments can be made in the short term. For example, when feeding livestock, changes can be made on the basis of the cost and quality of the feed, and performance of the animals, together with the value of their product. This can virtually be a form of day-to-day control. Production techniques can be changed in the long term if such a move appears to be justified.

Records of such items as stock numbers, receipt and disposal of goods, and utilisation of goods on the farm, all help management to control the business.

Suitably adjusted the profit and loss account prepared as a basis for the assessment of tax liability can be employed for management purposes to monitor the performance of the business as a whole. With the addition of other records the relative profitability of the various sectors of the business can be established.

Care has always to be taken to note that all records immediately become historical. Their future repeatability must be considered carefully before making any changes based upon them.

4.2. THE MAIN PHYSICAL RECORDS

4.2.1. Introduction

It is easy to keep too many records and waste time on those which will not be used, perhaps to the detriment of those that are of value. The first step is to ask if a particular record is legally necessary and if not to then assess the cost/benefit of keeping the record in terms of time, effort, and money.

This book of necessity restricts itself to the main examples rather than covering all records that are possible. Those dealing with labour and machinery are left to the appropriate chapters.

A good farm map can be a valuable asset. It is surprising how many of those in use are inaccurate, even though they have Ordnance Survey origins. This is because fences have been removed, areas have been reclaimed, or road works and building have taken place. Many farmers use maps as a basis for field records, some even entering details of fertiliser and seed use on pieces of paper stuck to the fields on the map. Such records are better than none at all. Although they may not be considered very permanent, the author has seen up to 10 years' data recorded on one map in this way. Good maps of farm drains can be very valuable when a drainage problem occurs.

The necessity for permanency and ease of retrieval are two significant criteria in the design of any record-keeping system. Visual display techniques can frequently be employed to great advantage, although some are time consuming to prepare. Computers may be employed more in the future to assist with storage, retrieval, and display.

4.2.2. Agricultural and Horticultural Return

This return enables appropriate government departments to prepare national agricultural statistics. The main return is taken in June. Information is required about total land area, rented or owned, together with details of respective crop areas including grass, the age of the sward, and whether the grass will be mown. Data for hay, straw, and silage stocks on hand have to be given.

Cattle, sheep, pig, and poultry numbers have to be recorded in great detail giving age, sex, and such aspects as whether for beef or dairy, if for breeding, and in certain cases whether pregnant.

Other things on the form include details of the labour force. The farmer has to state if he or his wife works on the farm and provide a break down of the staff employed by age, sex, and whether employed full time or part time.

4.2.3. Livestock Movement Book

In order to ease the problem of tracing stock which have been in contact with a notifiable disease, all movements of stock onto and off a farm must be recorded. Figure 24 shows the main headings of a livestock movement book. Permits may have to be obtained to move stock when serious outbreaks of disease such as foot and mouth disease occur in a region and sometimes may be refused. Licences have to be obtained in Britain to move pigs other than to a slaughterhouse, and a movement permit must be obtained to move sheep after the date for compulsory dipping against sheep scab has been reached.

Date	Type and breed of animal	Ear mark	Sex	Age	From	To
					Farm of departure or market	Auction mart or farm transferred to

FIG. 24. The livestock movement book.

4.2.4. Livestock Reconciliation Statements

Numbers of stock on most farms change frequently because of births, purchases, deaths, or sales. It is essential that good records are kept so that the number of stock at the end of the year can be reconciled with those at the start.

Example

At the start of a financial year a farmer had 160 cattle and 268 sheep. At the end of the year he had 175 cattle and 264 sheep. During the year he purchased 5 cattle and 25 sheep, 99 calves and 165 lambs were born, 3 calves and 12 lambs died, and 86 cattle plus 182 sheep were sold. Prepare a livestock reconciliation statement (Fig. 25).

Frequently farmers do not bother to break down cattle, sheep, or pigs into their respective age categories, but a common exception occurs on hill farms. Here the farmer may categorise his stock into ewes, wedders, hoggs, rams, and lambs, and for each category record numbers at start of year, bought, transferred in, and marked, on

Class of stock	Cattle	Sheep
No. at start Purchased Born	160 5 99	268 25 165
Total A	264	458
No. sold Died No. at end	86 3 175	182 12 264
Total B	264	458

FIG. 25. Livestock reconciliation statement.

	Year	19...					
	Cattle						
Month	Start of month	Births	Purchases	Total	Sales	Deaths	End of month
April	50	4	1	55	2	1	52
May	52	3	2	57	4	1	52

FIG. 26. Monthly stock summary and reconciliation.

the one hand, and numbers sold, transferred out, dead, unaccounted for, and at "end of year", on the other. The item "transferred in" in this case refers to hoggs, or possibly lambs, which have been transferred to the ewe flock after being mated. This data can then be used to calculate lambing percentages.

4.2.5. Recording Livestock Numbers

In order to supply the above information the farmer must undertake basic recording of livestock numbers. The type of record shown in Fig. 26 is recommended and should be completed each month.

In order to be able to allocate forage costs (i.e. for grass, silage, hay, kale, swedes, turnips, rape) to respective classes of stock using one of the techniques shown later in the book it is also necessary to complete the more detailed type of record shown in Fig. 27. The average monthly carry of each class of stock can be calculated if all the twelve monthly numbers for the class of stock are added together and divided by twelve. The farmer then has a much better knowledge of his livestock carry than if a figure is simply calculated from the average of the number on hand on the opening and closing

Year 19 . . .	Cattle										
				Other weaned cattle							
				Beef			Dairy				
Month	Dairy cows and (bulls)	Beef cows and (bulls)	Barley beef	2+ yrs	1–2 yrs	0–1 yr	2+ yrs	1–2 yrs	0–1 yr		
April	( )	( )									
May	( )	( )									
June ⟶											
February	( )	( )									
March	( )	( )									
Total											
Monthly average											

FIG. 27. Livestock carry record.

valuation days of the financial year. In fact some animals such as store bullocks purchased and finished over a short period may not even be on the farm on valuation day.

4.2.6. Feed Recording

Feed is usually the main cost involved in livestock production. Careful control of its use is essential. Feed records can be time consuming and many farmers fight shy of them, sometimes with attendant loss of efficiency and difficulty in allocating feed costs to respective classes of stock. The greater the number of different types of stock, and of feed, the bigger the problem.

A good record of all categories of feed coming onto, and produced on, the farm should be kept. If each one of these is a ready-mixed purchased feed to be fed only to one class of stock, then it appears to be a simple matter to cost each class of stock with the appropriate amount of its feed bought. Reconciliation of the quantity bought with the total which should have been fed, based on rations provided to stockmen, can often show discrepancies.

Recording by stockmen in the barn must be kept to a minimum to reduce errors. Where home mixing is practised the following technique can be tried. A board with large letters placed in the mixing room should show the ingredients for each respective mix, and each mix should be given a number or letter. The stockman should put the mix into the mixer, but before starting the motor he should record on a pad (suspended over the starter button, to remind him) the number of tonnes (and the number of the mix) to be mixed. He should add the date and his signature. Just four simple entries. In the farm office the quantities of each respective ingredient can be quickly calculated.

If each separate mix is put into a separate area of the store, and it can be guaranteed that it will be used for only one class of stock, no further record may be necessary. Frequently such guarantees are not available and some farmers require stockmen to record when any feed is drawn from the store, giving the class of stock, the name of the feed, number of bags, and total weight, again with signature and date.

Soya bean meal							Barl
In		Out			Running total		In
Date	kg	Date	Used by	kg	kg	Date	
1.12.84	In hand				500		
		2.12.84	D. cows	200	300		
		3.12.84	Sheep	150	150		
5.12.84	500				650		

FIG. 28. Office record of barn sheets and supplies.

Month	November							
Type of feed	Barley		Beet pulp		Bran		Total used	
Type of stock	50 kg	£	50 kg	£	50 kg		50 kg	£
Bullocks								
Sheep								
Pigs								
Total								
A. Amount on hand at start								
B. Purchased (invoiced)								
C. Home grain introduced								
Total A+B+C								
D. Amount used								
E. Amount on hand at end								
Total D+E								

Note: (A+B+C)=(D+E).

FIG. 29. Monthly feed allocation and reconciliation.

In the office these slips signed by the men together with delivery notes for feed should be recorded. Fig. 28 shows a possible record. The running-total column acts as a reminder to order supplies.

Once each month a physical check should be made of the actual quantities of feed in the barn. This should be reconciled with office records. The running total of Fig. 28 may be sufficient for some farmers but others will produce a full reconciliation (Fig. 29), introducing costs and allocation to stock. Costs should be entered net of discounts. Summaries can be made at the end of the financial year of all twelve monthly statements.

Discrepancies in feed records do occur. Some of the main reasons are as follows:

(i) feed dispensers or weigh scales are inaccurate;
(ii) loss of weight of grain in store;
(iii) stockmen overfeeding;
(iv) wastage, stealing;
(v) errors made by stockmen in barn records, some not recorded;
(vi) some feed fed to wrong class of stock;
(vii) errors in checking stocks and invoices for deliveries.

A scoop shaped like a pint beer glass is better than one shaped like a soup dish for dispensing concentrates which are not weighed each time. The first type has a smaller top surface area, so reducing variations because of the degree of height to which the concentrates are piled.

4.2.7. Field Records

A good field record book shows patterns of production over several years and may point to decisions which should be taken by management. Each field should preferably be recorded separately. The design should cater for entries of the name and size of the field. If it is being cropped, headings should be included for the name of the crop, variety, quantity of seed used, date sown; fertiliser applications including type, quantity, and dates; sprays including type, quantity, and dates; date and method of harvest; yield; plus a column for unusual and exceptional occurrences. For grass fields headings should include age of grass, date sown if this year, method of establishment, seeds mixture, quantity of seed, fertiliser application including type, quantity, and dates, if cut—type of product and estimated weights with dates.

4.2.8. Fertiliser Records

Fertiliser is a major cost factor in crop production. Many farmers take great care with their records to ensure correct use and accurate allocation of costs to respective

Field:	Date:
Fertiliser type:	Total area:
Total fertiliser:	Rate per ha:

FIG. 30. Staff fertiliser slip.

crops. Staff who are given the task of distributing the fertiliser can be given a slip such as that shown in Fig. 30 for each field they are about to work in. They should then hand the slip back into the farm office after the application together with comments, if any.

In the office it is recommended that an allocation and reconciliation sheet be compiled (Fig. 31).

Month: Year:

Fertiliser	22.11.11		34.5% N				Total	
Use on:	50 kg	£	50 kg	£	50 kg		50 kg	£
Barley								
Grass								
Potatoes								
Total used (A)								
On hand at start								
Purchased								
Total available								
On hand at end								
Total used (B)								
Discrepancy								

FIG. 31. Fertiliser use and reconciliation.

4.2.9. Individual Crop Records

It is important to allocate costs correctly to respective crops. The sheet shown in Fig. 32 is typical. As can be seen from several of these examples, financial data can be incorporated with physical data into the same record. In this case the object is to obtain the variable costs for each crop.

4.2.10. Other Records

Many other physical records can be kept to help management control. The above examples will serve to introduce some of the main types and others will be mentioned under appropriate sections later.

A farmer's diary can be particularly valuable to record appointments and to record major happenings each day irrespective of other records.

4.3. FINANCIAL RECORDS

4.3.1. Office Routine

A good office routine can greatly simplify the problem of keeping records, save time, enable full advantage to be taken of discounts, ensure that all money due is received,

Crop:				Year:					Area:		
Date	Fertiliser			Seed		Sprays			Contract work	Casual labour	Other variable costs
	Type	kg	£	kg	£	Type	l	£	£	£	£
Total											
Per ha											

FIG. 32. Variable crop cost record.

assist the farmer in the prevention of severe cash imbalances, ease planning problems, including those connected with taxation, and simplify the problem of presentation of documents to the accountant. There are a number of alternative procedures, and no one system will suit all farmers. It is important to find a system which works well and to stick to it rigidly.

A procedure similar to that depicted in Fig. 33 can be employed for purchases. The equipment required is three filing trays plus box files or a filing cabinet.

A good system for placing and controlling orders, or obtaining tenders for the supply of goods, and such things as buildings, is essential. Some farmers use order books with carbon duplicates and the latter serve as a record. Sometimes it is necessary to "follow up" or "chase up" orders to make sure that the goods are received on time. The frequent use of a telephone is recommended.

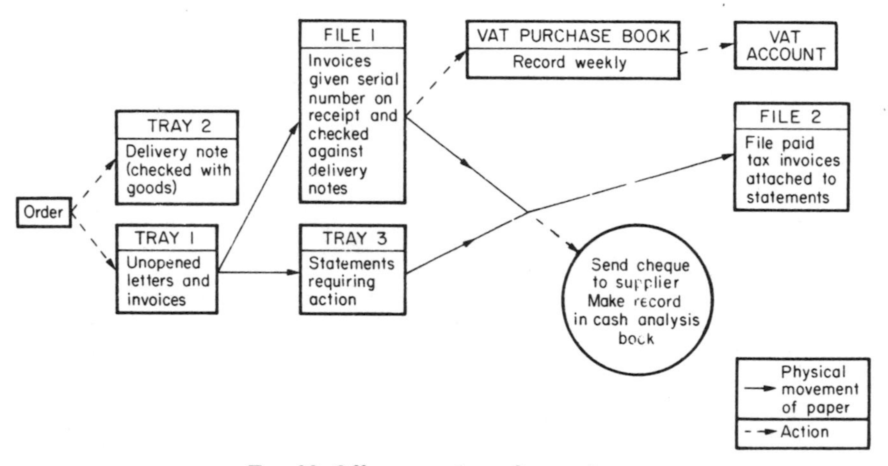

FIG. 33. Office procedures for purchases.

Delivery notes presented when goods arrive should be checked with the items received and put into tray 2 (Fig. 33). Unopened letters should be put into tray 1. Some of these contain invoices sent by suppliers. Invoices are the most important document in a purchasing routine and give full details of goods, prices, and VAT. They should be given a serial number by the farmer and checked against the appropriate delivery note. Once each week the VAT record should be completed. Until payment is made each invoice should be put into file 1.

Suppliers usually send a statement once each month. This states the number of the invoices to which it refers. Statements requiring action should be put into tray 3. A few firms issue joint invoice/statements although these are probably in the minority.

Within reason the farmer should select the time which is best to him before paying. There may be a discount for early payment. If this is 5% for payment before the end of the month it effectively represents 60% over the year. No farmer should be over anxious to pay, especially if he has a bank overdraft, but too long a delay will produce pressure from the supplier whose ultimate sanction is not to supply him again. It is not advisable to upset too many suppliers.

When the supplier is paid an entry should be made in the cash analysis book with the help of details on the invoices and those recorded on the cheque book stub. If payment is made by post a remittance advice note (Fig. 34) should be sent with the cheque to state who it is from and what it is for. A duplicate should be kept.

When payment is made to suppliers from whom several purchases have been made it is essential to pay an exact number of invoices rather than a round sum "on account", since it is necessary to know precisely what goods have been paid for and which bills have still to be paid.

Remittance advice

Date .... Our reference No. ....

From ....

....

....

Date	Reference	Amount
	Total	
	Less discount	
	Enclosed cheque No. .... for in settlement	

FIG. 34. Remittance Advice.

Invoices and statements should then be attached to each other and filed more permanently in file 2 in an order based, as far as possible, on the serial number given to each invoice. Since some statements contain items from several invoices, all of which may not have been received in sequence, a slight error may appear in the order. Sometimes a file for each major supplier may be started and the statement/invoices filed in this way.

A similar procedure can be operated for sales (Fig. 35). The diagram is basically self-explanatory, but one or two points must be made. Most goods sold by farmers have the price fixed after they have left the farm, e.g. stock sold through an auction or even fat cattle sold deadweight. The farmer does not therefore issue the purchaser with an invoice in these cases, but instead self-invoicing takes place. This means that the purchaser makes out the invoice. In the case of auctions, the auctioneer's clerks do the work.

In a few cases, such as when one farmer does contract work for another or sells him something, it will be necessary for the vendor to issue a tax invoice. It is always advisable for a farmer to make a note of contract work, and of sales on the day that the goods go off the farm, to ensure that ultimately they are all paid for. Whilst this is not so essential with organisations such as the Milk Marketing Boards it is particularly important when sales are made to individuals.

In some cases cheques for payment are not sent to the farmer. The Milk Boards arrange for direct credit of the farmer's bank account and then send him a statement. Cheques received by the farmer should be paid into the bank as quickly as possible.

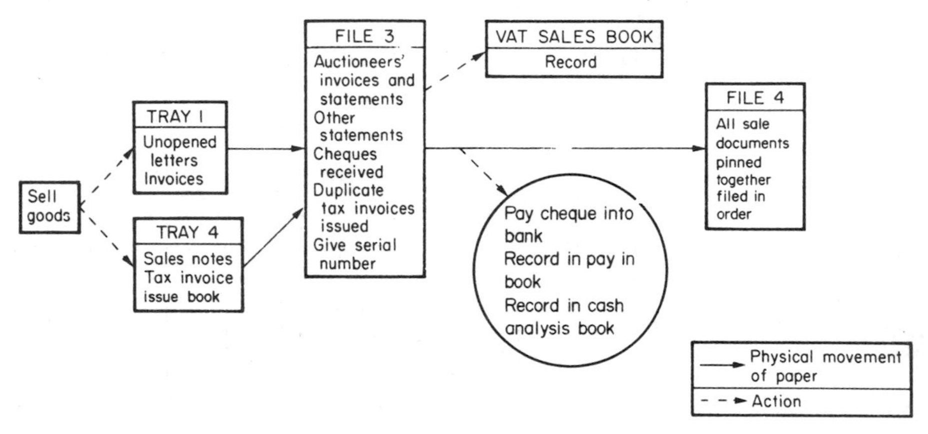

FIG. 35. Office procedures for sales.

Credit notes, usually printed in red, may be received when goods have been returned to a supplier or if a firm has been overpaid. The amounts on these should be entered in red on the cash analysis payments sheet, but when the analysis columns are totalled items in red should be deducted rather than added. Alternatively, if a similar item is subsequently purchased from the same firm the credit note can be attached to the invoice for that item and the difference between the two prices entered in the analysis sheet.

4.3.2. Cash Analysis

If a farmer is to establish the costs and outputs from the various sectors of his business the whole process will be so much simpler if at the outset allocation, or analysis, of items is made under appropriate headings. Two sets of sheets are required, one for payments with its analysis and the other for receipts with its analysis. Some people insist that only transactions which have taken place through a bank should be recorded since the bank statement is evidence of the passage of money. They suggest that if possible all payments should be made by cheque and that any cash receipts should be banked. Where cash has to be handled the use of a separate petty cash book is recommended. The basic concept of this advice is sound and should be followed. Alternatively, the Scottish Farm Business Records Book (SFRB) shows how cash transactions can be entered on analysis sheets.

					Total amount paid			Crop expenses					Livestock	
Date	Name and details	No. or quantity	Voucher No.	Contra items	By cash	By cash	VAT	Fertiliser and lime	Seed	Contract work	Casual labour	Other crop expenses	Feed bought	Cattle
				£	£	£	£	£	£	£	£	£	£	£
About 30 lines are drawn here so that entries can be made														
		Totals												

Continued from above

Expenses				General expenses						Other items		
Sheep	Vet med.	Milk and dairy	Other stock expenses	Regular labour	Machinery repairs, etc.	Fuel and electricity	Rent, rates building insurance	Repairs to property	Other expenses	Capital expenses	Private expenses	Cash drawn from bank Cash paid into bank
£	£	£	£	£	£	£	£	£	£	£	£	£

Note: In practice this is spread over two pages side by side. Space shortage prevents all headings being given here.

FIG. 36. Payment analysis headings (Scottish Farm Business Records Book).

Clearly someone unscrupulous enough to try to evade taxation on cash receipts may not wish to bank, or indeed record them.

A typical payments analysis sheet from the SFRB is shown in Fig. 36. The entry in the detail column should be brief since full details are shown on the invoice. Usually the name of the supplier and a brief description of the item is recorded. When cash is drawn from the bank the entry "cash" would be more appropriate, or for the costs of banking—"bank charges". The number/quantity column refers to the number or quantity of the items purchased. The voucher number should correspond to the serial number on the invoice or other voucher detailing the payment, but it is advisable to precede the number with the letter P for invoices which have been paid. The record should as far as possible follow the order of the voucher numbers. Farmers who record only transactions done through the bank may record the cheque number rather than the voucher number. If the item is a direct debit or a standing order, the letters DD or SO should be used.

The total amount paid should be entered under the "By cash" or "By cheque" heading appropriately, and then again under the correct analysis heading. Care should be taken to ensure that the figures are clear and that units, tens, and hundreds are under each other to facilitate correct addition of columns.

In the SFRB the analysis headings are printed, but in many other analysis sheets they are blank. In the latter case great care should be taken in choosing headings to ensure that the most useful are selected from the point of view of cost allocation. To facilitate more detailed analysis when columns are limited, the use of a code is sometimes advocated; e.g. for fertiliser it has been suggested that the letter G could be put against grass fertiliser and B against barley fertiliser. This is not very practical because all the fertiliser may be covered in one line for both grass and for barley, and in any case the amount purchased may not be the amount actually used on the crop. A separate record is therefore essential similar to those already shown in Sections 4.2.8. and 4.2.9.

Certain codes, however, can be useful. For example under "Other general expenses" T could be used for telephone, I for interest, and P for post.

The VAT column is included to enable the amount recorded as being paid to reconcile with the analysis of the payments. The entry as shown is not entirely suitable for completion of VAT records and it is preferable to keep a separate book for them.

Figure 37 shows the receipts analysis headings in the SFRB. Usually there are few problems with the number of columns, and a fairly detailed analysis can take place under respective crops and livestock. Voucher numbers should be prefixed with an R for receipts.

4.3.3. Contra Accounts

A typical example of a contra transaction is for a farmer to obtain £600 worth of dairy cow concentrates from a merchant and at the same time he sells the firm £360 worth of barley. The merchant then demands payment of £240 and a cheque for this amount is sent. Another example is for the farmer to buy £250 worth of spring wheat seed and sell the same firm £1000 worth of last year's wheat. In this case the farmer would receive a cheque for £750.

It is clear that such procedures require careful recording if the full costs and returns

Date	Name and details	No. or quantity	Voucher No.	Contra items	Total amount received		VAT Charged or repaid	Crops			Livestock		
					Kept in cash	Paid to bank		Wheat	Barley	Potatoes	Cattle	Milk	Sheep wool
				£	£	£	£	£	£	£	£	£	£

About 30 lines are drawn here across the page so that entries can be made

		Totals											

Continued from above

Miscellaneous	Capital grants	Machinery equipment vehicles sold	Private receipts	Cash drawn from bank
				Cash paid into bank
£	£	£	£	£

Note: In practice this is spread over two pages side by side. Shortage of space prevents all headings being shown.

FIG. 37. Receipts analysis headings (Scottish Farm Business Record Book).

Payments analysis

Date	Name and details	No. or quantity	Voucher No.	Contra items	Total amount paid		Feed	Seed
					By cash	By cheque		
3 Nov.	J. Wall dairy cubes @ £120/t *Less* credit 4 t barley @ £90/t	5	P 102 P 102	£360		£240	£600	
5 Feb.	Smith's grain: wheat seed	1	R 72	£250				£250

Receipts analysis

Date	Name and details	No. or quantity	Voucher No.	Contra items	Total amount received		Wheat	Barley
					Kept in cash	Paid into bank		
3 Nov.	J. Wall credit for barley @ £90/t	4	P 102	£360				£360
5 Feb.	Smith's grain: wheat @ £100/t *Less* seed wheat @ £250	10	R 72 R 72	£250		£750	£1000	

FIG. 38. Entry of contra account items in the cash analysis sheets.

Payments analysis

Date	Name and details	No. or quantity	Voucher No.	Contra items	Total amount paid		Labour	Private expenses	Cash drawn from bank
					By cash	By cheque			Cash paid into bank
3 Dec.	Self cash drawn					£400			£400
5 Dec.	Wages				£250		£250		
6 Dec.	Wife—housekeeping				£150			£150	
7 Dec.	Cash for calves J. Brown				£320				£320

Receipts analysis

Date	Name and details	No. or quantity	Voucher No.	Contra items	Total amount received		Cattle	Potatoes	Cash drawn from bank
					Kept in cash	Paid into bank			Cash paid into bank
3 Dec.	Self cash drawn				£400				£400
7 Dec.	Cash for calves J. Brown		R 68		£40	£320	£360		£320
8 Dec.	Cash for potatoes		R 69		£10			£10	

FIG. 39. Cash transactions.

are to be established at a later date by the farmer and the accountant. Figure 38 illustrates how this could be done. The figure entered in the amount columns must be the actual net amount paid or received, and that under the "Analysis" column must be the full value of the goods bought or sold. The contra entry is effectively the figure which represents the value of the items sold, if this is less than the value of the items bought; or the value of the items purchased, if this is less than the value of those sold.

4.3.4. Cash Transactions

If cash is drawn from the bank a farmer following the SFRB form of analysis should follow the procedure shown in Fig. 39. The amount on the cheque used to draw cash should be entered in the analysis column headed "Cash drawn from bank or Cash paid into Bank" on both the payments and the receipts analysis sheets. As the cash received is spent, the amount should be entered on the payments sheet under the "Total amount paid by cash" column and under the appropriate analysis heading (see 3 Dec., Fig. 39).

When cash is received a voucher number should be entered on the invoice for the transaction and this, plus the amount received, should be entered on the receipts analysis sheet. If it is paid into the bank it should be recorded as shown in Fig. 39 for the receipt from the calves, and the same amount recorded in the payments analysis sheet. If the cash is not banked it should be treated as shown by the receipt for the potatoes.

4.3.5. Alternative Cash System

Not all farmers follow the SFRB technique. The alternative is to practise a petty cash system. When a cheque is cashed to provide petty cash the amount is recorded on the payments analysis sheet as shown in Fig. 40.

Payments analysis

Date	Details	Contra	Amount	Petty cash	Labour	Private
3 Dec.	Petty cash, wages, housekeeping		£460	£60	£250	£150

FIG. 40. Cashing cheques for petty cash.

Notice that compared to the SFRB sheet the "Amount" column heading is different; there is no reference to the bank *per se*, a column for petty cash is introduced and the one cheque may be used to draw cash for wages, private expenses, as well as cash. A separate petty cash book is kept to record miscellaneous transactions as shown in Fig. 41. More column headings can be added and the bottom line can be used to obtain page totals and for checking accuracy. The amount under receipts less the amount under expenses should equal the balance in hand. When a new page is started the balance in hand is entered as "brought forward" and put in the "Amount" column under receipts, and in the "Balance in hand" column.

4.3.6. Discounts

The best procedure for dealing with these is to reduce each item covered by a statement by the percentage discount given. The reduced amount is entered in the "Total amount paid" column and the figures put into the "Analysis" columns reflect the same discount.

4.3.7. Cross-checking "Analysis" Columns

The figures in each column on any one page of the analysis sheets should be totalled. These figures should be carried forward and entered in the same columns on the next sheet used, provided it is for the same financial year. The totals for all the "Analysis" columns on the payments sheet should equal the total in the "Total amount paid" columns plus the total of the "Contra" column. On the receipt sheet the totals should equal the total of the "Total amount received" column. It can be seen that each of the totals reflects the cumulative position since the financial year started.

			Expenses			Receipts			
Date	Detail	Cheque or voucher No.	Amount (£)	Stamps (£)	Petrol (£)	Amount (£)	Eggs (£)	Potatoes (£)	Balance in hand (£)
3 Dec.	From bank					60			60
7 Dec.	Eggs—at gate					5	5		65
8 Dec	Stamps		8	8					57
9 Dec.	Petrol		20		20				37
10 Dec.	Potatoes—at gate					6		6	43
Check totals			£28	£8	£20	£71	£5	£6	£43

FIG. 41. Petty cash book.

4.3.8. Outstanding Accounts

If a farmer is owed money by a person or merchant, that person or merchant is a debtor as far as the farmer is concerned. If the farmer owes money to someone, that someone is his creditor. At the end of a financial year a farm business usually has both debtors and creditors. Lists should be made of each category and subsequently great care must be taken to ensure that the correct costs and returns are debited or credited to the year in which the goods were bought or sold.

At the financial year end the columns on the analysis sheets should be ruled off and the figures in each column totalled.

When the outstanding accounts are eventually settled, their details should be entered on the analysis sheets for the new financial year but they should be identified

Payments analysis

	Total amount paid (£)	Fertilisers (£)	Seeds (£)	Feed (£)	Wages (£)	Rent (£)
Note: Creditors at end of previous year (PY)		300	—	1,000	—	—
Financial year in question	⟩	⟩	⟩	⟩	⟩	⟩
Totals of columns for year	57,500	7,000	500	36,000	12,000	2,000
Deduct creditors at start of year		300	—	1,000	—	—
		6,700	500	35,000	12,000	2,000
Add creditors at end of this year		600	150	1,200	—	—
		7,300	650	36,200	12,000	2,000

Receipts analysis

	Total amount received (£)	Wheat (£)	Cattle (£)	Milk (£)	Sheep (£)
Note: Debtors at end of previous year (PY)		400	—	4,200	100
Financial year in question	⟫	⟫	⟫	⟫	⟫
Totals of columns for year	70,000	3,500	4,000	61,000	1,500
Deduct creditors at end of previous year, i.e. start of this year		400	—	4,200	100
		3,100	4,000	56,800	1,400
Add debtors at end of this year		350	360	7,000	—
		3,450	4,360	63,800	1,400

⟩ Represents all the lines of figures recorded during the year for items purchased and paid for.

⟫ Represents all the lines of figures for all items sold for which money was received.

FIG. 42. Treatment of outstanding accounts.

by the code PY (previous year). The total of the PY items for both receipts and payments should be obtained. Figure 42 (which has the number of headings reduced to help illustration) shows how the PY items should be treated to ensure allocation to the financial year in which the goods were actually bought or sold. For example, consider the payments analysis sheet. At the end of the previous year the business owed £300 for fertilisers and £1000 for feed which had already been received. These were actually paid for in the current year in question. Since the farmer is only interested in the data for the year in which the goods were purchased the £300 and the £1000 should be deducted from the appropriate analysis columns. At the end of the financial year currently in question the business owed £600 for fertiliser, £150 for seeds, and £1200 for feed. These are actually paid for in the next financial year but must be extracted from next year's sheets and charged against the current financial year if a true picture is to be obtained.

This procedure of entering the items in one year, extracting them, and entering them in another may seem unnecessarily complicated. It is necessary for two reasons. First it is essential that the money entered in the "Total amount paid" or "Total amount received" columns is a true record of when the financial transactions do actually take place in order to facilitate an accurate reconciliation with the bank balance (Note from Fig. 42 that when the year end adjustments were made to the analysis columns no amendment was made to the "Total amount paid" or "Total amount received" columns.) The second reason has already been stated, that of ensuring that the correct data is obtained for the year in which the goods were actually bought or sold. Hence the necessity to transfer information from one year to the other.

The totals for the "Analysis" columns suitably amended for debtors and creditors at the beginning and end of the year can then be employed in the construction of the profit and loss account.

4.4. BANKING AND RELATED RECORDS

4.4.1. The Current Account

Farmers use a current bank account for most financial transactions. Some may have one account for the farm and another for their own private finances. The current account enables them to sign cheques to pay bills and provides a convenient method of keeping money received.

The person or firm to whom a cheque is made out is the "payee". If a cheque is not crossed the payee can sign it on the back and take it to the bank of issue and obtain cash for it. A cheque crossed with two parallel lines, usually with "& Co." in between, has to be paid into a bank. If the words "Not negotiable" or "A/C payee only" are written on it, then the cheque can only be paid into the payee's account. In the interests of security the latter procedure is frequently recommended.

If a farmer has paid for an item by cheque and he is subsequently dissatisfied with the item it is possible in certain circumstances to stop payment by contacting his own bank and so prevent the payee receiving the money. There is very little time, however, since it must be done before the cheque has been paid in and cleared.

Bankers now issue cheque cards to clients. Each provides evidence of the particular client's signature and carries a number. If this number is put on the back of a cheque the client's bank guarantees the payee the amount on the cheque provided that this does not exceed a sum printed on the front of the card. However, by putting the card number on the cheque the client foregoes the right to cancel or stop it.

4.4.2. Standing Orders, Direct Debits

If a farmer regularly pays a sum of money to a company or body, say monthly or quarterly, he can save time by signing a banker's order, sometimes called a standing order. This instructs the bank to automatically make these payments for him until such time as he specifies.

Alternatively he may sign a direct debit form. This instructs the bank to pay the amount currently being demanded by the particular organisation. One example is for annual membership of the Automobile Association, which may increase from year to year.

4.4.3. Credit Transfers

The farmer may receive some money directly into his account, without receiving a cheque, by bank giro credit transfer. An example is the payment for milk by the Milk Marketing Boards. The farmer is informed of the amount transferred by his particular Board, and his account is automatically credited with this amount.

The farmer can make payments by credit transfer himself. He can go to a bank, fill in a form giving the payee's name, the name of the latter's bank and account number, the amount to be paid, sign the form, and pay over the amount. The payee's bank will automatically be credited.

4.4.4. Paying Money into a Bank

For many items the farmer receives cheques as payment. When he takes these to the bank to put into his account he fills in a paying-in slip and retains a copy of the basic details on the counterfoil attached to this slip. He is advised to request a book of paying-in slips rather than use individual forms because the counterfoils remain stapled together in order of "paying-in" and form a useful record.

4.4.5. Bills of Exchange

Some farmers use bills of exchange to pay for goods such as feed and fertiliser. These are orders which fix a future date of payment for a specific amount. For example, most fertiliser firms give discounts for out of season sales and deliveries because it suits their business arrangements. A farmer may wish to take advantage of this discount but may not want to pay for the goods immediately. He therefore asks the firm to take a bill of exchange which legally binds him to pay for the goods at a specific date in the future (Fig. 43). In practice the firm produces the bill of exchange, gets the farmer to sign it, and presents it for payment to the farmer's bank on the due date. The farmer is notified by the firm shortly before this is done to remind him of his commitment.

The term "bearer" on the bill is necessary because bills can be sold. For example, if the firm subsequently found that it required the money before the due date it might

Bill No.	Date ....
Accepted payable at: Name and address of farmer's bank: .... Signed (by farmer)	Pay Richardson's Fertiliser Company or bearer on .... the sum of £ .... (£) for value received to Name of farmer .... Address .... Signed .... Credit Controller Richardson's Fertiliser Company

FIG. 43. A bill of exchange.

sell the bill to a broker who would pay them less than its full value and make his profit by redeeming it in full on the due date.

4.4.6. Overdrafts and Loans

A farmer may negotiate overdraft facilities with his bank if he thinks he will not have sufficient funds in his account to cover his expenditure. He will have to pay interest on any money borrowed when he goes into "the red". In Britain this is usually related to the Bank of England base rate, a figure which varies with the current cost of borrowing money, and is used to control money supplies. Normally the overdraft rate is $2\frac{1}{2}$–$4\frac{1}{2}$% above the base rate. If a farmer has excess funds in his current account, which does not pay him interest, he may transfer some to a deposit account which pays taxable interest at about 2–4% below base rate.

When a farmer requires a more permanent source of extra finance for a purchase he may negotiate a bank loan rather than an overdraft.

4.4.7. The Bank Statement

Every month by agreement with the bank the farmer should receive a bank statement for his current account. This itemises each transaction which took place during the month including the date, and either a very brief description of the transaction or the number of the cheque which the farmer signed. In the later case he can check back to the cheque book stub or other record of payment. In addition the amount to be credited or debited for each transaction is stated and the current balance of the account shown.

4.4.8. Bank Reconciliation

It is advisable to reconcile the cash analysis book with the bank statement. To do this each entry on the bank statement should be checked in turn with the relevant entry in the cash analysis book. Both should be ticked if they agree. There may be certain items on the statement which are not recorded on the analysis sheets. These include standing orders, direct debits, and bank interest charges, all of which should now be entered on the analysis sheets. Frequently a sum recorded on the credit side of the bank statement does not correspond to an entry in the receipts analysis sheet. A check on the paying-in book may reveal that several cheques were paid in together and the bank has simply recorded the total.

Ideally by this stage each entry on both the bank statement and the analysis sheets will have been cross-referenced and ticked. In practice there may still be discrepancies: The reasons for this include:

(a) payments which have been made to the bank by credit transfer and which by this time have not been entered on the farm's analysis sheets;
(b) cheques received by the farmer, entered on the analysis sheets, but not yet banked, or if banked have not been credited to the account;
(c) cheques paid by the farmer, entered on the farm's payment analysis sheet, but not yet presented to the bank by the payee;
(d) errors on the analysis sheets such as incorrect addition and mistakes in entry.

In the case of (a) the entry should be made on the analysis sheet; and for both (b) and (c) the bank statement should be corrected. Mistakes in entry should be rectified.

	£	£
Farm cash analysis		
Balance	+950	
Deduct bank interest		−370
	+950	−370
Revised balance	+580	
Bank statement		
Balance on statement		−170
Add cheque not cleared	+800	
Deduct cheque not presented		−50
	+800	−220
Revised balance	+£580	

FIG. 44. Bank reconciliation.

The bank reconciliation statement can now be prepared. For example, a farmer calculates from his records that his bank balance should be in his favour by £950 but his bank statement shows that he is overdrawn by £170.

Examination shows that a cheque for £800, which he entered into the analysis sheet, was paid into the bank after the bank statement was completed. The bank statement showed interest charges of £370 which he had not entered on his payments sheet. A cheque for £50, which he signed to buy some equipment and entered on the payments sheet, had not been presented by the payee for payment. The reconciliation statement for his case is shown in Fig. 44.

The £580 is the figure which will be shown in the closing balance sheet.

4.5. VALUE ADDED TAX

4.5.1. Background to VAT

Value added tax (VAT) is a tax on sales of certain goods and services and is used throughout EEC countries. All businesses whose annual turnover is above a certain figure have to be registered with the Customs and Excise and are given a registration number to quote on transactions.

The tax is levied each time goods change hands or services are supplied in such a way that in most cases the ultimate customer eventually pays the tax and not the intermediaries. When a business buys goods it is charged VAT, effectively as an input tax, and when it sells them it has to charge its customers VAT, an output tax. Each business is obliged to produce monthly or 3-monthly VAT returns. From these it is possible to obtain the total of the VAT inputs and the total of the VAT outputs. The business then either pays, if the output total is the greater, or receives, if the input total is larger, the difference between the two totals to, or from, the Customs and Excise.

The VAT is charged as a percentage of the value of the goods or service. Goods and services are listed in categories. Some are classified exempt from tax, others are given a zero rating, and yet others are taxed at the standard rate or percentage. Petrol and certain luxury goods are given a higher rate and cars put into a special category. Items like vehicle licences and grants or subsidies are outwith VAT.

The following is a selection of items associated with agriculture according to category:

Exempt

Land and building sales	Wages
Letting of land, rent	Discounts
Bank interest and charges	Insurances

Zero Rated

Sales and purchases of most classes of stock and most classes of their produce

Grain, hay, potato sales	Grass let, hogg wintering
Animal feed, seeds	Electricity, tractor diesel

Standard Rated

Wool sales	Fertiliser
Contract work	Sprays
Vet services	Haulage
Farm implements	Telephone
Property repairs	Implement repairs

Suppliers of goods which are exempt do not have to charge VAT as an output tax but they cannot claim VAT refunds on their purchases. Those selling zero-rated goods also do not charge VAT on these sales but can claim input tax paid on goods they buy. This means that because farmers sell zero-rated produce, but buy many standard-rate items, they usually receive VAT refunds from the Customs and Excise.

Since this tax system was introduced there have been modifications especially to the rates of VAT. It is one of the means by which governments attempt to control the direction of the economy of the country.

4.5.2. VAT 100

VAT records have to be kept by registered producers and the form VAT 100 has to be completed and sent to the Customs and Excise within a month of the end of a given tax period. These periods are quarterly but a farmer who expects to have refunds regularly can apply to be assessed monthly and so obtain the money back quicker.

4.5.3. VAT Records

To facilitate records the invoices for goods or services supplied by any person registered for VAT must show that person's VAT number, the date, both the supplier's and the customer's name and address, a description and number of items supplied, amount payable excluding VAT, discount, and the VAT charged.

A farmer is best advised to keep a separate purchases and sales book for VAT. This will simplify the completion of VAT 100, speed up the reclaim for tax if the farmer delays paying his bills, and avoid errors when he rounds off the sum he pays to a supplier.

An example of a purchases record is shown in Fig. 45a, and an example of a sales record is shown in Fig. 45b.

Date 19 . . .	Supplier and particulars	Voucher No.	Invoice total (£)	Deductible input tax	Total taxable inputs	Other items
4 Sept.	J. Cod, fish meal	P44	200	—	200	—
6	A. Bolt, fencing material	P45	345	45	300	—
20	P. Plant, fertiliser	P46	3000	391	2609	—
26	A. Butcher, household	P47	50	—	—	50
			3595	436	3109	50

FIG. 45a. Purchases record for VAT.

Date 19 . . .	Customer/item	Voucher No.	Sales invoice Total	VAT output tax	Outputs (excluding private)	Private, etc.
6 Sept.	Ring, 1 cow	R7	350		350	
8	Wool Board	R8	805	105	700	
12	A. Mann, barley	R9	900		900	
20	VAT repaid	R10	200			200
			2255	105	1950	200

FIG. 45b. Sales record for VAT.

Checks should be made to ensure that the total column equals the total of all other columns.

The VAT rate taken has been 15% for the standard-rated items. Thus the fertiliser invoice P46 is for a total of £3000 of which £2609 is for the fertiliser and £391 for VAT.

Invoices must by law be retained for at least 3 years so that a good filing system is essential. Discounts, if given, should be deducted before the VAT is charged.

4.5.4. VAT Accounts

At the end of each month, or quarter depending upon the period selected, the VAT sales and purchases must be totalled and entered on the VAT account. The latter can be similar to Part A of VAT 100 since it is used when filling in this form, but also acts as a record of the VAT 100 returns for the farmer. Figure 46 shows a system which is satisfactory.

Month	Total output tax	Tax deductible on inputs	Net tax payable	Net tax repayable
	£	£	£	£

FIG. 46. Headings for a VAT account.

4.6. VALUATIONS AND DEPRECIATION

4.6.1. The Objective of Valuations

Each farm business has a financial year and the profitability of the farm is assessed over this period. Valuations have to be completed before the profit can be established because farming is a dynamic process. Crops and animals will have grown during the year and it is not sufficient to consider sales alone because some of the produce resulting from the year's work may still be on the farm. Machinery, equipment, and buildings may have changed in value and account has to be taken of items such as feedingstuffs and fertiliser which have been bought and which appear in the farm expenses, but which have not been used.

A balance sheet is also completed on the last day of the financial year to show what the business is worth, i.e. its assets, and to show who owns these assets—the farmer, the bank, or the mortgage company. The assets such as stock and crop have to be valued for this purpose.

More often than not the financial year will differ from the calendar year. For many farmers the end of the year coincides with the last day of the month in which the anniversary of their entry to the farm occurs. Thus because of the difference between countries in farm entry dates, in England and Wales it is often 31 March or 30 September, and in Scotland 30 November or 31 May. A large number of farmers depart from this rule and select the end to coincide with a period when they are least busy and when it is easiest to undertake the valuation. For example, a hill farmer would not select the time when he was lambing.

When a farmer is giving up a farm to a new tenant there is usually an "outgoing valuation". Items valued depend upon the tenancy agreement, but can include equipment, stock, especially in the case of hill sheep farms, crop, unexhausted manurial residues, and improvements made to the farm. The outgoing farmer is then compensated, obtaining what is called his "tenant right". An ingoing valuation takes place before the new farmer enters and he may have to pay the value derived. The two valuations may be identical but could differ under some tenancy agreements. When a farm is sold, tenant right for growing crops, cultivations, and unexhausted manurial residues is usually payable by the purchaser to the vendor over and above the price of the farm.

4.6.2. The Annual Valuation

The annual valuation on the last day of a closing financial year becomes the opening valuation on the next day of the new financial year. It is as if the items included in the closing valuation are sold, on paper, to the next year. Farmers can undertake the valuation themselves although many employ a valuer.

The value of items for "taxation purposes" may differ from that for management, i.e. for the analysis of the performance of the business and for budgeting for change. Clearly any increase in value of an item registers as a gain, helps increase the profit, and so increases tax payable. If the item has not been sold before the tax is payable, tax on the "paper profit" element from it has to be paid for out of other funds. This could be particularly penal in a period of inflation when items such as breeding cattle, which are permanently on the farm, are gaining value. For this reason farmers should examine the short and long term economic implications of taking all possible steps to minimise increases in value when valuation takes place for taxation purposes. Accounts for management, however, must show a realistic up-to-date position to reduce the chances of fallacious conclusions being derived about current and possible future performances.

4.6.3. Livestock Valuation

Animals can be classed as "breeding stock", which are a "fixed asset" since they remain on the farm for a number of years, or "trading stock" such as beef stores, bacon pigs, or fattening lambs, which are sold off after a comparatively short period of time.

Whereas breeding males are normally valued individually for taxation purposes, other classes of stock are valued in groups. As a general rule the valuation for taxation purposes is the cost of production or market value, whichever is lower, but if the cost of production is not known it is permissible to take 60% of market value for cattle and 75% for other livestock. The cost of production is the purchase price, plus feed, labour, and any other costs incurred in keeping the stock up to the time of valuation.

4.6.4. Herd Basis for Valuing Livestock

At the start of a business, or when a new herd or flock is established, farmers can elect to the Inland Revenue to have breeding stock regarded as "capital assets". With the exception of replacement ewe hoggs on hill farms, this applies to adult stock only. All other animals must still be treated as trading stock, and changes in their value or cost of production must be reflected in the profit and loss account. After an election to be treated on the "herd basis", changes in breeding stock values are not included in this account unless numbers in the herd or flock increase. In fact under the herd basis the value per head of breeding stock remains the same until the unit is dispersed. However, where the herd basis is employed, sales of cull breeding animals do appear in the profit and loss account (henceforth referred to as the P & L account for short).

The significance of opting for the herd basis can be seen from an example. Farmer A has 80 dairy cows valued on a trading basis, which means their value is included in the P & L account at £200 each. He sells 20 at £300 each and replaces them with 20 home-reared heifers whose cost of production was £220 each. His output is £6400.

Relevant portion of account:

	£		£
Opening valuation: 80 @ £200 =	16,000	Sales 20 @ £300 =	6,000
		Closing valuation: 60 @ £200	
		20 @ £220	
Output	6,400	i.e. 80 @ £205 =	16,400
	£22,400		£22,400

Farmer B has the same size herd, 80 cows, but he opted for the herd basis and his cows are still valued at £120 each. He also obtains £6000 from the sale of 20 cows and like Farmer A he can put the cost of rearing his 20 replacements in the P & L account. However, when his heifers enter the herd they become valued at £120 each, a value which in fact is not included in the P & L account. This means that Farmer B's output is £6000 or £400 less than Farmer A, i.e. a reduction in taxable profit of £400.

If Farmer B purchases heifers which cause his herd to increase in size he may not be treated as favourably, from the point of view of taxation, as Farmer A. Under the herd basis the expenditure on stock to increase the herd is treated as a capital item and cannot be entered in the P & L account to reduce profit. Farmer A can buy extra heifers at, say, £500 each and value them at the closing valuation of £350, resulting in £150 to reduce profit.

When the herds are finally dispersed, Farmer B has a very significant advantage. Assuming that his cows are still valued at £120 each, and they sell for £400 each, the gain of £280 is neither included in the P & L account nor is it assessed for capital gains tax. By this time Farmer A's cows may be valued at £220 each and also sell for £400. The gain of £180 each is included in the final P & L account and taxed.

4.6.5. Valuing harvested Crops

Crops such as cereals and potatoes, which can be sold off the farm, may be valued at 85% of the market value less costs of marketing. This conservative valuation allows for losses in store. Alternatively the cost of production, including storage, can be used if this is lower.

It is difficult to assess the market value for silage since so little is sold, and care has to be taken with hay and straw because their value can fluctuate with the season. Many farmers have difficulty in establishing costs of production because of insufficient records, and use a "standard" value, which they obtain from their valuers or estimate, and retain the same value over a period of years, increasing it from time to time with inflation.

4.6.6. Valuing Growing Crops and Tillages

Growing crops and cultivations are usually valued at cost of production to date. Some farmers add contractor's charges for ploughing and drilling, etc., even if they do the work themselves, to the cost of seed, fertiliser, and spray used.

There appears to be great variation in the practice for valuing grass. With recently established fields the cost of production can be taken and in the subsequent years allowance made for deterioration. Many farmers do not value the older grass at all.

4.6.7. Valuing Unexhausted Residues

Residual values left after fertiliser applications and from dung, assessed from the values of feed fed to stock, can be obtained from tables published nationally for this purpose. Such a calculation is undertaken when a farm changes hands and obviously good records are necessary if the calculation is to be accurate.

It would be tedious to carry out such valuations each year for an on-going business so that generally for annual valuation purposes the same figure is carried forward unless there is a significant change in cropping policy. If a spring or early summer valuation takes place, manures applied for specific growing crops come under the heading of "growing crop".

4.6.8. Valuation of Stores

At the end of a year a farmer will usually have items on hand such as purchased stock feed, fertiliser, sprays, fuel, and machinery spares which have been paid for but not used. These must be valued at cost and then sold "on paper" to the next financial year so that the current year can be credited.

Care must be taken on certain farms where one year, at the valuation, the seed and fertiliser may all lie in store but the next year, because of better weather, they may be sown. A logical valuation policy must be adopted to ensure that this does not mask the true profits. If the farmer values recently applied seed and fertilisers, in full, in the growing crop valuation there is little problem, but for some crops some farmers keep the growing crop value constant. In the latter case it might be prudent to value recently applied materials as if they were still in store.

4.6.9 Valuation of Machinery, Equipment, Structures, and Improvements

The purchase of machinery, equipment, buildings, and improvements, is regarded as capital expenditure whereas money for something like fertiliser is trading expenditure. The latter will be quickly used up but a tractor bought this year should last several years. It might be regarded as unfair to charge the whole cost of an item like a tractor, and even more so a substantial building, against the financial year of purchase.

The value of the tractor will go down, that is it will depreciate, as the years go by because it wears out and becomes an older model. (Admittedly in times of very rapid inflation some farmers claim to sell tractors for as much as they paid for them but this is merely an illusion since the value of money has fallen meantime.) It is this depreciation element which is allowed as a cost in the P & L account.

4.6.10 Depreciation

Two methods of depreciation are practised. The "straight line" method and the "reducing balance" method. In general the former is used for buildings and the latter usually adopted for machinery.

The straight line method "writes off", i.e. it depreciates, an item in equal instalments over a given period of years. Farm buildings are usually written off over 10 years for management purposes. Thus if a building costs £20,000, £2000 is charged in each year's P & L account for 10 years. Expressed graphically (Fig. 47) this appears as a straight line. The written down value (WDV) of the building after 1 year is £18,000, after 2 years £16,000 and so on. If a grant has been received on the building the net cost after grant is depreciated.

For taxation purposes a building may be depreciated by 4% per annum, a figure fixed by fiscal regulations outwith the farmer's control. One reason why a higher rate is employed for management purposes is the speed of change which can take place in agricultural production policy. If the building is very adaptable to different enterprises then the need to write it off quickly is reduced. However, most buildings cannot be sold and if, for example, a specialist poultry house is put up and the enterprise is

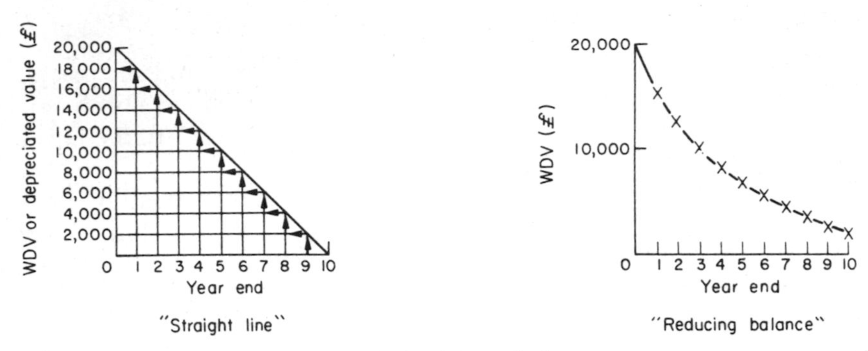

FIG. 47. Methods of depreciating assets.

abandoned after 5 years the building is still there. A common sense approach has therefore to be taken when fixing the depreciation rate for management purposes. If the period is too short the depreciation figure employed in budgets for the project may be so high that it puts the farmer off starting what might be a viable proposition. If it is too long the building might still be featuring in the farm's depreciation calculations after the project has been terminated and therefore effectively needs subsidising by activities which are still running on the farm. It also still appears as an asset in the balance sheet even though it is not being used.

An interesting point is that some farmers put minor extensions to buildings in their accounts as building repairs which means that the cost all appears in one year and is written off.

Farm machinery can be depreciated at 25% per annum on a reducing balance method for taxation purposes. However, for management purposes many people adopt 20% as the figure when considering the overall depreciation rate for the farm's machines. When undertaking budgets for the purchase of individual machines they may modify the figure. This matter is dealt with later in the book under the section on machinery.

4.6.11. Other Valuations for Management

If records are to be of any value to management they must illustrate a realistic picture. In the case of items such as stock and crop, valuations should approximate to market value less cost of marketing. For this reason management valuations frequently exceed those for taxation. Naturally for some growing crops, where it is difficult to assess market value, cost of production will have to be used even for management purposes.

4.7. PROFIT AND LOSS ACCOUNTS—INTRODUCING THE GROSS MARGIN CONCEPT

4.7.1. Introduction to P & L Accounts

P & L accounts are produced for two main purposes. The first is to obtain a statement of profit or loss for a business to serve as a basis from which liability for taxation can be derived. This will be covered in Section 7.3. dealing with taxation, but whilst the account must conform to legal fiscal requirements, the owner will be concerned to pay the minimum amount of tax, considering both long and short term implications. The second is to produce an account for management purposes which illustrates the performance of the business as accurately as possible so that sound judgements can be based upon it to help future improvements. The two types of account are not identical.

Many students are surprised and initially confused to find that in addition to the two main objectives there are several different ways of presenting the layout of an account, and also of expressing profit. The introduction of computer-based accounting may result in greater standardisation in future, but for the present the reader is advised to accept that it matters little that the layout varies from district to district, and from accountant to accountant, provided that the account serves one of the two functions outlined above, and provided that a common approach is used in formulating the account year after year so that one year's results can be compared with those of another.

Some confusion also arises over the term "trading account". In some industries a trading account is compiled which shows sales of goods less cost of purchases, with an appropriate adjustment for valuation of stocks on hand. The resultant figure is then transferred to a P & L account where wages, rents, fuel, and other overhead costs are deducted to show profit or loss. In agriculture it is usual to combine these two procedures and simply call the result a P & L account although more correctly it is a trading and P & L account.

4.7.2. Layout of a P & L Account

Figure 48 shows a typical layout and forms the basis for all the variants which can be found in the P & L accounts used in farming.

It can be seen that items in the opening valuation (OV) are assumed on paper to be bought from last year by this year and are added to the expenses. The items in the closing valuation (CV) are sold on paper to next year and the value added to the receipts. If receipts plus CV exceed expenses plus OV there is a profit. A loss occurs if the expenses plus OV are higher.

An absolutely vital point to note at this stage is that the P & L account does not reflect all the cash transactions of the year. Remember that although a machine or building may be bought and paid for within the year only part of the cost, the depreciation, is entered in the P & L account. Also a substantial loan might be repaid or received in the year but only interest on capital, and not the repayment or receipt of capital, can be entered in the P & L account. This means that a P & L account does not reflect the full capital position of a business.

Variations on the layout shown in Fig. 48 can be produced to advantage, but first it is necessary to understand one or two definitions and to introduce the gross margin concept.

4.7.3. Output Terms

(a) *Output* is a general term used to express the value of production of an enterprise (a sector of the farm such as a sheep flock) or of a whole farm. It is sometimes

P & L Account for year ending 31 March 19... for Church Farm (as basis for taxation purposes)

	£	£		£	£
Opening valuations			*Receipts*		
Livestock	89,380		Livestock	50,646	
Crops	26,784		Livestock products	99,100	
Stores	21,300	137,464	Crops	58,990	208,736
Stock purchases			*Closing valuation*		
Sheep	3,172	3,172	Livestock	88,680	
Variable costs			Crops	26,500	
			Stores	21,300	136,480
Feed	32,922		*Miscellaneous*		100
Vet. and medicines	3,110				
Stock sundries	5,380				
Seed	8,066				
Fertilisers	24,310				
Sprays	4,425				
Crops sundries	3,620				
Casual labour	4,500	86,333			
Fixed costs					
Labour	32,900				
Machinery	31,850				
Rent and rates	650				
Property repairs	6,400				
Building depreciation	2,000				
Interest	14,900				
Other overheads	3,250	91,950			
Subtotal		318,919			
PROFIT		26,397			
		345,316			345,316

FIG. 48. P & L account—Church Farm.

necessary to be more specific, and more definitive output terms are then employed.

(b) *Gross output* can be obtained from the following formula:
Gross output=(sales+closing valuation)−(purchases of livestock+purchases of any crop subsequently sold+opening valuation).

Note that the sales include sales off the farm plus the value of produce consumed in the farmhouse or by workers. If the farmer bought some barley which he fed to his stock the cost would not be entered in the gross output statement but would come under expenses. If he resold the barley both cost and sale price would be entered into the gross output calculation. When the gross output for a whole farm is calculated sundry revenue such as subsidies, grants, and money for contract work done for other farmers is included.

Any revenue included in the above formula is adjusted for debtors and creditors at the start and end of the accounting period.

Whilst the term "gross output" is a suitable way of measuring the output of a whole farm, when completing records for management purposes it is necessary to be more specific, especially when studying the performance of enterprises. If a farmer is looking at the various sections of his farm he frequently wants to assess how each enterprise would perform if it was standing on its own feet to see if he should change the area of land, or amount of other resources devoted to it. For example, he may grow barley which he could sell, but instead he feeds it to his dairy cows. The barley in this case is transferred from one enterprise to another, it is an "interdepartmental transfer". If the barley is not charged to the cows the farmer cannot obtain an accurate picture of the profitability of the cows or of the barley. A realistic figure, based on average market values, is therefore put on the barley, credited to the barley, and debited to the cows. The barley is a "transfer out" from the barley enterprise and a "transfer in" to the dairy herd.

This is purely a paper calculation for management purposes. The tax authorities are not interested because in effect the farmer has increased his output from the barley by the same amount as he increased the costs to his cows; result—zero difference in profit.

(c) *Enterprise outputs* are therefore calculated using the same basic formula employed to calculate gross output. However, the sales item must now include transfers out to other enterprises on the farm and the purchases of stock must include any transfers in from other activities on the farm.

4.7.4. The Gross Margin Concept

In crude terms:

$$\text{Profit} = \text{output} - \text{costs or inputs}$$

The costs of a farm include such items as feed, fertiliser, rent, labour, and machinery. When a farmer starts to examine the profitability of the enterprises in his business he is faced with the problem of how to allocate these costs. If he has kept good records he should be able to establish the cost of feed and fertiliser for each enterprise. However, the rent is for the whole farm—for buildings and land. To allocate it to enterprises is difficult and arbitrary. With extremely good labour records it might be possible to allocate the costs on a time basis, but time spent in maintaining the holding, such as repairs to fences, drains, roads, and buildings, may have to be allocated to enterprises on an arbitrary basis. Again, with good records the cost of machines might be allocated, in part accurately, and in part on an arbitrary basis. Consider a man operating a slurry tanker carting slurry from the pigs to a barley field. The farmer would have to decide how much of the man's time, the machinery cost, and the cost of maintaining the road should be charged to the barley and how much to the pigs.

The gross margin concept was introduced to circumvent this problem. Some costs such as those for feed can be allocated directly and reasonably accurately. In general the more animals the farmer has, the more feed he uses. Similarly the greater the number of hectares of barley a grower produces, the more seed and fertiliser he needs. In other words these costs tend to vary directly with the scale or size of the enterprise. The term "variable cost" was therefore devised to describe them although sometimes they are known as direct costs.

Variable costs are those direct costs which, given appropriate records, can be reasonably easily and accurately allocated to specific enterprises, and which tend to vary directly with the size or scale of an enterprise.

(It was seen in Chapter 3 that variable costs vary with output. This can give room for some confusion. For example, consider Farmer A and Farmer B. They use identical amounts of seed, fertiliser, sprays, and other variable costs per hectare for growing wheat. However, because of better land and better timing of operations, Farmer A obtains higher yields than Farmer B. Farmer A also stores his grain, unlike Farmer B, and realises a much higher price than Farmer B. Clearly the output achieved by Farmer A is higher than by Farmer B. This appears to suggest that variable costs do not vary directly with output. The real point is that given identical conditions of production, variable costs will vary with output. It is the differences in these conditions between wheat grown by the two farmers which causes the apparent anomaly with Chapter 3.)

Examples of variable costs:

Feed	Fertiliser	Casual labour
Seed	Sprays	Contract machinery
Vet. and medicines	Seeds	Specific dairy expenses
AI	Twine	Other specific livestock
Bedding	Silage additive	or crop expenses
Haulage		

Livestock purchases are not regarded as variable costs but are deducted when appropriate enterprise outputs are calculated. Casual labour and contract machinery are employed to undertake specific tasks, and it is therefore possible to allocate them directly to enterprises. Lime is best treated separately because its effects can be long lasting and a single crop should not be penalised.

It can be seen from Fig. 49 that when expressed graphically there is a tendency for variable costs to increase in a direct, linear pattern with increases in scale of the enterprise.

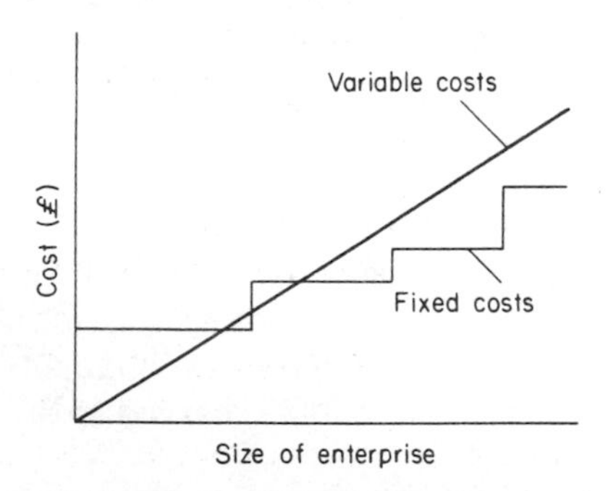

FIG. 49. Variable/fixed costs.

When the gross margin concept was devised it was decided that no attempt should be made to allocate certain costs such as rent and regular labour to each enterprise because of the difficulty involved and the arbitrary nature of the result. Instead these costs were to be put on one side and calculated for the whole farm. The wisdom of this strategy will be discussed later in Chapter 5 on analysis and Chapter 10 on planning.

P & L Account for year ending 31 March 19... for Church Farm

Opening valuation	£	£	*Receipts*	£	£
Livestock	89,380		Livestock	50,646	
Crops	26,784		Livestock products	99,100	
Stores	21,300	137,464	Crops	58,990	208,736
Expenses			*Closing valuation*		
Stock purchases	3,172		Livestock	88,680	
Feed	32,922		Crops	26,500*	
Vet. and medicines	3,110		Stores	21,300	136,480
Stock sundries	5,380		*Miscellaneous*		100
Seed	8,066				
Fertilisers	24,310				
Sprays	4,425				
Crop sundries	3,620				
Labour	37,400				
Machinery costs	31,850				
Rent and rates	650				
Property:					
depreciation	2,000				
maintenance	6,400				
Interest	14,900				
Other overheads	3,250	181,455			
Subtotal		318,919			
NET PROFIT		26,397		LOSS	-
		345,316			345,316

*Valuation of crop in ground £3,300.

FIG. 50. P & L account showing fixed and variable costs. (Note: If the phrase "As basis for taxation purposes" is written above a P & L account it clearly indicates the objective of the account and the nature of the profit calculated. Frequently such accounts do not include the value of any interdepartmental transfers that take place on the farm.)

Unfortunately the term "fixed costs" was established to describe these costs, although common and even overhead costs are sometimes used. The word "fixed" tends to imply that the costs remain static, which regretfully is not true, but compared with variable costs their changes are irregular and lumpy (Fig. 49). For example, consider a farm with dairy cows and cereals. The farmer employs one man plus himself. First the farmer intensifies his grass by using more fertiliser, keeps more cows, but still maintains the original area of cereals. His variable costs for feed and fertiliser would increase but he would probably be able to keep the extra cows with the same rent and same labour. A point would arise, however, where, if he increased his herd further, perhaps at the expense of some of the area of barley, he would have to ask his landlord for extra buildings, and he might have to employ an extra man. These additions would allow the farmer to continue increasing his herd, assuming he still has enough capital and land, until again the buildings, and/or labour, would be a

	Milk £	Cattle £	Sheep and wool £	Wheat £	Barley £	Straw £	Forage £	Potatoes £
Stock at beginning	-	78,600	10,780	18,700	2,884	100	1,800	-
Purchases*	-	-	3,172	-	-	-	-	-
Total B	-	78,600	13,952	18,700	2,884	100	1,800	-
Sold	98,000	35,266	16,030	18,700	15,100	2,430	-	22,760
Used by house workers	450	-	-	-	-	-	-	-
Stock at end	-	77,900	10,780	19,400	2,100	100	1,600	-
Total A	98,450	113,166	26,810	38,100	17,200	2,530	1,600	22,760
Gross output (A-B)	98,450	34,566	12,858	19,400	14,316	2,430	−200	22,760

*In the case of crop: only include items bought for re-sale.

FIG. 51. Gross output schedule for Church Farm.

constraint and the farmer would have to consider further increases in staff and/or buildings.

The distinction between fixed and variable costs is not always clearly marked and, indeed, in some industries the term "semi-variable cost" is used to describe such items as repairs to machinery, which are to some extent related to the size or scale of an enterprise.

One student described fixed costs as those costs which you have whether you produce anything or not, and variable costs as those costs which you have to incur if production is to take place. Whilst this statement is not entirely true there is a lot of truth in it. The landlord is not going to reduce his rent, the men are not going to refuse their wages, and the machinery is not going to stop depreciating simply because a farmer is not producing as much as he could. The real definition is as follows:

Fixed costs are those common or overhead costs which tend not to vary, except with large changes in the scale of production, and which are typically difficult to allocate except with the aid of complex records and in many cases arbitrary decisions.

In some cases the decision to treat a cost as a fixed cost rather than as a variable cost is related to the reasons why the costing is being carried out. A limited number of costs such as tractor fuel and machinery repairs are classified as fixed costs in the whole farm account because they are difficult to allocate, but in fact they would tend to vary with scale of production. When budgeting to establish the cost of one machine compared to another they would be treated as variable costs.

The gross margin of an enterprise can then be found from the following formula:

Gross margin = enterprise output − variable costs

Sometimes gross margins have been called by other names such as "contributions" or "gross profit". The latter term in particular should be avoided since it is used in widely different ways, both in agriculture and industrial business.

There is considerably more to the understanding of gross margins but this will be left to Sections 5.4. and 10.3. What has been covered so far is sufficient to examine some of the different presentations of P & L accounts.

P & L Account for year ended 31 March 19.., Church Farm (as basis for taxation purposes)

	£	£		£	£
To *variable crop costs*			By *Crops* GROSS OUTPUT		
Fertilisers and lime	24,310		Wheat	19,400	
Seeds	8,066		Barley	14,316	
Sprays	4,425		Potatoes	22,760	
Casual labour	4,500		Forage	−200	56,276
Sundries	3,620	44,921			
			By *Livestock* GROSS OUTPUT		
To *variable livestock costs*			Cattle	34,566	
Feed bought	32,922		Milk and dairy produce	98,450	
Vet. and medicines	3,110		Sheep and wool	12,858	145,874
Milk and dairy expenses	2,680				
Other livestock expenses	2,700	41,412	By *Miscellaneous*		
		86,333	Straw	2,430	2,430
To balance carried down GROSS MARGIN		118,247			
		204,580	GROSS OUTPUT FROM FARM		204,580
Fixed costs			By balance brought down GROSS MARGIN		118,247
To regular labour:			Wayleaves (e.g. telegraph poles)		100
Wages and National Insurance	30,600				
Wife's labour	2,300	32,900	Gain on sale of machinery		200
To machinery, equipment and vehicles					
Repairs and renewals	16,500		By private shares of:		
Fuel and electricity	1,500		Rent and rates of farmhouse	1,300	
Other machinery expenses	3,400		Fuel and electricity	375	
Depreciation	11,305	32,705	Car expenses	280	
To general overheads			Telephone	100	2,055
Rent and rates	1,950				
Insurances	1,280				
Repairs to property	2,400				
Fencing, draining	4,000				
Telephone	500				
Bank interest	6,000				
Loan interest	8,900				
Other general overheads	1,570				
Depreciation of improvements	2,000	28,600			
Total fixed costs		94,205			
Net profit carried to capital account*		26,397	Loss		-
		120,602			120,602

*Capital account in balance sheet.

FIG. 52. P & L account—Scottish farm business records type.

4.7.5. Different P & L Account Presentations

Some farmers amend the layout shown in Section 4.7.2. above to highlight both the variable and the fixed costs. They might present the P & L account as shown in Fig. 50.

One form of account developed in Scotland introduces the concept of a gross output schedule which is compiled before the actual P & L account is prepared. An example is shown in Fig. 51. It must be noted that it is possible to obtain a negative gross output. In the example shown in Fig. 51 for Church Farm the opening valuation of forage was higher than the closing valuation. Assuming that this was for hay and silage, the stock will have eaten all the food represented by the opening valuation and a high

Balance sheet - Church Farm as at 31 March 19...

Liabilities				Assets		
Last year £			This year £	Last year £		This year £
4,886	Sundry creditors		7,000	300	Cash in hand	450
38,000	Bank overdraft		30,000	-	Cash at bank	-
60,000	Bank loan		60,000	4,900	Sundry debtors	8,000
75,000	Mortgage on farm		74,000	21,300	Stores and sundries	21,300
				23,484	Crops in store	23,200
	Capital account			3,300	Crops in ground	3,300
	Last year's net worth	299,798		46,600	Trading livestock	45,900
	+ Profit for year	26,397		35,020	Machinery	38,265
	+ Capital introduced	1,000		42,780	Breeding livestock	42,780
		327,195		50,000	Permanent improvements	48,000
	− Private drawings	17,000		250,000	Heritable property	250,000
		310,195				
299,798	This year's net worth		310,195			
477,684			481,195	477,684		481,195

FIG. 53. Balance sheet—Church Farm.

proportion of the hay and silage grown during the year. In total they consumed more than was actually produced during the year, resulting in the negative output.

The gross outputs produced from the schedule are then transferred to the type of P & L account shown in Fig. 52.

4.8. THE BALANCE SHEET

4.8.1. The Function and Layout of the Balance Sheet

The main function of a balance sheet is to present a statement of a business's financial position on one particular day. The day selected is the last day of the financial year so that the P & L account and balance sheet can be presented together. The information provided can be used in several ways by the owner of the business or by a person who is asked to lend money to it. These uses will be clarified later when analysis of the balance sheet is discussed.

Although there are differences in the way balance sheets can be presented basically they all list the assets of the business in addition to the liabilities (Fig. 53). The assets include anything which the business owns including any valid claims on anything of value currently possessed by others. The liabilities include all the financial claims on the business. The term "balance sheet" is appropriate because the total on the assets side must equal the total of the liabilities.

The assets are owed to someone—the farmer, someone else, another business, or a bank. Technically the farmer is not the business but as an entrepreneur just an investor in it. Hopefully he is the main controller of it since he usually has most to gain or lose by its success or failure. Apart from cases where he has formed a limited company he can be charged to meet all the debts or be bankrupted. Sometimes as a condition of a loan to his business some pressure may be exerted upon him to follow a certain policy for the farm.

The farmer's stake in the business, or the portion of the total assets owned by him, is known as his net worth, net capital worth, or owner's equity. The people who have a claim on the business for the rest of the assets are the creditors. Thus it is usual to say that the total assets of the business less the claims of the creditors equals the farmer's net worth.

If the claims of the creditors are higher than the total assets of the business, and the only way to make the two sides of the account balance is to include an item called capital deficiency on the assets side, then the business is said to be insolvent as opposed to solvent.

A study of the assets side reveals that items are usually listed in order of liquidity. In the example shown (Fig. 53) the most liquid items are listed at the top and the least liquid, or fixed items, at the bottom. Thus cash in hand is fully liquid since it is immediately available, but progression down the list shows that it becomes more and more difficult to realise the cash for the particular asset without altering the business or, in the case of land, giving up the business.

The sundry debtors represent the money owed to the business for goods or services which it has supplied. In farming most of the organisations to which the product is sold, such as the Milk Marketing Boards, can be relied upon to pay the farmer fairly promptly, but there is always the possibility that debtors may not pay. Some businesses have gone bankrupt in spite of the fact that their balance sheet shows them to be solvent, simply because debts have not been received.

Usually the sundry creditors consist of the bills which have to be paid fairly shortly to feedingstuffs and fertiliser merchants, or other similar claims. The sundry debtors and sundry creditors can be obtained from the unpaid invoices in file 1 and sale notes in file 3 (see Section 4.3.1.).

The stores and sundries, sometimes known as deadstock, include items such as feed, fertiliser, and sprays on hand. Trading livestock represent store and fattening animals in particular, but also replacement youngstock which have not entered the breeding herd. Clearly if creditors had to be paid and money was short, these animals could be sold with less upset than the sale of the breeding stock.

Permanent improvements include items such as buildings, and these could be erected by a tenant (as opposed to a landlord) in some cases.

Usually the liabilities are listed in the order in which they will have to be met. Thus the mortgage is a much longer term loan than a bank overdraft.

A study of the capital account shows that the first figure entered is last year's net worth. This figure can be observed at the bottom of the liabilities column on last year's balance sheet. The profit is then added, and if any capital has been introduced from outside the business such as a legacy from a will, it is also added. Private drawings, including income tax, are then deducted to obtain this year's net worth.

4.9. MECHANISED ACCOUNTING AND COMPUTERS

4.9.1. Machine and Computer Accounting

Mechanised accounting, usually in conjunction with some outside agency or advisory service, is employed by some farmers to reduce the work involved in recording and analysing results. Typically each month basic information, recorded on specially printed forms, is sent to the agency which processes them and sends reports back to the farmer.

The system can avoid the necessity to prepare analysis sheets. Reports show the allocation of income and expenditure to respective enterprises both monthly and cumulatively. Fixed costs can be recorded in some detail so that they can be carefully examined. Bank transactions, whilst being recorded under headings, are detailed collectively on control cards which can be used to check bank statements. VAT records can be produced. At the end of the year P & L accounts, gross margin statements, and comparisons of actual results with budgeted data can be made available. Together the report cards form a valuable system for information storage, retrieval, and review.

Naturally the system costs money and there is still some time and care needed by the farmer when completing the initial forms. The work is kept to a minimum, however, by such practices as the use of cheque books which produce carbon copies so that the latter can be sent to the agency.

The key to the success of a mechanised accounting system lies in the coding of the items. Each transaction or record has to be coded correctly, normally by the farmer, so that it can be accurately dealt with at the accounting, or computer centre. If the coding is performed in a "slap happy" way then the final results will be worthless.

Computers have now made significant inroads into recording by outside agencies and also on many farms. There is almost a plethora of companies marketing computers and the associated software which can be used to record both physical and financial information. Such facilities, if employed correctly, can be an effective aid to the farmer. However, the volume of data which can be produced must not cloud the essential elements necessary to good management.

Invoice Date	Pay Date	Trader's Name or Number	Description of Contents of Invoice	Revenue or Cost Centre Code	Cheque No. or Pay In No.	VAT code	Invoice Total inc. VAT	VAT

Both care and time are necessary when feeding information into a computer. The above illustration is just one of the working sheets employed in connection with our own College Farm analysis programme and helps illustrate the point. All the data must be accurate but particular care must be taken to ensure that the revenue or cost centre code is allocated correctly.

CHAPTER 5

ANALYSIS AND APPRAISAL OF THE PERFORMANCE OF A FARM BUSINESS

CONTENTS

5.1. OBJECTIVES OF ANALYSIS AND ACCOUNT ADJUSTMENT

5.1.1. Objectives of Analysis

All records quickly become historical, but if they are carefully analysed they can not only be employed to measure the efficiency of performance of a business as a whole, but perhaps more significantly provide pointers for the future by highlighting current strengths which should possibly be exploited, or weaknesses which need eliminating or require modifications. Care has to be taken, however, to prevent false conclusions being derived. Both strengths and weaknesses can be of short term duration. Sound analysis therefore involves careful study of all the facets which contributed to a business's success or failure, and it also examines why changes took place with the objective of assessing the repeatability of each particular performance.

5.1.2. Adjustment of Accounts

One of the first objectives of analysis is to assess the true profitability of a particular farm. The profit compiled from the P & L account prepared for taxation purposes is influenced by procedures which are quite permissible under taxation regulations but which mask the real profit from the management point of view. The taxation account is therefore adjusted before it is used to assess the efficiency of the business. This is usually done so that the adjusted account is prepared in a form which is common to the area, say a college area in Scotland or a university region in England. A standardised form of preparation facilitates comparison of results between farms within the given area. There are virtually no differences between areas or regions in the principles involved in the adjustment, but minor variations in convention do occur.

When the adjustment is carried out it is essential to ensure that income which is not strictly a product of the business as a farm is removed. This includes wayleaves for electricity or oil pipelines, caravan lettings, shooting and fishing lets, timber sales, and private income or expenditure. Equally it is important to ensure that all genuine income of the farm is credited. For example anyone injudicious enough to have omitted receipts, paid for in cash, from the taxation account is only deluding himself if he omits them from the management account.

Figure 54 (p. 100) shows the account for Church Farm suitably adjusted for management purposes.

THE FORM OF PROFIT which has been calculated is a point which the reader must observe carefully. There are several different ways of expressing profit and it is therefore essential to ensure that it is clear which form is being used, especially if two farms are being compared.

NET FARM INCOME (NFI) *is the profit from an account suitably adjusted for management purposes which shows the return to the farmer and his wife for their manual work, their management, and the investment of their capital.* In other words no charge is entered into the account for the farmer's or his wife's work—manual or management.

MANAGEMENT AND INVESTMENT INCOME (MII) *is the profit from an account suitably adjusted for management purposes which shows the return to the farmer and his wife for their management and the investment of their capital in the business.* In this case a realistic charge is put into the account for the manual work of the farmer and for his wife if she works. This form of profit is possibly better for use in comparing profits between farmers than NFI because it allows for the fact that farmers differ in the amount of manual work which they do.

Where there is a paid manager the charge for his management is not included in the costs when calculating MII, but if he is a working manager a charge for his manual work is included.

The procedures for adjusting accounts are as follows.

Labour

It can be seen from the P & L account for Church Farm (Fig. 52), prepared for taxation purposes, that a charge of £2,300 has been included for the wife's work. Technically she must work before this procedure is undertaken, and some farmers actually pay the amount entered in the P & L account for the wife's manual labour into a separate bank account for her. In practice many farmers enter a figure for the wife's work which is equal to the current "wife's earned income allowance" whether she works or not. This reduces the profit of the farm, which in turn reduces the amount of tax the farmer pays, whilst the wife does not have to pay tax on earnings up to this level.

When compiling the NFI form of account the charge for the wife's labour is omitted irrespective of whether she works on not. When producing the MII type of account a figure for the real value of the wife's manual work is included in the costs. In the case of Church Farm the wife did not work.

No charge is included for the farmer's labour when calculating the NFI, but arguments can arise when establishing the cost of his manual labour to obtain the MII. Some farmers claim that they are worth significantly more than many of their men and some, in error, include time which could more justifiably be described as a charge to management. Most advisers have figures which are produced as standards and they apply these as realistically as possible to their client's accounts. The farmer who is establishing a figure for himself must endeavour to calculate a division of his time. Office work, going to market or the bank are clearly management, but one can debate the question of where to put late night visits which are alternated with the head dairyman to see if any of the cows are on heat. In truth there is little point in worrying too much about such fine decisions.

Where partnerships exist one partner and one wife can be entered as one farmer and his wife. A check must be made to ensure that a realistic charge has been entered for all other manual labour, including that for unpaid family and for partners.

The National Insurance must not contain the farmer's contribution. Charges for the workers board and lodgings can remain if provided by the farm, although it is usual to ensure that these are at the rates currently laid down by the Wages Board.

Interest

For management analysis purposes each business is regarded as being self-sufficient for capital. All interest charges on loans, overdrafts, and hire purchase are removed. Some businesses have to borrow more money than others simply because of the different financial positions of their owners. Removing interest charges prevents this fact from masking the true efficiency and allows more realistic comparability between businesses. Naturally no one forgets when money has been borrowed, and when considering changes to plans attention is focused upon the borrowing position. Some people do in fact state the interest charges below the figure for MII. This procedure is strongly recommended.

Rent

In management accounts all farmers are usually regarded as tenants although the system is not without its critics. In the case of tenant farmers the actual rent paid is employed, but for an owner-occupier a notional rent is charged. The latter can be fixed by reference to typical rents being paid for similar farms in the same district by farmers who have held their tenancies for a number of years roughly equivalent to the period that the owner-occupier in question has been in his farm.

The wide variation in rents can have a significant influence on profit levels. When comparing the profitability of farms it is advisable to examine the effect of the respective rents before assessing their relative efficiency.

Landlord's items

The P & L account for an owner-occupier prepared for taxation purposes contains certain items which can more realistically be described as a landlord's rather than a tenant's responsibility.

When this account is adjusted for management, and a notional rent is introduced, the landlord's items must be omitted. A commonsense approach is required. Even in the case of tenanted farms there are differences with reference to the costs which the landlords will meet. Some tenancies are on a full repairing lease with all repairs being the responsibility of the tenant.

Usually the owner-occupier account is adjusted to remove major property maintenance costs, substantial new building costs, and the charge for major drainage schemes, together with any stipends and building insurances. When fixing a notional rent, reference is frequently made to the responsibility for repairs of the tenant farmers used in the comparison.

Machinery

A check should be made to ensure that the items of new equipment, other than spare parts and loose tools, have not inadvertently been included under repair costs. Where new capital items, e.g. a new tractor, has been acquired then only the depreciation must appear in the P & L account.

Under fiscal regulations at the time of writing 25% of the written down value of a machine is included in the P & L account prepared for taxation. For management purposes many centres use 20% of the written down value, but deduct any gains in sale from this figure. A gain on sale occurs if the sale price is higher than the written down value. The selection of 20% for management purposes has to be viewed against not only the actual life of the machine, but also its replacement cost especially in times of inflation.

Private Shares

In many accounts prepared for taxation, private items appear in the costs, and to offset these the private shares are listed on the income side of the P & L account as shown in Fig. 52. The proportions of the total costs usually entered by accountants as being private shares are two-thirds of the rent and rates of the farmhouse, one-quarter of the fuel and electricity, one-fifth of the telephone bill, and one-fifth of the car costs. For management purposes it is essential that private shares are removed from the account and in the example for Church Farm (Fig. 54) they have been deducted from both sides.

Contra Accounts

Contra accounts should always be shown gross in a management account to ensure that full costs and income can be observed.

Valuations

Valuations must be examined to see if those used in the taxation account mask the real profit. There may have been some attempt to manipulate profit to even out or reduce taxation. For example, in a year with high profits goods in store may be undervalued. Clearly there is a risk that the next year will also have high profits; remember this year's closing valuation becomes next year's opening valuation, and if the goods are sold at much higher prices than the opening valuation they will contribute significantly to profits.

Valuation of quantities of goods on hand might be over- or understated. Both quantities and valuation prices might have been increased to impress a bank manager to encourage him to lend money. It is not unknown for expenditure and income to be switched from one year to another for similar reasons.

Valuations for management should represent conservative market value, and a common approach should be taken to opening and closing valuations. It will be of added advantage when calculating gross margins if they are in sufficient detail to foster a study of each separate enterprise.

Other Points

Milk sales should be recorded net of haulage and coresponsibility levy, but should be gross of milk recording and AI fees. The latter are entered under livestock expenses.

All private use of produce must be credited to the output of respective enterprises.

5.2. COMPARATIVE ACCOUNT ANALYSIS—WHOLE FARM

5.2.1. Efficiency Factors and Standards

Two basic methods of comparative analysis can be employed:

(i) whole farm analysis (incorporating balance sheet analysis);
(ii) gross margin analysis.

P & L Account for management purposes, Church Farm, for year ending 31 March 19...

To: *Variable costs*	£	£	By crops – *Gross output*	£	£
Fertilisers and lime	24,310		Wheat	19,400	
Seeds	8,066		Barley	24,000	
Contract work	-		Potatoes	22,760	
Casual labour	4,500		Forage	-200	65,960
Sprays	4,425		By livestock – *Gross output*		
Sundries	3,620	44,921	Cattle	34,566	
To: *Variable livestock costs*			Milk and dairy produce	98,450	
Feed bought	32,922		Sheep and wool	12,858	145,874
Home grown feed	9,684		By *miscellaneous*		
Vet. and medicines	3,110		Straw	2,910	2,910
Milk and dairy expenses	2,680				
Other livestock costs	3,180	51,576			
		96,497			
To balance carried down GROSS MARGIN		118,247			
		214,744	GROSS OUTPUT OF FARM		214,744
Fixed costs			Balance brought down		
To: Labour	30,600	30,600	GROSS MARGIN		118,247
To: Machinery, equipment, vehicles					
Repairs and renewals	16,220				
Fuel and electricity	1,125				
Other machinery expenses	3,400				
Depreciation	8,844	29,589			
To: General overheads					
Rent and rates	22,650				
Insurances	680				
Repairs to property	1,000				
Fencing and draining	1,400				
Telephone	400				
Other general overheads	1,570				
Depreciation on improvements	2,000	29,700			
Total fixed costs		89,889			
Net farm income		28,358	Loss		-
		118,247			118,247

FIG. 54. P & L account adjusted for management purposes—Church Farm.

Both involve the calculation of efficiency factors and comparison of the results with standards, and both lead to more detailed studies of the physical and husbandry factors of the farm. Efficiency factors can involve the measurement of a financial or physical performance. For example, MII per hectare for a farm or liveweight gain per pig per day. A standard is a figure employed as a yardstick to compare with the results calculated for a particular efficiency factor.

The principle of analysis is similar in concept to that adopted by a doctor when examining a patient. He studies the various symptoms shown, compares them with standards such as normal body temperature, relates his findings to one another, and makes his diagnosis or appraisal. Sometimes the symptoms are obvious and a quick diagnosis can be made, but other cases necessitate thorough investigation. Too quick a diagnosis can, however, be dangerous since the symptoms of one disease can sometimes mask another which might be more insidious.

Equally in farm business analysis it is important to relate symptoms to one another and to make a sufficiently thorough examination to ensure that the real causes of problems or factors of success have been established.

Standards selected for comparison come from a variety of sources. Previous year's results from the same farm can be employed to see if there has been any change. If the results differ a check must be made to establish the reasons. New technical innovations, a particularly favourable or unfavourable season, or a period of favourable prices may have been responsible.

Various bodies ranging from commercial firms to universities and colleges publish standards. The important thing is to compare "like with like". Standards from similar farms in the same area are infinitely preferable to national standards. Sometimes standards for specific efficiency factors represent the averages in performance for a similar group of farms. On other occasions the organisations publishing standards divide each group of farms according to their profitability and produce results for the respective efficiency factors for the most profitable 25%, the least profitable 25%, and the mid-50% sections of each group.

NOTES ON THE ADJUSTMENT OF THE P & L ACCOUNT FOR CHURCH FARM (Fig. 54)

1. Check that a common approach has been taken to opening and closing valuations. Result—no change necessary.
2. Increase output for barley to account for produce used on farm and increase feed costs by the same amount—effect on profit nil, but presents more realistic output and variable cost totals. Amount used £9,684.
3. Increase straw output by £480 to allow for bedding used by calves and increase other livestock expenses by the same amount. Effect on profit nil.
4. Remove wayleaves, i.e. non-farm income—reduces profit by £100.
5. Remove private shares from both sides of the account—effect on profit nil.
6. Remove wage for farmer's wife—increases profit by £2,300.
7. Adjust depreciation of machinery to 20% depreciation from the 25% employed previously. New depreciation £9,044, increases profit by £2,261.
8. Remove gain on sale of machinery from income and also from depreciation—effect on profit nil.
9. Add notional rent of £22,000—reduces profit.
10. Remove £600 property insurance—assuming it to be landlord's share.
11. Remove repairs to property and enter a more realistic tenant's repairs figure of £1,000—increases proft by £1,400.
12. Remove charges for fencing, draining and put in realistic tenant's figure of £1,400—increases profit by £2,600.
13. Remove both bank and loan interest—increases profit by £14,900.
14. Consider if depreciation on improvements to property would be for tenant's or landlord's structure: decision taken to assume that it was a tenant improvement—leave at £2,000.
15. Assess value of farmer's manual labour—decision £6,800.
16. Calculate MII by deducting value of farmer's labour from NFI. Answer: £28,358−£6,800=£21,558.

Target figures set in advance of operations are used by some people as their standards.

5.2.2. Step 1—Analysis of Profit

The first step in analysis is to assess the efficiency of performance of the business as a whole by an examination of its profit. This is usually done by calculating the NFI and the MII per hectare and comparing the results with appropriate standards. The owner of the business then has to decide if the performance attained meets his objectives and justifies the management involved, plus the capital invested, with all the attendant risks.

The return on tenant's capital provides a means of assessing whether the capital investment is justified. This is given by the formula:

$$\text{Return on tenant's capital} = \frac{\text{MII}}{\text{Tenant's capital}} \times 100$$

The tenant's capital can be obtained by reference to the balance sheet. Most farmers want to see their stake in the business increasing from year to year. Reference to the capital account in the balance sheet and the following formula can give the percentage growth in net worth:

$$\frac{\text{Profit for year} - \text{Private drawings}}{\text{Opening net worth}} \times 100$$

Ideally the percentage growth should at least keep pace with inflation. Obviously excessive private drawings relative to the profit can be an obstacle.

It can be seen from Fig. 55 that profit = output − costs. Clearly if profit is inadequate it may be a result of insufficient output or excessive costs, or a combination of the two. It is therefore necessary to examine both outputs and costs and to relate one to the other.

5.2.3. Step 2—Examination of Outputs

The gross output per hectare can be calculated and compared with standards. Alternatively the net output per hectare can be employed for this assessment.

Net output = gross output − (cost of purchased feed + purchased seed). The deduction of the cost of those items which have been grown on other farms produces, in net output, a measure which in many ways more accurately reflects the true productivity of the farm land than gross output. For example, a farmer who buys a large proportion of his stock feed might be expected to be more intensively stocked and have a higher gross output than a farmer who buys very little feed. Net output is considered by many to be a better basis on which to compare such cases.

Clearly farmers with large and intensive pig, poultry, barley beef, or veal calf units, the so-called concrete enterprises, can be expected to have very high outputs per unit area. To avoid their influence it is common practice to calculate the gross or net outputs per hectare from crops and grazing livestock, thus omitting those enterprises which do not directly use the land.

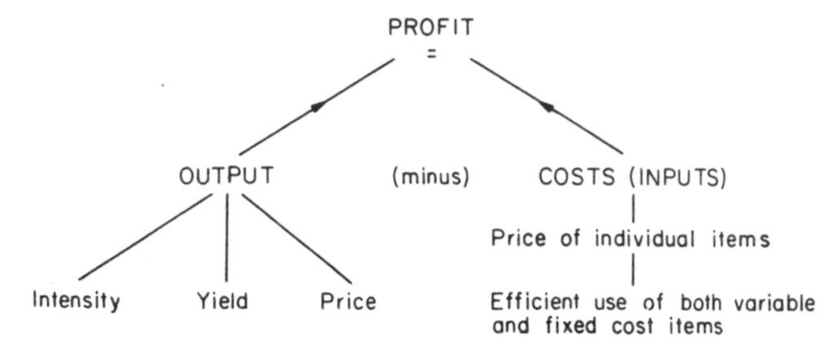

FIG. 55. The profit formula.

The output of a farm is a reflection of the intensity of the farming system and its enterprises, the yields of the crops and stock, and the prices received for the goods (Fig. 55).

A. Intensity

Intensity is influenced by three main factors:

(i) the output character of the farm's enterprises;
(ii) the density of stocking;
(iii) turnover efficiency.

(i) *The output character* of an enterprise refers to the level of output which might typically be expected from that enterprise. For example, dairy cows usually have a high output per hectare compared to single suckler cows. Thus, if when comparing two farms with similar land types one has predominantly dairy cows and the other has a commercial, as opposed to pedigree, single suckler unit, the first farm could be expected to have the highest output.

Some authorities measure the output character of a farm by using standard outputs. They list the area devoted to each enterprise on the farm and multiply these figures by the standard outputs for the respective enterprise. The standard outputs are a reflection of what could be expected with typical yields and prices. The total standard output for the farm is then compared to other farms to obtain a measure of the level of intensity.

(ii) *The density of stocking* is frequently measured by employing the livestock unit system. Tables such as that shown in Fig. 56 are compiled to relate all animals to a standard livestock unit (LU) which is one dairy cow. Reader's should not be surprised to see different figures used in different publications although some attempt at standardisation has been made in MAFF publication *Definition of Terms used in Agricultural Business Management*, which relates stock in terms of their metabolisable energy requirements to a 625 kg Friesian cow producing a calf and 4,500 litres of milk per annum. Figure 56 has been based on their data.

The system is used not only to measure the density of stocking but also to allocate the variable costs of forage production and the forage area to respective classes of stock. Although it involves a slight digression it is a convenient point to illustrate the full use of the technique by reference to Church Farm.

The stocking of the farm on an average carry basis is obtained by taking the average of 12-monthly counts of respective livestock numbers. The appropriate LUs are then entered and the total LUs obtained (Fig. 57).

Stock	Livestock units	Stock	Livestock units
Dairy cows	1.00	Ewes and ewe replacements	
Beef cows	0.75	(excludes suckling lambs):	
Bulls	0.65	Light weight	0.06
Barley beef	0.47	Medium weight	0.08
Other cattle:		Heavy weight	0.11
0–12 months	0.34	Lambs:	
12–24 months	0.65	Birth to store	0.04
24+ months	0.80	Birth to fat	0.04
		Birth to hoggets	0.08
		Purchased stores	0.04
		Rams	0.08

FIG. 56. Table for conversion of stock numbers to livestock units.

Average stock carry		Livestock units	Total	
Dairy cows:	100	×1		100.00
Dairy young stock:				
0–1 year	36	×0.34	12.24	
1–2 year	35	×0.65	22.75	
2+ year	17	×0.80	13.60	48.59
18–20 month beef:				
0–1 year	40	×0.34	13.60	
1–2 year	26	×0.65	16.90	30.50
Ewes	200	×0.11	22.00	
Lambs	300	×0.04	12.00	
Rams	6	×0.08	0.48	34.48
				213.57

FIG. 57. LU calculation—Church Farm.

The cropping of Church Farm is:

	Hectares
Grass	100
Maincrop potatoes	10
Winter wheat	30
Spring barley	60
	200

The total number of LUs is divided by the forage area on the farm to obtain the measure of stocking density. The forage area includes the grass used for grazing and conservation and such items as kale, rape, swedes, but not homegrown feed grain crops.

$$\text{Church Farm stocking density} = \frac{213.57}{100} = 2.14 \text{ LU per forage hectare}$$

The area of forage that each type of stock uses can be obtained by dividing the total LUs attributable to each class of stock by the stocking density:

	Hectares
Dairy cows	100 ÷ 2.14 = 46.72
Dairy young stock	48.59 ÷ 2.14 = 22.70
Beef	30.50 ÷ 2.14 = 14.25
Sheep	34.48 ÷ 2.14 = 16.11
	99.78

The addition of the respective areas gives a cross-check on the arithmetic, but slight errors due to rounding-off can occur.

The total forage cost on Church Farm was £19,221. This can be allocated by first dividing the forage costs by the total number of grazing LUs and then multiplying the results by the total number of LUs for each class of stock:

$$\text{Forage cost per LU} = \frac{£19,221}{213.57} = £90$$

Allocation:

Dairy cows	100 × 90 =	£9,000
Dairy youngstock	48.59 × 90 =	£4,373
Beef cattle	30.50 × 90 =	£2,745
Sheep	34.48 × 90 =	£3,103
		£19,221

Doubts about the accuracy of such allocations have been expressed by some people. In theory adjustments should be made for high-yielding dairy cows so that animals giving above 4,500 l are allocated more than 1.00 LU. In practice few people do this and in fact on the author's college farm cows giving 8,000–9,000 l probably eat less forage in the form of silage during the winter than animals producing 4,500 l. This raises some doubts about the justification of any adjustment in LUs according to yield. Practical experience on the same farm suggests there could be a question about the comparability of ewes and lambs with dairy cows on the basis of the new MAFF system. The latter amended the old system which assumed that five ewes and their lambs were equivalent to one cow. A lot depends upon the system of management. It could be argued that ewes spend a considerable part of their time acting as scavengers on most farms, especially during the winter, and that the lower allocation of LUs to sheep is justified.

As with most management practices, intelligence has to be used in the application and interpretation.

The cow equivalent grazing day (CEGD) system can produce more accurate allocations of forage, but it is time consuming. Like the LU system it relies on the comparison of all stock to one cow. Detailed grazing records have to be kept which register the number of animals in terms of cow equivalents in particular fields, together with the number of days the stock are in these fields. For example, assume that 80 cows are in field A for 4 days and 200 ewes plus 300 lambs are in field B for 12 days, the total CEGDs for these situations are:

	Stock	Cow equivalents	Days	CEGD	
Field A	100 cows	100 × 1 = 100	4		400
Field B	200 ewes	200 × 0.11 = 22	12	264	
	300 lambs	300 × 0.04 = 12	12	144	408

Winter fodder is allocated accurately to stock and 1 tonne of silage consumed is equivalent to 24 CEGD, whereas 1 tonne of hay is equivalent to 75 CEGD.

At the end of the year the total number of CEGDs for each class of stock is calculated and the forage area and costs are allocated in proportions relative to the respective totals. Alternatively, sectors of the farm can be considered, and from detailed records of forage costs for each sector, together with those for stock using the forage from these areas, even more accurate allocations can be made.

(iii) *Turnover efficiency* of both crops and animals can influence outputs and, depending upon the costs involved, can also affect the return on capital employed. Examples: reducing the time taken to fatten stock to increase throughputs; improving the calving or the farrowing index to improve production per year; following early potatoes by a catch crop to get most out of the land. Selling winter wheat in the autumn, rather than storing it, and using the money to purchase store bullocks to fatten over the winter, illustrate how turnover of capital can increase outputs.

B Yields

Yields from both stock and crop have a significant influence on output. For example, consider the effect on output of an increase in the yield of wheat of 0.5 t/ha at current prices. If this is achieved by better husbandry, which may cost nothing, the impact on profit is significant. Yields of stock and crop and performance figures such as liveweight gain per day should be examined and compared with standards.

This is an opportune time to remind the reader of the care needed in calculating averages whether it be for yields, prices, or any other factor. Consider a farm which has the following yields for wheat:

Field	Size of field (ha)	Yield per hectare (tonne)	Total yield of field (tonne)
A	20	4.5	90
B	30	5.6	168
C	40	6.2	248
D	10	5.8	58
	100	22.1	564

A common mistake is to say that the average yield is 22.1 t divided by 4 fields equals 5.52 t. In fact it is 564 t divided by 100 ha equals 5.64 t/ha, i.e. a weighted average should be used.

C. Prices

Prices are the other main factor influencing output. They should be examined to see if they are the source of strength or weakness for the business. Marketing policy may warrant examination including timing, type of market, and type and quality of product. The significance of prices can be illustrated by the fact that an increase of 0.1p per litre of milk would increase Church Farm's income by £583.

5.2.4. Step 3—Examination of Costs

Having examined outputs the first ideas for improvements may start to crystallise, but costs and use of resources must be studied before any decisions are taken.

The first step is to calculate the cost per hectare for each of the following and then to compare the results with standards:

(i) *Labour costs* including casual and unpaid labour. Check that the standards include a charge for the manual work of the farmer and his wife.
(ii) *Machinery costs* including fuel, electricity, repairs, vehicle licences, depreciation, and contractor's charges. Each of these may also be examined individually on a cost per hectare basis.
(iii) *Rent*
(iv) *Other fixed costs* studied collectively or individually.
(v) *Fertiliser costs.*

5.2.5. Step 4—Relating Costs to Outputs

Great care is necessary when interpreting these results. On some farms labour can substitute for machinery and on others, which are more highly mechanised, the reverse situation can occur. The interelationship between the machinery and the labour costs must be examined.

Such results are also little guide to the efficiency with which the respective items are used, and it is important to establish the result achieved in comparison to the money spent. This can be done in part by relating outputs to costs. The following should be calculated and compared to standards:

Gross output or net output/£100 labour costs,
Gross output or net output/£100 machinery costs,

and to partially overcome the problem of labour/machinery substitution:

Labour + machinery costs per hectare,
Gross output or net output/£100 labour + machinery costs.

Generally speaking the gross output should be at least five times the labour costs and net output at least four times the labour charge. More efficient farms can produce much better figures.

The gross output should be seven or eight times the machinery costs, depending upon the type of farm, with the net output five times. More efficient farms are one point higher in each case. It must be noted that these are generalisations, and appropriate standards should be employed.

Unfortunately, expressing gross output or net output per £100 labour and/or machinery costs still has limitations. A high output can be achieved with extremely good husbandry and could mask a very inefficient use of labour or machinery. Conversely, the efficient use of labour or machinery could mask poor output and still produce a reasonable figure for output per £100 labour or machinery.

Each of the figures, gross output per hectare, labour costs per hectare, gross output per £100 labour, etc., should be examined individually and then carefully related to one another.

If any of the results are unsatisfactory it may be necessary to study resource use in more detail. For example, if machinery costs appear to be high, detailed analysis may be warranted. This may focus attention on specific items such as the replacement policy. Techniques employed to study labour and machinery use will be detailed in the appropriate chapter.

Although appearing rather detailed for whole farm analysis it is common practice to work out other efficiency factors to measure the cost/effectiveness of what are called the variable costs. The following list illustrates this point:

(i) gross output per £100 feed costs for pigs or poultry;
(ii) margin of the value of milk produced over concentrate costs;
(iii) feed conversion factors;
(iv) cost of feed per kilogram of liveweight gain.

It is also common practice to calculate the gross margin per hectare for the whole farm even where there is insufficient recording of the variable costs to permit full gross margin analysis.

5.2.6. Step 5—The Appraisal

The final step is to produce a full appraisal of the business. This is facilitated if the calculations for the efficiency factors are listed together with standards as shown for

Efficiency factor	Church Farm £	Top 25% £	Average £
NFI per ha	142	156	112
MII per ha	108	115	65
Gross margin per ha	591	625	540
Gross output per ha	1,073	1,130	970
Labour costs per ha*	210	197	175
Machinery costs per ha	148	156	164
Rent and rates per ha	113	117	98
Other fixed costs per ha	35	40	38
Gross output per £100 labour	512	573	554
Gross output per £100 machinery	726	724	591
Gross output per £100 labour and machinery	300	320	286
Fertiliser costs per ha	122	128	90
Return on tenant's capital %	9.68	11	1
Stocking LU per ha	2.14	2.4	2.0

*Including £6,800 for farmer's manual labour.

FIG. 58. Efficiency Factors and Standards—Church Farm

Church Farm in Fig. 58. Two standards columns are listed; one shows the average results for a sample of similar farms together with data from the top 25% of these farms in terms of profitability.

For the exercise it is assumed that the objective for Church Farm is to generate a level of profit which will provide the farmer with a good standard of living, give a return on capital which is comparable with that being achieved by the best group of farmers, and obtain performances which maintain or improve the future viability of the business. The farmer feels, however, that he does not wish to increase his contribution to the manual labour.

Present performance figures show that the business can be compared with others in the top 25%, although there are obviously some in the group with better results. The failure of Church Farm to completely match the gross output per hectare for this group must be related to its relatively inferior stock carry. It appears to be using a similar amount of fertiliser to the others in the group, therefore the issue of "fertiliser application practice" and "grassland utilisation management" could be raised. But note that the farm is "mixed cropping and livestock". Before final conclusions are reached it would be necessary to compare actual use of fertiliser on grass.

The labour costs on Church Farm are higher than others in the top category but it would be essential to study the proportion of land devoted to labour intensive activities, such as potatoes, between this and the other farms.

Machinery costs are slightly below the figure shown by the top set but Church Farm uses casual labour to pick potatoes and others may have machines.

A check on Church Farm and on the standard farms shows the following percentage contribution to gross output from respective enterprises:

	Church Farm (%)	Top 25% (%)	Average (%)
Milk	46	47	40
Cattle	16	15	14
Sheep	6	5	7
Potatoes	11	12	8
Cereals	20	20	30
Miscellaneous	1	1	1

Church Farm compares very much with the top group, whereas the average group appear to derive a lower percentage of their output from high output character enterprises such as dairy cows and potatoes.

Generally the results for Church Farm are well above those for the average group, notably in terms of MII and return on tenant's capital. This does not mean that the farmer should be complacent. These results are just for one year and are historical. It is the future which counts.

It is clear from the appraisal that several points need further examination. If sufficient data was not available to enable detailed gross margin analysis to take place, attempts would be made to assess those factor's contributing to output such as yields and prices from any information which could be found. In this case the matter is left until gross margin analysis is covered.

It is common under whole farm analysis to make some reference to the balance sheet, and indeed it is necessary to refer to the balance sheet in order to calculate the tenant's capital on which to base the return on capital.

5.3 BALANCE SHEET ANALYSIS

5.3.1. Tenant's Capital (Operating Capital)

Tenant's capital consists of those assets which would be common to a business irrespective of it being tenanted or owner-occupied. In most cases it is obtained by calculating the average of the opening and the closing valuations of the following items:

(i) stores and sundries;
(ii) crops in store;
(iii) crop in ground;
(iv) trading livestock;
(v) machinery;
(vi) breeding livestock;
(vii) permanent improvements.

Generally speaking this can be described as a satisfactory practice although it is conceivable that a case could arise where a high percentage of the stock and crop had been sold and at the time of the valuations the proceeds were not reinvested in the business. To avoid this some organisations practise a second method and arrange for valuations to be made each month and take the average of these. This contrasts with the third, and crudest method, which is to simply take the closing valuations for the tenancy items listed above.

None of the methods fully considers the day-to-day working capital involved which can vary substantially from business to business. It will usually be highest on farms without a regular monthly income, and comparatively lower than might be thought for dairy farms where there is a monthly milk cheque.

5.3.2. Return on Tenant's Capital

The formula for the calculation of the return on tenant's capital was given in Section 5.2.2. Anything which influences the MII or the tenant's capital will obviously affect the result.

Valuation changes can very much influence output and therefore the MII. Any change recorded must be realistic, as must the notional rent attributed to an owner-occupied farm and the value of the manual labour of the farmer and his wife.

A major limitation of calculating the return on capital from most balance sheets is that the asset values for many items are entered at below realistic values with the result that the return on capital is falsely high. An attempt can be made to calculate more realistic values, and care must be taken when comparing returns on capital to check the approach used in calculating the capital involved.

5.3.3. Growth in Net Worth

The net worth is the claim that the owner has against the business after all other legitimate claims have been met. A review of the balance sheets for several years should, as suggested earlier, show a growth in net worth which at least keeps pace with inflation. Care has to be taken, however, to allow for any injections of additional personal capital which have been made. Examination of private drawings in relation to profits points to the amount of capital which has been left in the business for growth and development.

The calculation of the growth in net worth for Church Farm indicates the care that is necessary in interpretation (see Fig. 53). From the business point of view it appears to be £310,195−£299,798=£10,397, which represents a growth of 3.47% over last year's figure. However, £1,000 was introduced from external sources and borrowing was reduced by £9,000 effectively increasing the percentage of the liabilities represented by the farmer's stake or equity. These points must be considered when comparing the growth in net worth with previous years and with standards.

Another point which must be remembered is that this farm is owner-occupied and no increase in value of the land has been entered into the calculation. In spite of this, the land value influenced the opening net worth, and hence the divisor in the calculation of the percentage growth in net worth. This could make the task of increasing the percentage growth more difficult than for tenant farmers, even allowing for factors such as rent. Again it calls for care in interpretation, or for a management form of balance sheet with up-to-date values, and with all farmers placed on the same footing with reference to ownership. In practice few people undertake this procedure.

5.3.4. Current and Fixed Assets

Current assets consist of cash in hand or cash at bank and those items which could be converted into cash within a relatively short space of time. They include:

(i) cash in hand;
(ii) cash at bank;
(iii) sundry debtors;

(iv) stores and sundries;
(v) crops in store;
(vi) crops in ground;
(vii) trading livestock.

Any investments outwith the farm made by the business which could be cashed in the short term would also be included.

Fixed assets are longer term investments in the farm and comprise:

(i) breeding livestock;
(ii) machinery;
(iii) tenant's fixed capital;
(iv) heritable property.

Tenants fixed capital refers to items such as building improvements or extensions.

The current assets of Church Farm are £102,150 and the fixed assets £379,045.

5.3.5 Liquid Assets

Liquid assets are those liquid funds such as cash in hand or cash at bank and near liquid assets which include sundry debtors. The total for Church Farm is £8,450.

5.3.6. Current Liabilities

Current liabilities are claims on the business which have to be met in the short term, normally within a year. They usually consist of bank overdraft and sundry creditors. The current liabilities for Church Farm are £37,000.

5.3.7. Balance Sheet Ratios

A study of the relationship between the respective categories of assets and liabilities gives some measure of the ability of the business to meet the claims which could be made against it in the short, intermediate, and to some extent long term periods. This information is of value not only to the farmer when planning but also to a person, such as a bank manager, who is considering the viability of the business with the prospect of lending it money.

Fixed assets tie up capital for a long time, and in order to allow the business to remain solvent any borrowing for them should generally be from long term sources. Current assets can be financed from short term borrowing in part, but it is desirable that a proportion is derived from long term sources to avoid the possibility that some current assets have to be prematurely realised to finance short term claims.

5.3.8. Short and Medium Term Ratios

$$\text{Current ratio} = \frac{\text{Current assets}}{\text{Current liabilities}}$$

$$\text{Church Farm data} = \frac{102{,}150}{37{,}000} = 2.76:1$$

This measures the intermediate solvency of the business. In the case of Church Farm it shows that it is reasonably placed to meet short term liabilities and the situation has improved from the previous year when the ratio was 2.33:1. If all the current liabilities had to be met at once, few of the current assets would have to be realised. Had the figure been less than 2:1 it could indicate a more risky situation although the acceptable ratio always depends upon how near to sale the respective assets are.

$$\text{Liquidity ratio} = \frac{\text{Liquid assets}}{\text{Current liabilities}}$$

$$\text{Church Farm data} = \frac{8{,}450}{37{,}000} = 0.23:1$$

Frequently a 1:1 ratio is said to be desirable, but this needs considerable qualification. In the case of Church Farm there is £23,200 worth of crop in store ready for sale. If this is considered in relation to the current liabilities, the 0.23:1 ratio gives no major concern for worry. If the assets had not included such readily saleable items, the extent of the bank overdraft facilities would be examined, creditors would be studied to see how long payment to them could be delayed, debtors would be approached for payment, or some of the assets would have to be sold.

If liquid assets are excessively high relative to current liabilities, it might indicate that some capital is being under-utilised.

5.3.9. Working Capital

Accountants frequently talk of the term working capital which is current assets minus current liabilities. For Church Farm it is £65,150.

A high figure signifies that a large proportion of the current assets have been financed by capital which does not have to be paid back quickly. Some farms need more working capital than others. If there is a long interval between expenditure and sale of goods produced the requirement will tend to be higher than if the period is short. This is exemplified by the case of costs incurred in keeping cows for suckler beef production, where there is a long interval, and cows kept for milk production, where the milk cheque is received quickly, facilitating prompt payment of feed and other bills.

If a business is able to obtain credit over an extended period at reasonable cost it can ease the need for working capital. For example, some merchants allow credit for cereal seeds and fertilisers until the grain is harvested and can be sold.

The ratio of working capital to fixed assets can be calculated to see if investment in fixed assets is impairing the ability of the business to pay current production expenses. Equally if calculated from two or three succeeding balance sheets it can show if fixed assets are having to be depleted in order to maintain sufficient working capital. This can be serious since it prejudices future production.

5.3.10. Long Term Aspects

The relationship between the owner's net worth and the capital owed to outside agencies is very important from several points of view.

(i) *The net capital ratio* is one method of making this comparison.

$$\text{Net capital ratio} = \frac{\text{Total assets}}{\text{Total liabilities} - \text{Net worth}}$$

	Church Farm (£)	Contrast Farm (£)
Total assets/liabilities	481,195	481,000
Net worth	310,195	100,000
Liabilities—net worth	171,000	381,000
Net capital ratio	2.81 : 1	1.26 : 1

It can be seen that if Church Farm experienced a difficult period and made a loss, or the profit was below what the owner needed for living expenses and taxation, it would be less serious than for Contrast Farm where the net worth is only equal to a small part of the total assets. Naturally if Church Farm made losses for a number of years it would also be very serious, but the owner would have more time to change the policy than the owner of Contrast Farm. In addition someone asked to lend money would be much happier to consider the request from Church Farm than from the other business. Banks and other sources of capital tend to look at balance sheets from several years to see the trend in net worth in relation to other liabilities before lending money.

(ii) *Capital gearing* measures the ratio of the capital borrowed on a long term basis to the owner's net worth.

	Church Farm (£)	Contrast Farm (£)
Long term loan—A	134,000	300,000
Net worth—B	310,195	100,000
Gearing (ratio A : B)	1 : 2.3	3 : 1

Contrast Farm is more highly geared than Church Farm because the ratio of capital borrowed on a long term basis to net worth is higher. This can have some advantage in a highly profitable situation, but when returns are low it can create problems. Much depends upon the interest rates being paid on the borrowed capital compared to the return made on all the capital invested in the business. Consider the situation shown at the top of the next page.

Clearly in all cases Church Farm has the highest surplus because of the enormous interest cost Contrast Farm incurs, but in terms of return on owner's capital Contrast Farm is superior provided that the return on total capital employed is greater than the interest paid on the long term capital.

	Assuming return is 12% of total capital employed		Assuming return is 5% of total capital employed	
	Church Farm (£)	Contrast Farm (£)	Church Farm (£)	Contrast Farm (£)
Total return	57,743	57,732	24,060	24,050
Interest on long term loan at 9%	12,060	27,000	12,060	27,000
Surplus left	45,683	30,782	12,000	−2,950
Surplus as % of owner's capital	14.72	30.73	3.87	Negative

There are two points to be made. First, that the risks in having a high percentage of borrowed capital must never be forgotten. Second, that if the risks are not too high it can pay handsomely to farm on borrowed money if this can be used to generate a higher return than the interest which has to be paid on it, since the surplus can be kept by the person taking the risk of borrowing. It may be necessary for some businesses to become more highly geared in order to make a good profit.

(iii) *The ratio of times covered* expresses the number of times by which the fixed interest charges of the business are exceeded by its profits before deduction of tax and interest. A ratio of at least 5:1 is sometimes recommended because at this level the profits could decline by 80% before the interest charges would equal the profit. In times of high interest rates many have to accept poorer ratios.

On Church Farm the ratio was just over 4.6:1 and unless profits increase it would perhaps still be short of 5:1 next year.

5.3.11. Church Farm—Conclusion

Most financial sources would examine several balance sheets, rather than the one permitted by space in this book, before lending money. On the surface the ratio of times covered suggests caution, but if additional borrowed capital created significantly higher profits this ratio could alter. The farm is low-geared and a significant proportion of the assets are owned by the proprietor. They could be used to pay off existing loans plus substantial new ones if the need occurred.

It is highly possible that if the management of Church Farm was considered to be of a sufficiently high standard by a lending agency and suitable plans for the future were forthcoming, the business could attract both long and short term loans. Bank managers and others would therefore look further than the balance sheets before making their decisions.

5.4. GROSS MARGIN ANALYSIS

5.4.1. Introduction

Gross margin analysis is used to study the performance of the enterprises, i.e. the productive components, of a farm to obtain information about the business's strengths and weaknesses in greater depth than can be derived by whole farm analysis. The technique can only be adopted if sufficient records have been kept to facilitate the allocation of variable costs and items which contribute to the outputs.

Essentially gross margins measure the efficiency with which variable cost inputs are converted into outputs. Maximum outputs do not always produce the highest gross margins especially if they are associated with very high variable costs. Moderate outputs achieved with low variable costs can produce satisfactory results.

The usual procedure is to calculate the gross margin per head for each group of stock and at the same time establish the gross margin per forage hectare used by the respective group of animals. For arable crops the gross margin per hectare is obtained.

When the gross margins have been calculated they can be compared with last year's results on the same farm, with figures which were budgeted for the respective enterprises or with standards from similar farms.

The procedure can conveniently be divided into a series of steps.

5.4.2. Steps in Gross Margin Analysis

Step I

The first step is to decide the activities for which gross margins will be calculated. For example it is usual practice to divide wheat into winter wheat and spring wheat, dairy herds are split into dairy cows and dairy replacements, pigs may be separated into breeding units and finishing units, and cattle for beef production are split into different categories according to the finishing system adopted. Crops which could be sold including barley that is consumed on the farm, are treated separately but grass, hay, silage, and other forage crops are usually not differentiated, unless intended for sale, but are allocated as variable costs to the stock consuming them.

Step II

This step entails checking records for accuracy. The value of good farm records will be realised in the next steps when allocating various items to respective activities.

Feed allocation frequently presents most problems. Veterinary surgeons will usually give a break down of their costs according to class of stock.

Records of interdepartmental transfers are essential. In the last chapter it was mentioned that it is imperative that each activity for which a gross margin is to be calculated should stand on its own and must neither be subsidised by, or provide subsidy for, any other activity. If this is not rigidly adhered to, a false picture of the efficiency of the respective activities will arise and negate the value of any findings.

The example of barley grown on the farm was mentioned before, together with the need to transfer it at real market values. Other examples include calves born to dairy cows which are credited to their dams, but any transferred for rearing are debited to the dairy young stock activity or the beef unit. Home-reared dairy heifers entering the milking herd are credited to the dairy young stock at realistic market values and debited to the dairy cows. There are many other examples all of which require records.

Remember that all this is done so that sufficient gross margins can be calculated to check whether it is worth growing feed barley, if it pays to rear dairy stock, especially if some are sold, or which beef system is most profitable.

Step III

The calculation of the gross margins for each activity is undertaken in turn. First it is necessary to establish the respective enterprise outputs.

Sales must include all interdepartmental transfers. For example, milk transferred to the calves must be credited to the dairy cows but only if it was suitable for sale. Any subsidies received should be included.

Valuations must be in sufficient detail to facilitate calculations for respective activities. Changes in value per unit of animal or crop between the opening valuation and the closing valuation assessments must be treated with care. If there is a genuine difference in quality or, in the case of stock, in size, the different valuations are acceptable. If the difference is more attributable to monetary inflation or deflation, i.e. for reasons which are not a true reflection of the management efficiency and are not repeatable to the same degree, then it is recommended that changes are omitted. For example, if dairy cows are each valued at OV at £300 and at CV at £350, £350 should be taken in each case. If this is not done then comparisons between the years will be less valid and, as suggested above, the issue will be clouded by whether the business has operated in a period of fluctuating monetary values.

The alternative to this is to enter the valuation change and state at the end of the calculation the portion of the gross margin attributable to changes in value of money.

Possible exceptions can be considered. If the herd in which the cows had a CV of £350 had substantially increased in size with the introduction of a large number of heifers at £600 each, the drop in values to £350 each would have a significant impact on the gross margin. In this case a commonsense approach would be required and some increase in valuation would be justified or, alternatively, it would be necessary to interpret the gross margin prepared with care.

Purchases must include interdepartmental transfers of stock.

Step IV

Next the variable costs must be established for each activity. Interdepartmental transfers, such as feed barley, must be charged. The problem of allocating forage costs raises some questions, but the methods outlined under whole farm analysis can be employed. Frequently the gross margins for stock using forage are first calculated before deduction of forage costs and then after a charge has been included.

In any form of comparative analysis it is essential that like is compared with like. Care must be taken that the items included in gross margins which are to be compared are similar. For example both Farmers A and B have identical cereal enterprise outputs and the same seed, fertiliser, and spray costs. However, Farmer A employs a contractor to combine harvest his grain and bale his straw, and casual labour also assists with carting the grain. These items are all variable costs. Farmer B does all this work with his own machinery and staff. Farmer A will therefore have a much lower gross margin than Farmer B even though they are basically equally efficient.

This does not negate the value of gross margins, but illustrates the care which is necessary in interpretation.

Step V

In this step the gross margins are compared with standards and an appraisal of performance is undertaken. Great care is necessary. There can be significant fluctuations in gross margins from year to year due to weather, prices changes, and many other factors, some outside the farmer's control. When using them as a basis for budgeting they are normalised. This means that a typical performance over recent years is taken rather than results at the extreme ends of the spectrum. Changes in husbandry and management may, however, have produced long term repeatable performances. All reasons for changes should be examined to see if they can be exploited further or removed if adverse.

Avoid the old student trap. Having established the gross margins per hectare some then deduct the average fixed cost per hectare to obtain net profit. Obviously this is unrealistic because, for example, potatoes contribute significantly more per hectare to fixed costs than sheep.

5.4.3. Cropping Year Basis

Figure 59 illustrates why it is advisable to calculate gross margins for crops on a cropping year basis. It shows that a single crop of winter wheat can overlap three financial years. Consider year II crop. This is sown in the autumn of year I, harvested in year II, stored and sold in year III. Assume that 30 ha, 40 ha, and 50 ha were harvested in years, I, II, III respectively. The financial data in year II would refer to 30 ha sold, 40 ha fertilised, sprayed, and harvested and 50 ha seed—establishment costs. It is necessary therefore to select a harvest year and to extract the establishment costs

Year I	Financial year II		Financial year III		
March	April	March	April	March	April
	Year I Crop sold				
Year II Crop sown	Year II Crop harvested		Year II Crop sold		
		Year III Crop sown	Year III Crop harvested		

FIG. 59. Gross margins—cropping year

from the previous year, sale value from the next year, and consider these together with the variable costs of the harvest year if a meaningful answer is to be obtained.

Care has also to be taken when calculating the gross margin per head from growing and finishing animals. The data from within a financial year may refer to different numbers of animals at respective ages, e.g. there may be 30 calves 0–6 months, 20 calves 6–12 months, and 25 calves 1–2 years. When the numbers are identical the actual gross margin of the older bunch cannot be ascertained without looking at a previous financial year because costs may have changed.

5.4.4. Church Farm—Gross Margins

The following data was collected from Church Farm:

(i) Sales off Farm

	£		£		£
Milk	98,000	Lambs	13,500	Wheat	18,700
Cull cows (22)	7,128	Wool	650	Straw:	
Calves (14)	868	Cull ewes	1,280	Barley	1,680
Dairy heifers (11)	6,600	Potatoes	22,760	Wheat	750
Beef cattle (39)	20,670	Barley	15,100		

Plus: £450 worth of milk to house and workers
£600 ewe premium

(ii) Interdepartmental Transfers

	£
Calves from dairy herd to dairy young stock (37)	2,220
Calves from dairy herd to beef unit (42)	2,940
Dairy heifers from dairy young stock to dairy herd (25)	15,000
Milk from dairy cows to dairy young stock	450
Milk from dairy cows to beef	500

Barley produced from:	Last year's harvest £	This year's harvest £
to dairy cows	—	5,800 (56t)
to dairy young stock	518 (5t)	1,036 (10t)
to beef	518 (5t)	1,346 (13t)
to sheep	—	466 (4.5t)
Straw: to dairy young stock	—	480 (24t)

(iii) Purchase of Stock

Gimmer sheep (42) £2,772 Annual Share Tups Bought £400.

(iv) Variable Costs—usage (adjusted for stocks on hand)

	Fertilisers (£)	Seeds (£)	Sprays (£)	Sundries (£)	Casual labour (£)
Forage	15,600	1,236	855	1,330	—
Potatoes	1,440	2,250	390	1,900	4,500
Wheat	2,950	1,400	1,500	150	—
Barley	4,320	3,180	1,680	240	—
	24,310	8,066	4,425	3,620	4,500

	Purchased feed £	Vet. and medicines £	Miscellaneous £
Dairy cows	24,000	1,600	3,200
Dairy young stock	3,506	500	820
Beef	3,816	360	880
Sheep	1,600	650	480
	32,922	3,110	5,380

(v) Valuations

Opening valuation (£)		Closing valuation (£)
32,000	Dairy cows	32,000
24,900	Dairy young stock	24,000
21,700	Beef	21,900
10,780	Sheep	10,780
18,700	Wheat	19,400
2,884	Barley	2,100
1,800	Silage and hay	1,600
100	Straw	100

The records are now used to calculate the gross margins.

Forage Costs

It is recommended that where grazing livestock are involved the first step should be to allocate the forage costs. It can be seen from the above records that the total cost for forage of fertiliser, seeds, sprays and sundries was £19,021. There was a decrease from £1,800 to £1,600 in the value of the forage valuations which in effect means that £200 worth of winter fodder was consumed above the amount produced in the year. The total forage cost was therefore £19,221. This was allocated to each class of stock earlier in the chapter, on page 105.

Gross Margin for dairy herd (100 cows on average)

	£	£	£
Output:			
Milk: sold off farm	98,000		
to farmhouse	450		
to calves	950	99,400	
Calves: sold (14)	868		
to dairy young stock (37)	2,220		
to beef (42)	2,940	6,028	
Cull cows (22)		7,128	112,556
Less: Heifers transferred in (25)			15,000
		OUTPUT	97,556
Variable costs:			
Purchased concentrates	24,000		
Home grown barley	5,800	29,800	
Vet. and medicines		1,600	
Miscellaneous and AI		3,200	34,600
Gross margin before forage costs			62,956
Forage costs for cows			9,000
Gross margin after forage			53,956
Gross margin per cow (÷100)			540
Forage area allocated earlier in chapter to cows			46.72 ha
Gross margin per forage hectare from cows	$\frac{53,956}{46.72}$	=	£1,155

Gross Margin for dairy young stock

	£	£	
Enterprise output:			
Closing valuation	24,000		
Sales off farm (11)	6,600		
Transfers to dairy herd (25)	15,000		45,600
Less: Calves from dairy herd (37)	2,220		
Opening valuation	24,900		27,120
		OUTPUT	18,480
Variable costs:			
Purchased concentrates	3,506		
Home grown barley	1,554		
Milk from dairy herd	450		
Vet. and medicines	500		
Miscellaneous	820		
Straw	480		7,310
Gross margin before forage costs			11,170
Forage costs			4,373
Gross margin after forage			6,797

Forage area allocated to dairy young stock 22.7 ha

Gross margin per forage hectare = $\frac{6,797}{22.7}$ = £299

Gross Margin from Beef

	£		£
Enterprise output:			
Closing valuation	21,900		
Sales (39)	20,670		42,570
Less: Calves from dairy herd (42)	2,940		
Opening valuation	21,700		24,640
		OUTPUT	17,930
Variable costs:			
Purchased concentrates	3,816		
Homegrown barley	1,864		
Milk from dairy herd	500		
Vet. and medicines	360		
Miscellaneous	880		7,420
Gross margin before forage			10,510
Forage cost			2,745
Gross margin after forage			7,765
Forage area allocated to beef			14.25ha
Gross margin per forage hectare			£545

Gross Margin for Sheep (average carry 200 ewes)

	£		£
Output:			
Closing valuation	10,780		
Lambs sold	13,500		
Cull ewes	1,280		
Wool	650		
Ewe premium	600		26,810
Less: Gimmers bought (42)	2,772		
Tups bought	400		
Opening valuation	10,780		13,952
		OUTPUT	12,858
Variable costs:			
Purchased concentrates	1,600		
Homegrown barley	466		
Vet. and medicines	650		
Miscellaneous	480		3,196
Gross margin before forage			9,662
Forage costs allocated			3,103
Gross margin after forage			6,559
Gross margin per ewe (÷200)			£32.8
Forage area allocated			16.11ha
Gross margin per forage hectare			407

Gross Margin for Wheat (30 ha)

Care is necessary to ensure that only data relevant to the crop harvest year being studied is employed. Thus the farmer notes from his records that the £18,700 under sales for wheat refers to grain from the previous year. This year's crop is all in store and is shown in the Closing Valuation at £19,400. The cost of seed and part of the fertiliser for this harvest year has been extracted from the previous financial year; £1,590 for seed and £790 for fertiliser. Of the £2,950 shown as fertiliser expenditure for the current year for wheat £2,210 was for this year's crop and £740 for the establishment of next year's. All the sprays and sundries were used on this year's crop.

	£	£
Output:		
Grain in store	19,400	
Straw sold	750	20,150
Variable costs:		
Seed	1,590	
Fertiliser	3,000	
Sprays	1,500	
Sundries	150	6,240
Gross margin		13,910
Gross margin per hectare		464

Gross Margin for Barley (60 ha)

A detailed study of the records shows that of the £15,100 recorded for barley sales £1,848 was for the previous year's crop and £13,252 for this year's. Care has to be taken to include only that quantity of barley fed to stock which was produced during the current year. The variable costs shown earlier are all for this year's crop.

	£	£	£
Output:			
Sales	13,252		
CV in store	2,100		
Fed to stock	8,648	24,000	
Straw: sold	1,680		
to DYS	480	2,160	26,160
Variable costs:			
Fertiliser		4,320	
Seed		3,180	
Sprays		1,680	
Sundries		240	9,420
Gross margin			16,740
Gross margin per hectare			279

Gross Margin for Potatoes (10 ha)

	£	£
Output:		
Sales		22,760
Variable costs:		
Fertilisers	1,440	
Seeds	2,250	
Sprays	390	
Sundries	1,900	
Casual labour	4,500	10,480
Gross margin		12,280
Gross margin per hectare		1,228

Appraisal

A detailed appraisal of each enterprise should now be undertaken. Much of the information given in Chapter 9 will be of assistance in this context, and only a brief treatment will be given here. It can be seen from Fig. 60 that Church Farm has above average gross margins per hectare for most activities, with the exception of barley. The farm falls below the "High Standard" figures in every case, but is not far behind in many instances. Results from previous years would establish if this was a normal picture. It is part of the task of the detailed appraisal of the activities to see why performances were above or below average and to see if gross margins could be increased to advantage. There are, however, further points which must be made about gross margins.

Activity	Church Farm	High standard	Average standard
Gross margins per hectare (£)			
Dairy cows	1,155	1,250	800
Dairy youngstock	299	400	220
Beef	545	600	480
Sheep	407	540	350
Potatoes	1,228	1,300	900
Wheat	464	530	445
Barley	279	540	297
Farm average	590	635	495
Gross margins per head (£)			
Dairy cows	540	580	440
Ewes	32.8	39	29

FIG. 60. Summary of Gross Margins

5.4.5. Evaluation of the Gross Margin Concept

Gross margins can be employed by agricultural businesses to very great advantage, but the basic simplicity belies the fact that if they are not fully understood they can be dangerous. The variation in gross margins of individual activities from year to year and the need to check that a common approach has been made to the items included in their calculation, before comparison with previous results and standards, has already been mentioned. Their requirement for detailed records is offset by the value they have in measuring the efficiency of sectors of the business. It has also been claimed that they have more value for inter-farm comparison than whole farm efficiency factors because they are less influenced by farm size and type. Caution is still necessary, however, because differences in the fixed cost area between farms can affect gross margins. If one farmer has invested in storage facilities for his wheat he should be able to obtain a better price than someone who sells at harvest; the farmer with a fully environmentally controlled pig building might obtain better growth rates than a farmer with a cold draughty unit. In each case the first farmer has increased his fixed costs in order to improve his gross margins.

Gross margin analysis provides very useful data for planning but again caution is necessary. A list of gross margins per hectare for the activities on a farm might suggest that the farmer should increase those activities with the highest gross margins at the expense of those with the lowest. In fact this is one of the early stages in gross margin planning, but for the unwary it has pitfalls. It assumes that the increase in scale of an activity will result in a linear increase in gross margin. This may or may not be true. Figure 61 illustrates that a change in scale may sometimes result in the gross margin per unit of activity, per head or per hectare, decreasing. Many factors may be responsible.

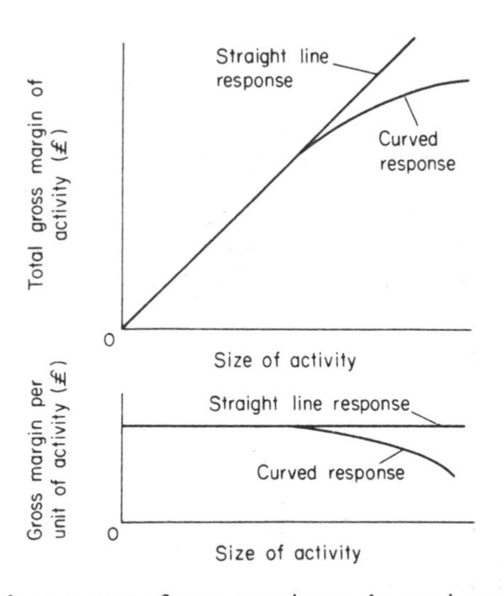

FIG. 61. Possible responses of gross margins to changes in scale of activity.

Management and labour may become so stretched that both the increased sector of the activity and the original sector receive less individual attention, and timeliness of operations becomes a problem. Expansion of an activity may necessitate it being undertaken on poorer land on the farm with attendant reductions in yield. Beneficial interrelationships between activities may be removed. One enterprise, albeit with a low gross margin, may be contributing indirectly to the output of another, e.g. sheep on a grass break may be contributing to the fertility of an arable farm.

It must not be thought from this that expansion always results in a decrease in gross margins per unit—far from it. Sometimes they improve because the size of the activity is increased to a point where it justifies (a) bigger or better machinery, giving better timeliness of operations, (b) improved buildings facilitating easier and improved stock management, and (c) greater attention of management to husbandry and marketing aspects. In addition the quantities of produce sold and inputs bought may foster improved product prices and quantity discounts.

Great care is therefore essential when predicting what will happen in the future. If an error is made in calculating the gross margin of one unit of activity it will be multiplied in the plan by the number of units of that activity selected. Since future performances are frequently difficult to assess it is often suggested that the sensitivity of plans is tested to see what price and quantity changes they could stand before giving room for concern.

Gross margins are frequently criticised because they neglect fixed costs. In fact they do the reverse because at the planning stage they focus attention on them by emphasising the need to consider their overall influence on business performance.

Protagonists of full costing, the time-consuming technique which attempts to allocate fixed costs to activities even if the result is arbitrary, suggest that it is infinitely better than gross margin analysis. This too has significant dangers. Consider the case of a farmer who, having fully costed his farm, finds that there was little profit from his potatoes. He stops growing them and increases the area of cereals and finds that he is making less profit than before. This is because he omitted to recognise that the potatoes were absorbing many of the fixed costs which remained even after they had been abandoned. Properly undertaken the gross margin planning technique would help avoid such mistakes.

In contradiction to the view normally held it has been suggested that farmers have more control over fixed costs than over variable costs. This provocative idea can have an element of truth in some situations. If a farmer is intent on maximising his gross margin from wheat he possibly has little choice in the variable cost levels he must employ to attain this goal. He can, however, exercise his judgement on the machinery and labour he uses. This hypothesis does not fully hold water with all activities. The dairy farmer has to exercise a great deal of judgement about the level of his variable cost inputs and it is hard to see how any of them can do much about rents or the mortgage.

Much of this section appears critical of gross margins but if all the points are considered with care it must be reiterated that they can be a valuable aid to analysis and planning. However, before reaching the stage where changes in farm policy are considered, it is necessary to look at some of the background issues influencing planning decisions.

CHAPTER 6

AGRICULTURAL LAND

CONTENTS

6.1. LAND AS A BASIC RESOURCE

6.1.1. Land Use and Classification

Land is a primary agricultural resource but it varies enormously in quality. This variation has a major influence on the type of farming practiced and upon farm values. In some cases land is capable of improvement but the cost/benefit of such action requires careful evaluation. Drainage, soil moisture deficit, fencing, accessibility and lime status are important considerations. The whole spectrum of fertility is a major factor for each farmer to consider.

The United Kingdom has about 18.7 million ha devoted to agriculture. Over one-third of this is rough grazing, about a quarter is permanent pasture and just under 40% is cultivated or in temporary grass, albeit that some of the latter has been in grass for many years. In times when high agricultural production has been a major priority land has been regarded as a particularly scarce resource. Worries have often been expressed about its loss for building, or for roads and for forestry. The problems created by food surpluses in the EEC in the mid-1980s reversed this concern in many people's minds and uses for land, other than for farming, were being considered.

Farms in remote areas, with poor soils and other natural difficulties, are usually the first to suffer when agriculture faces a depression. Farm businesses in these areas have then to receive increased consideration for government financial support on environmental and social grounds if they are to survive.

In 1987 the British Government published a document entitled "Alternative Land Use and the Rural Economy". One aim was to persuade farmers, especially in marginal areas, to take land out of agricultural production. For example it promised relaxation of planning controls for building, although green belts and other areas such as those of outstanding natural beauty were excluded. More money was promised for farmers to plant woodland and also to foster ancillary businesses on farms. The document also announced that increased research would be undertaken on alternative and less intensive cropping and livestock techniques.

Three main methods are adopted to classify land in Britain. These are (i) the MAFF/ADAS Agricultural Land Classification for England and Wales; (ii) the Land Use Capability Classification of the Macaulay Institute for Soil Research (MISR) and (iii) the Department of Agricultural and Fisheries for Scotland (DAFS) classification. The broad comparison between the three presented here is from the Scottish Agricultural Colleges advisory manual.

MAFF	MISR LUC	DAFS	Brief description
1	1	A+	Highest quality land with very minor or no limitations to agricultural use and capable of growing a wide range of crops with consistently high yields.
2	2	A	Very good fertile land with minor limitations which exclude it from A+. Capable of growing most crops with consistently high yields.
3a	3.1	B+	Good fertile land with moderate limitations. With a high level of management it is capable of growing above average crops of cereals, roots and grass.
3b	3.2	B	Medium quality land with moderate limitations. Capable of growing average crops of oats, barley and grass, also roots except on carse land.
3c	3.3		
4	4	B−	Poor quality land with moderately severe limitations, being only capable of growing crops, including some cereals, for the winter maintenance of stock.
4	5 Improved	C	Only suitable for improved pasture; cultivated only to enable reseeding to take place.
5	5 unimproved 6 7	D	Land used as rough grazing.

6.1.2. Structure of Agricultural Holdings

The twentieth century has witnessed an enormous change in the structure of British farms as they have progressively increased in size and the number of farmers has been reduced. Britain does not have the type of legislation which exists in Denmark to restrict farm size and the average area of holding is significantly above that for any other country in the EEC. Many countries in Europe have a very large number of farms which do not have sufficient work for even one man.

There have been several surveys to assess the optimum size of farm for maximum efficiency, but they have met the problem of differences in such things as the potential for production of the land and changes in the economic factors influencing profitability. Certainly between the 1950s and the 1980s many smaller farmers in Britain found the financial competition from larger farms too severe. Some took advantage of EEC schemes which encouraged voluntary amalgamation, co-operation and retirement or resettlement. Many received compensation for giving up milk production and not all found it easy to survive after the compensation was exhausted.

Forecasting the future for the size of farm businesses in times of overproduction and economic difficulty for agriculture is not easy. Political uncertainties increase the problem. There are those who consider that special small farms should be created, or that at least some form of financial support should be given to allow the younger generation to enter farming.

It is most probable that economic efficiency will be the main factor determining farm size in the immediate future. Low or economic use of fixed costs will be important. Family farms supplying their own labour can have low fixed costs, especially if the farm and tenancy capital have been passed on to the next generation without significant taxes at death. Many such farms will survive. There is, however, a strong possibility that, given no political intervention, farm sizes will continue to increase because many large businesses will be well organised and efficient in terms of both production and marketing. This does not mean that all large farms will remain viable but generally their owners will be better placed than those on smaller holdings to devote time to research new techniques and even to pay for advice.

6.2. LAND VALUES

6.2.1. Sale Prices

Land prices have fluctuated since records commenced. Table 3 shows the average prices for land sales in England and Wales.

Table 3. *Land Prices*

	£/ha		£/ha		£/ha
1820s	55	1960	198	1980	3,039
1870s	120	1965	403	1981	3,162
1900s	50	1970	494	1982	3,098
1920s	72	1972	554	1983	3,321
1930s	63	1975	1,213	1984	3,496
1945	91	1977	1,291	1985	3,586
1955	135	1979	2,316	1986	

Although the profitability of farming has had a major influence on land values there have been times, notably in the 1970s and early 1980s when the price paid was well above its immediate agricultural value. During this period land was bought as an investment which provided a sound hedge against inflation, with good prospects of long term capital appreciation in terms of the real purchasing value of money.

Many reasons other than inflation and capital growth were advanced for the price increase of the seventies. Purchases by financial institutions, such as insurance companies and pension funds, were often blamed. The Northfield report published in 1979 found that the institutions only owned about 1.2% of all agricultural land in Great Britain, although in the years 1975–8 they had bought 10% of the land offered for sale.

It seems certain, however, that the main reason for the increase in land values was that farmers stimulated the demand. Many added more land to their existing units, in some cases to spread costs over a greater area, or to provide for sons, but also as an investment with growth potential and favourable taxation benefits. Some farmers clearly opted to pay interest on money borrowed to finance farm purchase after considering their taxation position. Adding more land to a farm bought at comparatively low prices several years earlier still resulted in the average price per hectare of the enlarged farm appearing attractive. Entry to the Common Market had resulted in the prospect of higher product prices and it was observed that prices of land on the Continent were higher than in Britain.

The availability and cost of both mortgages and working capital, relative to inflation, the shortage of land and its use for other purposes, together with the money supply from farmers who sold their holdings very advantageously for development, all helped fuel the increase in prices paid for farms.

In the mid-eighties the increase in land prices was arrested and in some cases values declined. There was extreme concern about EEC policy following the introduction of milk quotas and statements that it could not continue to devote such a high proportion of its annual expenditure to support an agricultural industry which was producing food surplus to demand. The adverse weather conditions of 1985 together with the failure of output values to keep pace with cost increases created financial problems for many farmers. Some who had taken too great a risk and obtained large mortgages at fixed interest rates to buy their farms, or who had borrowed heavily to finance building improvements, and even to buy stock and machinery, found themselves insolvent.

Alarming as the British experience was for some it was not on the scale experienced in some parts of America, especially in the mid-west. In common with many of their British counterparts farmers in this area had paid high prices for land in the seventies but when the world prices for maize and soya fell in the eighties, in some cases below cost of production, a large number of these farmers could no longer service their loans. Their main asset, their land, dropped in value, often to less than the loan still outstanding on the purchase price. Not only did a significant number of farmers go bankrupt but many banks which had provided them with their loans had to close.

Estimating the future for land prices in such difficult times is fraught with danger, especially as the issue is very much influenced by political decisions. The quality of the land, coupled with any other comparative advantages it may have for the production of products offering economic returns for production and marketing efficiency, will be very important. There are many people who consider that top quality land will remain a good investment.

It is a bigger problem forecasting the future value of farms with soils of low capability but which, because of the support structures of the EEC in the 1970s and early 1980s, produced satisfactory returns to tillage crops. In fact it was these farms which were blamed by some for the surpluses of cereals since they considered that they were better suited to grass and livestock production. When it was recognised that problems also existed with surpluses of livestock products there were those who instigated the idea that a "set aside" programme, or payment for not growing crops, should be adopted on such land. They claimed that cropping should be the provence of farmers on good land.

The attitude of politicians to agricultural support on socio-economic grounds; their final decisions on how to control over-production; interest rates; taxation and the general economic fortunes of Britain and the EEC will all play their part in influencing future land values.

6.2.2. Rents

The 1970s and early 1980s witnessed a significant reduction in the number of farms offered for rent in Britain. Since it was a relatively prosperous farming era demand forced up rental values for new tenancies. In some cases the percentage of potential profit, before deduction of rent, which had to be offered to secure a tenancy was over 40% and there were instances where it was above 60%. High rents put extra pressure on farmers to be efficient and 30% of profits is frequently considered a more realistic figure.

A major reason for the reluctance of landlords to let out any land which became vacant was that they could enjoy a more favourable taxation situation if they took the land "in hand" and farmed it themselves rather than let it out again. There were also the Agricultural Holdings Acts which gave relatives the right to succession so that once a farm had been let it would be unlikely that the landlord could regain the right to farm it for a considerable period. If the land had then to be sold, as was sometimes the case when taxes had to be paid at the landlord's death, it had to be offered subject to tenancy rather than with vacant possession. Farms sold under the latter heading fetch higher prices because purchasers can farm the land themselves.

There have been some amendments both to the way in which landlords are taxed and in the rights to succession of a tenancy, but these have not been sufficient to significantly increase the number of farms to rent. However, when the financial fortunes of farming declined the number of applicants submitting for tenancies fell in many areas.

It used to be recommended that 2.5–4% of the vacant possession value of a farm should be offered as rent. When tenancies were in high demand it was frequently necessary to pay more than this in order to beat the competition. However, in any economic climate it is important to prepare careful budgets to calculate what rent a new business can stand. In times of difficulty it might also be prudent to establish how sensitive a farm would be, at the rent offered, to a reduction in profits before deduction of rent.

6.2.3. Assessing the Value of Agricultural Land

(i) Land and Climate

When assessing the value of land it is important to consider its cropping and stocking potential. Particular attention must be paid to soil type and a soil auger can be a useful tool. The productivity and range of crops which could be grown, suitability for given classes of stock, and level of fertility should be studied. It may be possible to cash in on fertility. Many farms which have been in grass used for finishing beef and sheep can grow excellent crops of barley if the pH is corrected. If the fertility is not high the cost of putting it right relative to the price being asked for the land should be checked. Previous cropping is therefore relevant.

Drainage, rainfall, aspect, slope, prevailing winds, fences, hedges, shelter, length of season, weed problems, depth of soil, stones, roads, size and shape of fields, public rights of way and electricity pylons should all feature in the examination.

Well-drained fields facing south may warm up early in the Spring or allow stock to remain out late in the season without poaching the land. Slope, shape and size of fields can influence the use of machines.

The layout of the farms must be considered. Farms within a ring fence are much easier to manage than those which are scattered. Checks should be made on access to fields by dairy cows, the necessity to cross main roads and on roads to fields for collecting crops or moving stock.

In hill areas the amount of in-bye or lower land and its quality, or potential for improvement, can be very important. Accessibility to the hill and ease with which fencing, liming, slagging and draining can take place must also be considered.

(ii) Buildings and Related Factors

New buildings are extremely expensive. The type, layout and state of repair of those in existence can greatly influence the price, especially on stock farms, or where crop storage is considered. If new buildings might have to be erected, study local planning regulations and look at the slope of the land to assess the possible cost of foundations. Note the distance of the buildings from the public highway and maintenance requirement of the farm road. Check on the fixtures which have to be taken over.

A very important item in the future on stock farms will be the regulations with reference to slurry disposal. The system on the farm should be examined and the cost of any amendments considered. Three-phase electricity and a good mains water supply to the steading and fields can be useful assets.

If it is a rented farm the amount of maintenance and repairs to be undertaken by the landlord and his willingness to erect new buildings may influence the rent.

The importance of the farm house is a subjective factor. In areas close to cities or in beauty spots it may be a major feature for some purchasers. Distance to shops and schools, suitability for bed and breakfast, central heating and state of maintenance can be crucial to other buyers.

(iii) Other Factors

Proximity to suitable markets can play a major role in determining farm policy and profits. Farmers on the islands off the coast of Scotland know the significance of this when they buy hay or concentrates from the mainland or market their stock. Sugar-beet factories, markets for specific classes of stock or horticultural produce can be important in other areas.

Farm cottages, distance to schools, shops and sources of entertainment, travelling shops, bus routes and competition from other industries can influence the ease with which staff are attracted. The supply of casual labour in the locality is relevant in some cases.

Quotas, the ease with which milk can be collected in winter, situation with reference to farm gate sales, sporting facilities, mineral rights and trespassers are other points to note.

(iv) First Impressions

The time of year when the farm is walked and the climate on the day can sometimes give a misleading impression. Untidyness, weeds, old machinery, wood piles and mud can do the same. Bad fences and many other features can be tolerated if the price of the farm is right. Always consider the cost of correcting deficiencies and the significance of the deficiencies to immediate production requirements. Other people may be put off by a challenge.

Some people recommend a field by field assessment as the farm is walked. After the visit is over these values can be averaged on a per hectare basis and employed to determine the final bid.

The considerable cost of the ingoing must be appreciated especially by the farmer who is short of money. This is the charge for such things as unexhausted manurial residues, seeds and crops in ground, fixed equipment and tenant-right on buildings.

Valuing land at times when it is increasing rapidly in price is even more difficult because it is then necessary to assess the premium which others might be prepared to pay for the investment potential.

6.3. LAND TENURE

6.3.1. Systems of Land Tenure

Owner-occupancy occurs where the farmer owns the land which he farms. A tenancy exists where a legal contract between the owner of the land and the person who farms it has been drawn up to establish the terms of the lease. Parliamentary Acts restrict the terms of such contracts in Britain.

Lease-back arrangements arise when a farm is purchased, in many cases by an institution, and it is then leased back to the farmer for him to continue farming, but now as a tenant.

Communal ownership takes place when land is publicly owned and communal tenure occurs when people have communal rights to use land.

In Britain there are several examples of variations from the straightforward owner-occupier or tenancy situations. There are specific grazing rights for New Forest, and many other examples of common grazing lands. In certain areas, particularly in Cumbria, sheep graze the fells which are not fenced, the farmers relying on the hefting instincts of sheep to prevent them from straying too far. Villagers in some parts have common rights to the village green.

In Scotland crofters are the tenants of small parcels of what, in relative terms, they would describe as good land, and they can graze stock on a large area of poor land. They usually only rent the land and provide their own houses.

6.3.2. Ownership and Tenancy

The advantages of the tenant:

(i) A tenancy may be the only way into farming for someone with limited capital. Since all capital can be employed in tenancy items it is possible to consider a larger farm than would be the case if land was to be purchased. This may allow the tenant to generate capital and to ultimately purchase a larger farm.
(ii) The tenant can have greater flexibility in the use of his capital because less of his money is in fixed assets.
(iii) In some cases, but by no means all, the landlord will pay for major improvements such as additional buildings.
(iv) Some landlords are understanding in fixing rent levels for a new entrant, especially if there is a chance that he may improve the farm. Economic pressures on landlords have made this less and less possible.

The advantages of the owner-occupier:

(i) The primary advantage in times of inflation can be the increase in value of the land, but the converse is true if land values fall. Land may have speculative value.
(ii) Land is a good security for a loan.
(iii) Provided that mortgage charges can be met there is a high degree of security of tenure.
(iv) There is no rent to pay.
(v) Ownership of land can confer prestige, create great satisfaction and promote interest in farm improvement. Sporting interests may be fostered.
(vi) In the United Kingdom owner-occupancy confers special taxation benefits both in relation to profits and also to transfer of capital during life and at death.

Disadvantages of the tenant:

(i) In some countries there is a lack of security of tenure and the landlord may be able to influence the policy for the farm. This disadvantage does not apply in the United Kingdom where various Agricultural Holdings Acts have given the tenant increasing security of tenure with the exception of one or two special circumstances and, whilst there are conditions about the state of the farm at the end of the tenancy, essentially the tenant can select his cropping policy.
(ii) Rent reviews occur but there are controls in the United Kingdom to govern their frequency and to ensure that they are fair.
(iii) The tenant forgoes many advantages of the owner-occupier and in particular the increase in value of land. However, where a tenant can pass on his tenancy to his successors when he dies he can do so without tax being levied on the land.

This means that in this context the tenant who increases the size of farm has the advantage over an owner-occupier because the expansion does not attract the same tax liability.

Clearly one or two of the points listed here under disadvantages could justifiably have been put under advantages as far as the British tenant is concerned.

Disadvantages of the owner-occupier:

(i) There is a danger of investing in fixed assets at the expense of working capital.
(ii) Mortgage interest and principal repayments may reduce the generation of capital which is necessary for the development of the business.
(iii) There are worries about Inheritance Tax and the possibility of passing on the farm intact to children.
(iv) The problems of land management are added to those of farm management.

6.3.3. Rent Amendments

Rents can be reviewed at any time by mutual consent of landlord and tenant. If one party does not agree then the terms of the lease agreement come into force. This may specify the interval between changes. Where the agreement does not make such provision, one party can refer to arbitration. The minimum interval between rent amendments is 3 years. A landlord can issue a tenant with a notice of reference to arbitration, but this facility is also open to the tenant.

The Agricultural Holdings Act 1984 for England and Wales and the Agricultural Holdings (Amendment) (Scotland) Act 1983 introduced new guidelines to arbibrators when fixing rents. These Acts essentially established the basis of what might be described as a more just system since arbitrators, when deciding a new rent, could ignore comparisons with unrealistically high rents being paid by some people for new tenancies.

6.3.4. Security of Tenure

Tenants have protection under the terms of the various Acts of Parliament. There are slight differences between the regulations for England and Wales, where the Agricultural Land Tribunals deal with disputes and Scotland, where the Scottish Land Court is responsible.

In all three countries a landlord can give a tenant notice to quit. This must be in writing and for a date at least twelve months ahead. The tenant can serve a counter-notice but this must be within one month. This right of the tenant can, however, be withheld for several reasons. These include cases where the tenant is bankrupt; where a certificate of bad husbandry has been granted against the tenant; where the tenant has failed to pay rent or, having been given a written demand to rectify a breach of the tenancy agreement has failed to do so even though it could be remedied; the land is required for certain public uses; or where the landlord's interest in the farm is severely prejudiced by a non-remedial breach of the tenancy agreement by the tenant.

The Agricultural Land Tribunals and the Land Court will only find in favour of the landlord in limited circumstances. The main reasons are for bad husbandry; where the landlord can prove greater hardship to himself by the tenancy continuing, than to the tenant; and in the interest of good estate management.

6.3.5. Succession to a Lease

Different Acts exist for England and Wales and for Scotland which determine succession to a farm tenancy. The Agricultural Holdings Act 1984 abolished the statutory right to succession in the case of "new" tenancies in England and Wales unless the tenancy was obtained under the provisions of the Agriculture (Miscellaneous Provisions) Act 1976 or the landlord and tenant formed a contract to the 1976 Act. Tenancies contracted under the 1976 Act restricted tenure to three generations, including the present occupier. The Agricultural Holdings Act 1986 consolidated the legislation of previous Holdings Acts but also gave some amendment to detail.

Details relating to succession in Scotland can be found in the Agriculture Act 1958, the Succession (Scotland) Act 1964, the Agriculture (Miscellaneous Provisions) Act 1968 and the Agricultural Holdings (Amendment) (Scotland) Act 1983. Essentially up to 1984 there was no limit in Scotland to the number of generations that might qualify for succession. The 1983 Act established that for "new" tenancies created after 1 January 1984, the landlord could serve notice to quit at the time the relative claimed succession. It then became necessary for the relative to prove to the Land Court that he had sufficient financial resources and relevant training and experience to farm the holding with reasonable efficiency. The Land Court was also empowered to grant the notice to quit (i) if the holding of the agricultural unit of which it formed part was not a two-man unit and the landlord intended to use the land to amalgamate with other specified land within 2 years of the termination of the tenancy; (ii) if the successor already occupied agricultural land which was a two-man unit and distant from the new holding.

6.3.6. Compensation for Tenants

Where a tenancy comes to an end, specific rules govern the compensation which the outgoing tenant can claim from the landlord. This can conveniently be treated under two separate headings.

First, there is disturbance compensation. If a tenancy is terminated when the tenant cannot be held responsible, one year's rent, and in certain cases 2 years' rent, must be paid to the tenant by the landlord. In addition a further sum equal to four times the annual rent may be payable to assist the tenant to reorganise his affairs. Again, legal advice may be necessary to clarify the position.

Second, where a tenancy comes to an end the tenant may be able to claim against the landlord for improvements made during the tenancy and for tenant right. The amount of the compensation which might be payable under this heading will probably reflect the increase in value of the farm as an agricultural holding attributable to the improvement. In some cases the landlord must have given written consent prior to the improvement taking place. If this agreement covers compensation then the agreement determines the compensation. In a case where the landlord refuses to agree to an improvement the tenant can appeal to the Agricultural Land Tribunal or Land Court.

Short term improvements, which could be of value to the new tenant such as lime applications or unexhausted manurial residues, also attract compensation. Their value is calculated according to specific rules. The same applies to growing crops.

The landlord will probably charge a new tenant an ingoing to defray at least some of his costs of paying compensation for improvements. It is also possible for the landlord to claim against the outgoing tenant for dilapidations and damage.

6.3.7. Farming Partnerships

Partnerships will be covered in greater detail in the next chapter, but here they are considered in relation to land use.

Many different forms of farming partnerships exist. Perhaps the most common is father and son or some other form of family arrangement. Tax planning may be a factor promoting the establishment of this type.

Non-family partnerships include cases where the owner of the land and the person who actually farms it provide a share of the working capital. Profits are split according to an agreed procedure which usually reflects the amount of capital invested and the management load for each partner. In some cases the partner who does not own the land has very little security because he is not covered by the Tenancy Acts. He may not get any compensation, other than his share of the investment, if the arrangement is terminated.

The benefit he does get is a chance to farm and the possibility of generating some capital. The owner gets a manager who has a stake in the business, and who should therefore be motivated. He also derives tax benefits because profits are taxed as earned income. In addition there are benefits associated with inheritance tax. Sometimes agreements can be found where greater security and compensation rights are given to the farmer partner.

Another example of a partnerhip is where an owner of a farm with vacant possession and an existing farmer with another farm agree that the owner pays for all the variable costs of fertiliser, sprays, and seeds for the vacant farm, and the farmer provides labour and machinery. Again profits are split according to a mutual agreement.

Other variants of these non-family type partnerships exist. Many give little security to the partner who does not own the land.

CHAPTER 7

CAPITAL AND TAXATION

CONTENTS

7.1. THE USE OF CAPITAL

7.1.1. Capital as a Resource

Capital is often considered to be the most limiting resource in agriculture. However, there are many cases where this requires qualification. Frequent examples can be found where farmers have adequate capital but they are unable to purchase extra land near to their existing holding because none is for sale. Sometimes farmers have to give up certain enterprises because insufficient skilled labour is available. Many dairy farmers have gone out of milk production for this reason. Shortage of labour in hill areas has forced other farmers to amend their systems of management.

When considering capital it is important to remember that it has two main dimensions—quantity and time. This is particularly relevant in relation to borrowed money. There is little point in borrowing adequate funds for a long term project if the money has to be paid back before the project has sufficient time to generate the capital for the repayment. It is therefore vital to understand the time scale involved in any investment.

Reference back to the study of balance sheets will show that the main division of capital invested in the assets is into *landlord's capital* and into *tenant's capital*. If these two types of capital are further examined, the time scale of any borrowing necessary to finance them can be ascertained. Equally if the farmer's own capital is involved it is possible to establish how long it will be "locked up". The landlord's capital is invested in land and buildings and must clearly be from long term sources.

The tenant's capital can be divided into sections. Some of it, such as that in breeding livestock and machinery, was earlier classified as fixed because it was considered to be a fairly long term investment. Breeding livestock generates income by effectively multiplying the money employed in the variable costs of feed, etc., which they require. It is therefore possible to borrow capital from medium term sources to finance both these and, because of the length of its life, to purchase machinery. In fact money borrowed for any depreciating asset such as machinery must, in the interests of good management, be paid back before the end of its life.

Whereas long term credit is usually borrowed for at least ten years and generally twenty or more years, medium term capital has to be repaid within three to ten years. In many cases modernisation and even extension of buildings by tenants is undertaken using medium term capital.

A farmer also requires capital to finance the day-to-day operations associated with production. In other words he requires working capital to buy feedingstuffs, fertiliser, seeds, and to pay such things as labour, repairs, and veterinary bills. Short term capital, to be repaid in 2, or slightly more than 2, years, can be borrowed for this.

Figure 62 illustrates how funds are deployed.

7.1.2. Making Capital Work

All capital costs money. If it is the farmer's own and he has invested it in the farm business there is an opportunity cost of not investing it elsewhere, such as a building society or in government securities. If the capital has been borrowed there will be interest charges in addition to the obligation to repay the loan. In either case capital should be made to work to make more money so that the farmer can cover his costs, has enough money to live on, and can build up his business. The risks that will be taken to do this will largely depend upon the attitudes of the particular farmer, although in some cases, if it is borrowed money, the lender may have some influence.

Some enterprises quickly produce a return on the capital invested in them whilst others lock capital up for long periods without any return. Veal calves and broiler chickens have such short production cycles that the money they require can be used

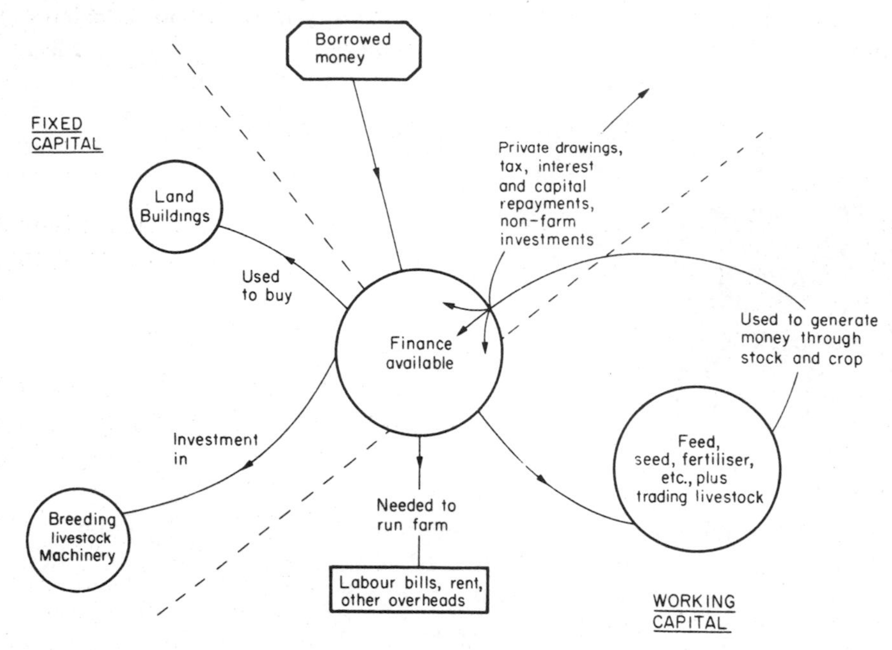

FIG. 62. Deployment of farm funds.

more than once in the year. Dairy farming locks large amounts of capital up in cows but they are quick to produce a return from the working capital. Suckled beef calf production, however, takes much longer than dairying to produce income. Investment in machinery and such items as grain storage bins can only give returns if they are effectively used to make a contribution to appropriate enterprises.

The salient point is that before any major capital investment takes place a very careful appraisal should be undertaken to establish how much capital should be used and where it should be allocated in the best interests of the business. Budgeting and other techniques can be used in this context, but the main discussion on investment appraisal is left to a later chapter so that examples from Church Farm can be employed to demonstrate the techniques.

When major policy decisions for capital investment have been made and a plan is in operation, it is still necessary for the manager to control capital on a day-to-day basis. It is important that the amount of capital left lying idle is kept to a minimum and that funds which are working are giving the best returns possible. There must be enough money to pay the bills, but it might be possible to find sufficient cash for these out of income rather than leave large sums in the bank.

Much will depend upon the nature of the business and the pattern of income and expenditure during the year. Hill farms only have receipts at certain times. These are mainly derived from the lamb or suckler calf sales, the wool cheque, and the Hill Livestock Compensatory Allowances. The hill farmer therefore has to manage his funds very carefully because he has to leave sufficient money to last between the times of the year when income is received in order to have money to live on and to provide the working capital to pay for labour and other items necessary for the day-to-day running of the farm.

In contrast the dairy farmer has a regular monthly milk cheque and may find it easier than the hill farmer to pay bills out of income.

7.1.3. Managing Cash Using Cash Flows

(i) Cash Flows

Cash flows can be used to assist the management of capital and to help ensure that it is put to effective use. A cash flow records the movements of cash coming into and going out of the business.

A cash inflow is money coming in or being received.

A cash outflow is money going out or being spent.

A *cash flow statement* should be produced for each business. This should show, in detail, the projected cash flows for defined periods of time. In many cases periods of three months are selected, but where there is rapid turnover of money, or the financial state of the business is fairly critical, intervals of one month are chosen.

The total cash inflows for the period are added together, as are the total cash outflows. The difference between these two is the *net cash flow* for the period. This is positive if the inflows are greater and negative if the outflows are greater.

The position is easier to understand if all the transactions go through the bank. It will be possible to see the capital position at the bank at the end of the previous time period. This is then adjusted to take account of the net cash flow for the current period to give the new cumulative cash position. Figure 63 illustrates the point. If the cumulative cash flow at the end of the previous period was −£24,000 and the net cash flow for January–March was plus £7,000, the cumulative cash position at the end of March is −£17,000. During the next three months the net cash flow is −£4,000 and the cumulative cash flow at the end of June is −£21,000. At the end of the year the total cash flow can be ascertained. This was +£17,000, and since the overdraft at the start of the year was −£24,000, the cumulative position at the end of December is −£7,000.

(ii) Assessing Overdraft Requirements

If budgets are undertaken to show the cash flows in advance the possible trends in the cumulative cash position can be predicted. If these are negative they will indicate when overdrafts will be required together with their size. Suitable negotiations then have to take place with a bank manager to see if the money will be forthcoming or if another source of funds, or even a change of plan, is necessary.

Where the cumulative position is positive it shows the reserves which can be used to pay future bills. If these reserves are excessive it might be prudent to ensure that they are invested. The cash flow will therefore point to the times when money can be spent without embarrassing the financial position, and to times when it should be spent or invested to make maximum use of capital.

(iii) Preparation for Control

When a plan is in operation it is necessary to record the results and act accordingly

	Jan.–Mar.	Apr.–June	July–Sept.	Oct.–Dec.	Year
Cash outflows "listed" e.g. Machinery purchase Feed Labour					
Total outflow	−£9,000	−£14,000	−£6,000	−£7,000	−£36,000
Cash inflows "listed" e.g. Cattle Milk Wheat					
Total inflow	+£16,000	+£10,000	+£15,000	+£12,000	+£53,000
Net cash flow	+£7,000	−£4,000	+£9,000	+£5,000	+£17,000
Cumulative position*	−£17,000	−£21,000	−£12,000	−£7,000	−£7,000

*Assume cumulative position at end of previous year was −£24,000, i.e. bank overdraft.

FIG. 63. Cash flow statement.

to exploit strengths and remove weaknesses. Control of the cash flow is facilitated if, when the budget is prepared, room is left so that the actual figures which occur can be written beside the budgeted data. Thus the heading would be split as illustrated in Fig. 64. In the example shown it highlights the point that £3,000 more was spent than was budgeted. Detailed analysis of the items in the cash outflow will reveal where this actually occurred. It may not just be in one item.

Period / Item	January–March		
	Budget	Actual	
Cash outflows listed			
Total cash outflow	−£9,000	−£12,000	

FIG. 64. Cash flows for control.

(iv) Items Included in Cash Flows

To save space, Figs. 63 and 64 simply state "cash outflows listed" and "cash inflows listed", but see page 295. Under cash outflows each major item of expenditure is given a line. These items may include capital expenditure on stock, machinery, buildings, and improvements; feed, fertilisers, seeds, veterinary and medicines, miscellaneous livestock, and crop expenses; labour, rent, electricity, and repairs; interest, loan repayments; private drawings, taxation. Printed forms can be used or headings selected to suit the farm.

The cash inflows include statements of each main area of receipts, e.g. cattle, wheat, milk, plus an entry for any capital introduced from outside the business.

(v) Points to Note

When the plan is in operation all items of cash coming into and going out of the business must be recorded. Records must be accurate and none must be forgotten. Debtors and creditors should not be entered until the money is actually received or the bills are paid.

When capital expenditure takes place for items such as buildings and machinery the full amount of money spent is included when the cheque is signed. Depreciation on these items is not included in the cash flow because it is a notional item and no actual expenditure takes place. Valuations are not included in a cash flow. The purchase price of stock is included irrespective of whether the farm operates on the trading or the herd basis. Interdepartmental transfers are not entered in the overall farm cash flow.

Private drawings including standing orders and income tax paid from the farm account, are entered, and interest and loan repayments are included. The reduction in bank overdraft is taken care of by the cumulative position.

Capital introduced from outside the business is added.

(vi) Differences Between Profit and Loss Accounts and Cash Flows

The main differences are shown in Fig. 65.

The list does not entirely exhaust the possibilities but it is sufficient to show that there are fundamental differences between P & L accounts and cash flows.

Included in Profit and loss account and not cash flow	Included in cash flow and not P & L account
Depreciation Valuations Produce consumed in farm house Notional rent* Interdepartmental transfers* Payments in kind to labour (board, produce, meals supplied)	Capital expenditure Private drawings Income tax Loan repayments Capital introduced

* Management accounts only.

FIG. 65. Differences between P & L accounts and cash flows.

This demonstrates why a farmer can be making a satisfactory profit and still be in financial difficulties. For example, consider the farmer who spends £40,000 on machinery in one year. This all disappears from his bank account but only £8,000 might be put as depreciation in the P & L account. Such examples illustrate the importance of the cash flow.

(vii) Reconciling Profit with Cash

At the West of Scotland Agricultural College the document shown in Fig. 66 is used in advisory cases to illustrate to farmers where their cash came from during the year and where it went to.

(viii) Managing the Cash Flow

If the actual results of a cash flow are not working out as predicted in the budget, the manager may require to take some action. Where the financial position is weaker than predicted, and borrowing limits could be exhausted, it may be necessary to delay purchase of some items or sell produce earlier than intended. Care has to be taken, as far as possible, to ensure that such action does not prejudice the long term viability of the business.

Some managers deliberately "walk a tightrope" to keep money working. For example, an arable farmer may assess the amount of grain in store which he must sell to pay next month's labour and other bills. Tight financial management of this type can help increase returns on capital.

It has already been indicated above that where the cash flow is more favourable than expected, some form of amendment to the investment policy might be wise.

A good manager will always attempt to keep his debtors to a minimum, especially if he has an overdraft, because effectively he is financing the debtors. Had they paid their debts the money would have been in the farmer's bank with an attendant reduction in overdraft. Caution has also to be exercised to avoid keeping too many purchased goods in store earning nothing, but still contributing to the level of borrowing at the bank.

7.2. SOURCES AND USE OF CREDIT AND CAPITAL

7.2.1. Policy

Farmers differ in their willingness to be in debt and the risks that they will take in borrowing. Some believe that every business should make use of a certain amount of borrowed capital and others have little alternative but to borrow to survive.

A golden rule is to ensure that a loan can be serviced without prejudicing the future of the business or the standard of living of the owner. In other words it must be possible to pay both the interest and to repay the principal on time without seriously upsetting the cash flow position or causing assets to be realised before they should be in relation to the best interests of the farm. Although it is always dangerous to generalize, if a farmer has more than one-quarter of his income going out in debt repayment and interest charges, he can be in a serious position.

FARM NO. ______________

FARM TYPE ______________

SOURCE and DISTRIBUTION of CASH
STATISTICAL YEAR

YEAR END ______________

SOURCES OF CASH	£	£	£
Starting with net profit (for management purposes)			
Add Depreciation on machinery, equipment, vehicles and fixed capital			
Charge for board and lodging of employees			
Sales of machinery, equipment, vehicles buildings and land			
Grants received on capital expenditure			
Personal cash introduced into the business			
Decrease in valuation of livestock, crops, stores			
Decrease in sundry debtors			
Decrease in cash and bank balances (increase in overdraft)			
Increase in loans and bonds			
Increase in sundry creditors			
Deduct Produce consumed in the house			
Private share of expenditure			
TOTAL CASH "AVAILABLE" WITHIN YEAR			
DISTRIBUTION OF CASH			
Purchases of machinery, equipment and vehicles buildings and land			
Increase in valuation of livestock, crops, stores			
Increase in sundry debtors			
Decrease in overdraft			
Increase in cash and bank balances			
Decrease in loans and bonds			
Decrease in sundry creditors			
Personal drawings (including "wife's wage")			
Taxation			
Capital withdrawn			
Other drawings			
TOTAL CASH "DISTRIBUTED" WITHIN YEAR			

FIG. 66. Source and distribution of cash.

Cash flow budgets can help to assess both the need and the advisability of application for credit. Over optimism should be avoided when preparing these budgets because of the possibility that borrowing requirements become greater than borrowing facilities arranged.

If the source of credit is matched to the use to which it will be put so that long term capital is used for long term investment and medium term loans for medium term investments, the difficulties of early repayment will be minimised.

7.2.2. Long Term Credit

In England and Wales loans for farm purchase, or less often for large scale improvements, can be obtained from the *Agricultural Mortgage Corporation* (AMC) and in Scotland from *Scottish Agricultural Securities Corporation Ltd.* (SASC). The maximum mortgage obtainable is two-thirds of the value of the holding as assessed by the particular corporation. Usually it is less than this, and in practice it could be one-half their valuation or less. The actual figure may be related to the farmer's ability to service the loan. One method used to value the farm is to take twenty times the economic rent. The security is the farm itself and the deeds of the farm are held by the mortgage company.

Typically these loans are for between 20 and 30 years. They are usually serviced by half-yearly payments of interest and premium. If all payments are up to date the loan cannot be withdrawn before the end of the contract. The borrower has no automatic rights to repay it early but usually arrangements can be made, although a penalty may be incurred.

The interest charged on new loans reflects rates currently being charged for money. There are two main options. One is to accept a fixed interest rate in which case this rate of interest will be charged for the rest of the life of the loan. The other is to accept a variable interest rate in which case it can be amended at set periods—quarterly or half yearly.

Clearly when interest rates are low the fixed rate is the best option. When they are high the variable rates may be more attractive on the prospect of a fall.

The half-yearly payments are equal throughout the life of the loan except where the variable interest option has been selected. Initially they are mainly interest—which attracts tax relief, and little premium repayment—which does not. Progressively as the years pass the proportions are reversed.

An alternative method of repayment, particularly suitable for the younger man, is to take an endowment life assurance policy with profits for the size of the loan or part of it. When the policy matures, with profits, it is used to pay off the loan. This matter is considered again later under investment appraisal.

The Lands Improvement Company provides loans for up to 40 years for the net cost of improvements to property at an interest rate fixed at the time of the loan. An annuity charge (equivalent to an extra rent), actually paid half yearly, in addition to being the means of payment, acts as security for the loan. This is because it takes precedence over all other farm costs. The loan cannot be paid off early and if one farmer leaves the property the new farmer must take over the loan.

Private loans, usually from relatives and frequently at moderate interest rates, are another major source of long term capital. It is generally advisable to have a firm agreement between the two parties, particularly with reference to the terms of the repayment and the possibility of early recall.

Banks in the main supply short and medium term credit but they do sometimes provide long term loans, even for farm purchase.

Insurance companies provide assurance-linked loans for land or land improvement for periods of 10–25 years. This is a different situation to that outlined above in that the insurance company and not AMC or SASC provide the loan. It is, however, linked to a life assurance policy.

7.2.3. Medium Term Credit

Bank loans can be obtained for specific terms of up to 10 years or more. The interest can vary with the bank's base rate and is usually 2½–5% above it. One client may find that he is quoted a different rate to another. Interest is payable every 6 months. The security for the loan may be the land, guarantors (frequently these are relatives) or stocks and shares. The nature of the project for which the money is to be used and the personal standing of the client can be important.

The loan is not subject to recall before the end of the term but repayment is made at 6-month intervals for the life of the loan. It is important not to borrow more from this source than is required.

Bank overdrafts are in practice used more than bank loans by farmers as medium term finance. A major advantage is that interest is based on the amount borrowed on a day-to-day basis. As cheques are paid into the bank they help to reduce it. The amount of interest due is added up and paid quarterly. The rate can be 2½% or more above the base rate.

The amount of the overdraft limit has to be negotiated with the bank manager and there can be significant differences between bank manager's in the way that they view a particular client or his case. The limit can be reviewed annually or even more frequently and there is always the danger that the money could be recalled at short notice.

Effectively a bank overdraft is a facility to go on signing cheques even though there is no money in the bank. The limit is the amount to which the client can go into the "red". Unlike a bank loan it does not get entered as an inflow in the cash flow. There is no obligation on a client to borrow to the limit set by the bank manager so that borrowing can be adjusted to suit requirements provided that these are within the limits. This is important since borrowing can suit seasonal demands.

Agricultural Credit Corporation Ltd. (ACC). The ACC, which is government sponsored, will act as a guarantor for a farmer requiring a bank overdraft for farm improvements, but these must result in improved efficiency and profitability of the farm. The cost is normally 2½% above the rates charged by the banks for overdrafts and is therefore only suitable for someone who has insufficient security and background knowledge of agriculture to obtain a bank overdraft. Loans are usually for up to 10 years. Guarantees for loans to encourage efficient marketing are also available. *SASC and AMC* offer medium term loans but these cannot exceed 50% of the Corporation's value of the farm.

Milk Marketing Boards provide loans for purchase of bulk tanks or in some cases they purchase and then rent the tanks to farmers. Other bodies which supply medium term loans include the Highlands and Islands Development Board, which offers loans for such things as land reclamation in its area.

7.2.4. Short Term Credit

Banks overdrafts provide a very common source of short term credit. The way they operate was described under medium term credit, but in fact their main objective is to provide short term finance. There is a danger that overdrafts develop into what is termed a "hard core" loan. Instead of reducing them as it is the intention, the business comes to rely on them and they are carried on for a long time. The concern is that they can be withdrawn in times of credit restriction, and when interest rates are high they become costly.

Merchant credit is obtained by getting the agreement of a supplier that accounts can be settled at a future date which is after the time when payment is normally due. The rate of interest varies and may differ from farmer to farmer. In most cases it is above that currently being charged for bank overdrafts.

Repayment methods also vary. In some cases a bill of exchange is employed to guarantee payment on a specific date. In other cases a standing order progressively pays the sum due.

Some companies provide loans to purchase stock provided that feed is bought from them. This can highlight a danger of merchant credit because farmers become tied to the products offered by the particular firm.

Another way of obtaining credit is to simply not pay bills on time, but this is not entirely desirable and can be costly because of discounts lost.

Farmers must be careful to ensure that they do not overload themselves with merchant credit to the ultimate detriment of their business.

Auctioneer's credit can be available to some farmers. Frequently this is for store stock which must be sold back through the same auction. The farmer then receives a cheque for the difference between the purchase and sale price less the auctioneer's charges. Interest rates can be very favourable, but vary from customer to customer.

FMC (Meat) *Ltd.* provide unsecured loans for up to 12 months for the purchase of store animals and breeding ewes and for fattening home-bred stock. Finished animals must be sold to FMC. Interest is deducted from the sale price and is related to bank overdraft rates.

Hire purchase can be obtained for the purchase of new plant and machinery. In addition livestock can be bought and extensions to buildings financed in this way. It is very expensive compared to bank overdrafts.

The goods are not the property of the purchaser until the final payment is made.

Contract hire and leasing, although technically not sources of credit, do provide means of obtaining the use of machinery or even livestock.

7.2.5. Credit Worthiness

Lending agencies assess the credit worthiness of a client before making a loan. Most look for security. Where land is to be purchased the deeds to the farm may suffice, but guarantors and life assurance policies may be suitable in some cases.

The personal standing of the borrower can be especially important. Bank managers look for plans which are viable and for a person who could be expected to carry them out. An increasing number require projected cash flows to support other budgeted data.

Table 4. *Sources of credit*

Type	Main uses	Main sources
Long term	Farm purchase and improvement including building	AMD or SASC Relatives or private sources The Lands Improvement Company Insurance companies Banks
Medium term	Purchase of machinery, equipment, and breeding stock	Banks MMB or Scottish boards ACC
Short term	Working capital for seed, fertiliser, feed, labour, tending livestock	Banks Merchant credit Auctioneer's credit FMC Hire purchase

Bank managers have to assess their risk in lending money. Like farmers they differ in their attitudes to risk, and many examples can be quoted of applications to one bank being rejected and a similar application to a different bank being successful. Some banks have more money to lend than others and this can be a factor.

Previous balance sheets show trends in profitability, private drawings in relation to profits, and growth in net worth. The nature of the assets can be examined and related to the sources of credit already employed by the business. It is possible to tell a great deal from these data about the past performance of the business and about the farmer. The various ratios and capital gearing mentioned under the study of balance sheets may be calculated. Many bank managers prefer a client to own at least 50% of the assets although there are some notable exceptions especially if suitable security can be offered.

Projected balance sheets can be examined along with other budgeted data to establish the risks attached to making additional loans to the business.

7.2.6. Calculating the Cost of Capital

With the possible exception of loans from friends and relatives money which has been borrowed has to be paid back. During the time that it is borrowed the lender will expect interest. In some cases the Inland Revenue will allow interest as a charge to the business so that it reduces the tax payable.

The main factors influencing the cost of capital are the terms of its repayment, the conditions relating to payment of interest, and taxation savings.

Any borrower should endeavour to compare the cost of borrowing from different sources to establish which would be cheapest. There are many different systems for charging interest, but one common basis for comparison is to calculate the Annual Percentage Rate [APR] of interest for each source.

(i) Simple Interest

An example of simple interest is where £1,000 is borrowed at a "Flat Rate" interest of 14.5% for one year and at the end the borrower pays £1,145 to cover both the repayment of the loan and the interest. The APR is 14.5%.

(ii) Compound Interest

If none of the money or interest is paid at the end of the year then next year the 14.5% interest is charged on £1,145.

The formula for compound interest is:

$$A = P\,(1 + i)^n$$

where A = the compound interest; P = amount borrowed; i = interest rate; n = number of years.

(iii) Compounding Within a Year

If £1,000 is borrowed at 14.5% per half year, and no interest or loan principal is paid until the end of the year, the total payable at the year end is:

$$£1{,}000 + \left(£1{,}000 \times \frac{14.5}{100} = £145\right) + \left(£1{,}145 \times \frac{14.5}{100} = £166\right) = £1{,}311$$

Remember in this case the interest is 14.5% per half year or 31.1% per year; i.e. the APR is 31.1%.

The APR can be established using the following formula:

$$\text{APR} = (1 + i)^m - 1$$

where i = rate of interest per period in decimal and not percentage form
m = the number of times a year interest is charged.

For the above example where £1,000 is borrowed at 14.5% calculated half yearly this becomes:

$$\text{APR} = (1 + 0.145)^2 - 1 = 0.311 \text{ or } 31.1\%$$

(iv) Bank Overdrafts

The farmer is more likely through his bank overdraft to meet the situation where the interest is calculated quarterly. However, it is not as penal as the rate shown in (iii) above because allowance is made for the quarterly calculation by including the number four in the denominator. The interest is still higher than many people think.

Example: A farmer has an overdraft at 3% over the bank rate which stands at 12%, i.e. at 15% nominal rate.

In the first quarter his interest is:

$$\frac{£1,000}{4} \times \frac{15}{100} = £37.50$$

Assuming nothing is paid off, the overdraft for the second quarter is £1,037.5.

The interest on this is:

$$\frac{£1,037.5}{4} \times \frac{15}{100} = £38.90$$

For the third quarter the interest is £40.37 and for the fourth quarter it is £41.88.

The total interest for the four quarters is:

$$£37.50 + £38.90 + £40.37 + £41.88 = £158.65$$

In other words the interest is 15.86% and not the 15% which the bank manager would quote.

(v) Front Loading

Many loans from finance houses for hire purchase agreements and certain loans from banks involve "Front Loading", also known as "Add-on Interest". Some farmers buy cars, machinery, or household items using such a loan. Frequently farmers accept the source of finance suggested by their machinery dealer and do not check whether this is more expensive than the bank overdraft.

Assume £1,000 is borrowed for 24 months at 10% with payments to be made monthly. The lender starts off by adding the interest payable to the loan so that his records show an advance of £1,000+£100 (for year 1)+£100 (for year 2)=£1,200; (any compulsory credit insurance might be added to this).

The monthly charge is then calculated as follows:

$$\frac{£1,200}{24} = £50$$

The next step is to establish the APR. Typically the "large print" on the literature published by the lender will state the annual "Flat Rate" of interest because this is the lowest. Fortunately, in Britain at least, there is a requirement to mention the APR, even if only in "small print".

The APR can be established by first calculating the annual nominal rate of interest using the formula:

$$\frac{2\ mq \times 100}{p(n+1) + \frac{q}{3}(n-1)}$$

where m=number of payments made in one year
q=difference between amount lent and the total of the instalments
p=amount lent (purchase price−deposit)
n=total number of instalments

Once the nominal rate has been established this can be converted to an APR using the following formula:

$$APR=(1+\frac{i\%}{n})^n-1$$

where i% = nominal interest rate per annum
n = number of payments per annum.

For the example given above in which £1,000 is borrowed for 24 months at a flat rate of 10%.

$$\text{Nominal Rate}=\frac{2\times12\times200\times100}{1{,}000(24+1)+\frac{200}{3}(24-1)}=19.565\%$$

$$APR=\left(1+\frac{\frac{19.565}{100}}{12}\right)^{12}-1=0.214 \text{ or } 21.4\%$$

(vi) Repayment of "Front Loaded" Loans

Compared to a bank overdraft, which is simply reduced if suitable funds are generated by a business, early repayment of a loan which is "Front Loaded" is not so straightforward. This is because The Rule of 78 applies. This states that the amount of interest due on the loan falls as the principal left to repay decreases. It must be appreciated that each monthly repayment contains an interest element and a repayment element. Since the monthly repayments are identical this means that the amount of principal repaid each month increases, whilst the interest declines.

Where a loan is repayable over a year the rule is that the interest is allocated in portions of 78ths. In month 1 it is 12÷78, in month 2 it is 11÷78; in month 3 it is 10÷78, and so on until in the final month it is 1÷78. The total 78 derives from 12+11+10+9+8 etc.

If the loan is for 2 years and not 1 year instead of using 78ths the interest has to be apportioned in 300 ths which is 24+23+22 etc. Where the loan is for 3 years the portions are derived using 666ths.

If £1,000 is borrowed for a year at 10% with monthly repayments the loan plus interest is £1,100 and the monthly repayment

$$\frac{£1{,}100}{12}=£91.67.$$

The situation created because of Rule 78 is shown in Table 5.

Table 5. *Application of Rule of 78*

Month	Total Repayable		Interest		Principal	
	Payment £	Balance £	Payment £	Balance £	Payment £	Balance £
0	—	1,100.0	—	100.00	—	1,000.00
1	91.67	1,008.33	$100 \times \frac{12}{78} = 15.38$	84.62	76.29	923.71
2	91.67	916.866	$100 \times \frac{11}{78} = 14.10$	70.52	77.57	846.14
3	91.67	824.99	$100 \times \frac{10}{78} = 12.82$	57.70	78.85	767.29
4	91.67	733.32	$100 \times \frac{9}{78} = 11.54$	46.16	80.13	687.16
			etc. up to month 12			
12	91.67	0	$100 \times \frac{1}{78} = 1.28$	0	90.39	0

The borrower might think that if he wished to repay the full loan after the end of month 1 only £1,000−£91.67=£908.33 would be sufficient. In practice he has to pay £962.17+£91.67=£1,053.84, and £53.84 over this period gives an APR of about 65%.

The reason for this is that the Consumer Credit Rebate Rules allow the lender to charge for setting up the agreement. If the loan was originally given for less than 5 years 2 extra months can be added to the term of the loan, and if it was for more than 5 years 1 extra month can be added. The £962.17 is obtained by adding the outstanding principal of £923.71 to the interest portion of the next month's payment plus the next 2 months interest i.e. £14.10+£12.82+£11.54.

[Note: Other systems for calculating the cost of borrowed money are covered in Chapter 11 under Investment Appraisal]

(vii) Taxation

Only the interest portion of the finance charges associated with borrowing are allowed as a cost when calculating profits and the item for which the loan is used must be one allowable by the Inland Revenue.

Assume a farmer has a loan of £1,000 and his interest at 14% in one year is £140. If his profit is such that his highest taxation band is 27% then the relief is £140×27%=£37.8 and effectively the loan has only cost him 10.22%. If his highest tax band is 50% then the relief is £70 and effectively the loan has only cost him 7%.

7.2.7. Other Sources of Money

(i) Self-generated

Over 60% of the capital invested in agriculture in Britain is self-generated. In other words it is invested out of profits. Taxation reduces the amount that is available and in some cases influences the way in which it is invested.

(ii) Grants—general

Government grants are available for certain items of expenditure. Over the years both the rates of grant and items eligible have changed. They have frequently been higher for farmers in Less Favoured Areas than those in standard areas. At the time of writing there were two main Agricultural Improvement Schemes, one national and the other EC, the former being financed by the UK Exchequer and the latter receiving support from the European Farm Fund (FEOGA).

The important point for any farmer wishing to avail himself of grants of this type is that he has to find a proportion of the capital himself. In doing so he must be careful not to "overstretch" his business, and he must be fully aware of the financial risks involved. He may decide to borrow a large proportion of his share and he should consider not only the effect on current cash availability as a result of interest and capital repayments, but also the possible change in interest rates and even the future for the enterprise associated with the investment.

There are also regional grants such as those operated through the Highlands and Islands Development Board, grants for forestry and tree-planting, for tourism, and a range of other grants.

(iii) Production Grants

These include the Hill Compensatory Allowances which are payable to farmers who keep breeding hill cattle and sheep on land in less favoured areas. In effect these are direct contributions to income and enable many hill farms to remain viable. Variable premium schemes for beef and sheep were also in operation at the time of writing to give support to beef and lamb incomes and there was also a suckler cow premium scheme.

(iv) Government Loans

In some countries, but not in Britain, government sponsored loans have been made available to farmers at cheap interest rates.

(v) Sale of Farm on Leaseback

Some farmers have generated working capital by selling their farm on a leaseback arrangement. Anyone doing this loses all the benefits of ownership and must balance this against the possibility of making a higher return on his capital in the foreseeable future, either on his existing farm alone or with additional land which he might be able to obtain. Long term implications must be carefully considered.

(vi) Borrowing from Sources Outwith the United Kingdom

When interest rates are high and/or loans are difficult to obtain it may be easier to get suitable loans from a bank outwith the United Kingdom, e.g. loans have been obtained from banks in Switzerland, West Germany and Canada.

A risk taken with any such loans is that if the value of the pound falls in relation to the currency in which the loan was obtained, then the real cost of interest and repayment could be much greater than was foreseen. Conversely if the pound rises in value then the real cost of interest and repayment could be less than expected.

7.3. TAXATION

7.3.1. Use of an Accountant

Taxation is a very specialist subject and both the fiscal regulations and levels of tax charged can vary frequently. Every farmer should therefore employ an accountant who will ensure that all allowances are obtained to minimise the liability to taxation. Expert advice on other tax matters is also valuable, in particular with reference to Inheritance Tax.

This book will therefore only present a brief summary of taxation. VAT has already been discussed elsewhere, but Income Tax, Capital Gains Tax and Inheritance Tax are all important.

7.3.2. Income Tax Schedules

Income or profits are taxed under the regulations specified by Schedules as follows:

(i) Schedule A

Landlords who obtain profits from letting land or property, whether for agriculture or for other purposes, are taxed under this schedule.

(ii) Schedule B

This schedule applies to profits from woodlands managed on a commercial basis unless the owner elects to be assessed under Schedule D, Case I.

(iii) Schedule D, Case I

Farm profits are taxed under this heading which applies to income from trades or businesses.

(iv) Schedule D, Case II

This applies to income from professions and vocations.

(v) Schedule E

Income from employment and notably Pay As You Earn PAYE is taxed under this heading.

7.3.3. Income Tax Allowances

Wage and salary earners are given allowances specified in the budget by the Chancellor of the Exchequer. Any income up to the total of these allowances is tax free. A married man gets a higher rate than a single person or a wife who has earned income. Once the allowances have been established there may be deductions for tax unpaid in previous years and for such things as interest payments received from investments which have not already had tax deducted from them. The net figure is the allowance to be set against pay. For those people paying tax under Schedule E this is converted to a code by the Inland Revenue who notify both the employee and employer. The code reflects the earnings on which the employee does not pay tax. It normally consists of a number followed by a letter e.g. 360 H means the allowance is £3,600. The letter H usually indicates that it is a married man and the letter L that it is a single person.

Any earnings above the level of the allowance attract tax at rates specified by the Chancellor of the Exchequer. Usually the first part of additional earnings is taxed at a comparatively low level and as annual income rises the rate becomes progressively higher. For example at the time of writing the first £17,900 of income above allowances was taxed at 27%; the next £2,500 at 40%; the next £5,000 at 45%; the next £7,900 at 50% and so on. These are the so-called "tax bands" or "tax brackets".

7.3.4. Income Tax and the Farmer

(i) General Points

The profits of a farm business are attached to the farmer and not the farm. If he has more than one farm they are all treated together for the purposes of trade so that losses of one can be set against the profits of another. Any expenditure which can be used in compilation of taxable profits must be "wholly and exclusively laid out for the purposes of the trade".

(ii) Allowable Expenditure

Farmhouse: Up to one-third of the cost of repairs, rates, and services is normally allowed, but this is at the discretion of the local Inspector of Taxes.

Farm Cottages: Expenses of maintaining farm cottages are allowed if occupied by farm employees. This does not apply if the cottage is occupied by a partner.

Workers: The cost of workers' wages, including the cost of their keep if boarded, is allowed. A farmer cannot charge for his own wage but a normal wage can be allowed for a spouse or members of the family if they work. This effectively means that the wife's wages, up to the level of the wife's earned income allowance can be used to reduce the taxable profit. In practice this should be examined to establish the level at which she would be required to pay National Insurance and if necessary to avoid such payment a lower wage might be entered.

The employer's share of National Insurance for his employees is classed as tax deductable.

Interest: Any interest on overdrafts or loans directly employed in the farm business can be treated as an expense. With hire purchase only the interest portion is allowable as an expense whereas the capital repayment part is eligible for writing down allowances.

NFU Subscription: Seven eighths of the National Farmers Union Subscription is allowable.

Travel Expenses: The cost of travel associated with the business is chargeable as an expense.

Other Expenses: Insurance associated with the business; maintenance of property and fences; hire of equipment; subscriptions appropriate to the business; and expenditure involved with the welfare of employees are all allowable expenses.

(iii) Expenditure Not Allowed

Frequently farmers pay for private items out of the farm's bank account. However, living expenses, private drawings, tax payments and the cost of life insurance policies must not be included in the P & L Account. Any farm produce eaten by the farmer and his family must be added to the income side of the account. Whilst the cost of cleaning out ditches is allowable, new drainage systems would usually be treated as capital expense and not allowed.

(iv) "Agreed" Profits

The profits of a farm have to be "agreed" before an assessment for tax is made. Many accountants prepare P & L Accounts for farmers which have depreciation rates that vary from one type of machine to another, and include a depreciation on buildings which is not appropriate to a taxation account. It is surprising that accountants do not present accounts in a standardised form. The depreciation costs must be added back to the profit and writing down allowances deducted from it in their place.

(v) Agricultural Buildings

An annual writing down allowance of 4% of the cost, net of any grant, for a new building can be employed to reduce the profit for taxation purposes. A system of balancing adjustments can be invoked if it is a "short life" building, or if it is demolished or destroyed.

(vi) Writing Down Allowances—Machinery

Machinery is treated on a "pool" basis and a writing down allowance of 25% of the "pool" value can be claimed to reduce the taxable profit. Capital expenditure on new machines and the value of machines which are sold must feature in the calculation.

Example

Assume that the Written Down Value (WDV) of a farmer's machinery pool brought forward from last year is £18,000. During the year he buys a machine for £6,000 and sells another for £2,000.

Year A

	£
WDV of pool brought forward	18,000
Sales	2,000
	16,000
Additions	6,000
	22,000
Writing down allowance @ 25%	5,500
Written down value of pool	16,500

Assume that next year he sells a machine for £1,000 but does not buy anything.

Year B

	£
WDV of pool brought forward	16,500
Sales	1,000
	15,500
Writing down allowance @ 25%	3,875
Written down value of pool	11,625

If the farmer's receipts from sales of machines in a given year exceed the pool value then the excess is raised as a charge to tax.

(vii) When is Tax Paid by the Farmer?

The farmer pays tax on the profit shown by the account ending in the previous taxation year. Half of the tax due is paid on January 1 and half the following July 1. This means that if his financial year ends on March 31 he pays half the tax due in January and the other half in July next year. However, if his financial year ends on May 31 the first payment of tax on the profits is not January next year, but January of the following year. It is fundamental that this is borne in mind when managing the cash flow.

When a new business is established tax is assessed as follows:

Year 1: Actual profits Year 2: First 12 months Year 3: Preceding year's profits	An election can be made to have tax based on actual profits for years two and three, but such an election must be made within 7 years of the end of the 2nd year of assessment.

Example

Assume a business commenced on 1 October 1990 and the financial year selected was 1 January to 31 December. The tax would be based as follows:

1st tax year: Profit from 1 October 1990–5 April 1991
2nd tax year: Profit from 1 October 1990–30 September 1991
3rd tax year: Profit from 1 January 1991–31 December 1991

(viii) Relief for Fluctuating Profits

With the exception of the opening and closing years of a business a farmer can average the profits for two consecutive years for the purposes of tax assessment provided that the difference in profits is at least 30% of the higher of the two profits. There is also some relief where this figure is between 25% and 30%. This helps the farmer spread the tax burden and may help reduce the total tax paid if the top level of his income falls in lower tax bands.

(ix) Calculation of the Tax Payable

Assume that for Year A in the example shown in (vi) above the accountant calculated the profit to be £26,000 but used a depreciation on machinery of £4,000. This depreciation is added back and the writing down allowance is deducted to arrive at the agreed profit.

	£
Profit calculated by accountant	26,000
Depreciation added back	4,000
	30,000
Writing down allowance	5,500
	24,500

In common with his employees the farmer can claim allowances for being married and possibly other allowances. Assume that his total allowances are £4,200. This reduces the amount on which he pays tax to £24,500−£4,200=£20,300.

At the rates pertaining when this book was written the tax would be as follows:

	£	£
First	17,900 × 27% =	4,833
Next	2,400 × 40% =	960
	Total Tax =	5,793

(x) Higher Rates of Tax and Income Received Net of Tax

Farmers, like private individuals, may receive income from investments outside the farm which has already had tax at the basic rate deducted from it. For example assume that a farmer whose income puts him in the middle of the 40% "tax bracket" received £80 interest from £1,000 invested in a building society.

In reality the £80 is only 73% of the interest payable because the standard rate of 27% tax would have already been deducted. Therefore the full interest £x is obtained as follows:

$$£x \frac{73}{100} = £80$$

$$£x = \frac{80 \times 100}{73} = £109.59$$

The farmer has to pay tax at 40% on this:

$$109.59 \times \frac{40}{100} = £43.84$$

However, the building society has already paid £109.59 × 27% = £29.59. He therefore has a further £43.84 − £29.59 = £14.25 to pay.

Many people receive dividends from investments in shares of public companies and attached to the cheque is a statement of tax credit. This is the amount of tax deducted at the basic rate and if the recipient is in a higher "tax bracket" it must be treated as described for the building society interest.

7.3.5. Capital Gains Tax

(i) CGT and Exemptions

When the disposal by sale or gift of an asset results in a gain over the acquisition price capital gains tax (CGT) may be payable. There are some exemptions. These include the principal private residence such as the farmhouse, certain life assurance policies, and moveable objects with a life of 50 years or less. Livestock are not subject to CGT where the farmer has elected to be treated under the herd basis (if on a trading basis income tax would be charged on any profit).

Each person is allowed to make a gain annually up to a limit specified by the

Chancellor of the Exchequer which is exempt. Losses can be employed to offset gains. Transfers of assets at death are not subject to CGT, but gifts other than between husband and wife are taken into account.

(ii) Transfer of Business Assets

If the recipient of a gift of business assets agrees with the donor then the CGT can be held over. This means that no tax is paid at the time of the gift but the recipient accepts that the gift has been received at current value less the CGT that would have been payable. The payment is effectively postponed or even abandoned if the recipient dies whilst still owning the asset.

(iii) "Roll-over" Relief

If business assets such as farm land are sold and the proceeds used to purchase new business assets then CGT can be deferred ("rolled-over"). The gain on the original asset is deducted from the cost of the new one. There are time limits and at the time of writing the new asset must have been acquired within a period of 1 year before to 3 years after the sale of the original asset.

"Roll-over" relief can be obtained where, say part of a farm is sold and the money is used to undertake capital improvements on the land still left.

(iv) Indexation

Only gains made since 6 April 1965 are subject to CGT. As a result of the 1982 Finance Act indexation to allow for inflation has been possible. If the asset was obtained before 6 April 1982 (1 April 1982 for companies) the indexation is related to its value at 31.3.82, but only if this is greater than the purchase price. If the asset was acquired after this time then indexation is possible from the time it was acquired to the time of its disposal.

In effect indexation means that the cost of the asset can be multiplied by the increase in the retail price index between March 1982, or when it was acquired if later than this, up to the point of disposal. This reduces the gain or may even give rise to a loss.

The Chancellor of the Exchequer specifies the rate of tax which is payable. It is due on 1 December following the year of assessment or, if later, 30 days after the assessment is received.

(v) Retirement Provisions

Provided a person with a family business satisfies certain conditions, upon retirement over the age of 60, relief can be obtained on the value of business assets up to a sum specified by the Chancellor of the Exchequer. One main condition is that the business must have been owned for 10 years, although there is proportional relief if the period is shorter.

7.3.6. Inheritance Tax

(i) Rates of Tax

This is a tax on transfers of capital at death or on gifts made within 7 years of the death of the donor. The tax is based on the loss to the donor and not the benefit to the donee. The rates of tax are amended from time to time but those applicable when this was written illustrate the principle.

Slice of cumulative chargeable transfer		*Cumulative total*	*% on slice*	*Cumulative total tax*
	£	£		£
The first	90,000	90,000	Nil	Nil
The next	50,000	140,000	30	15,000
The next	80,000	220,000	45	47,000
The next	110,000	330,000	50	102,000
Above	330,000		60	

Gifts made within 7 years of death were taxed at a tapered rate as follows:

Years between gift and death	*% of full charge* at death rate
0–3	100
3–4	80
4–5	60
5–6	40
6–7	20

Thus if a person died leaving £200,000 which was subject to inheritance tax the tax would be:

Slice £	Rate of tax %	Actual tax £
0– 90,000	0	0
90,001–140,000	30	15,000
140,001–200,000	45	27,000
		42,000

However, had the person given this money away 5 years before death the tax would have been £16,800.

(ii) Exemptions

There are several important exemptions:

(a) *Transfers between spouses* These are exempt from Inheritance Tax [N.B. if a husband leaves his estate to his wife she has this inheritance in addition to any money of her own to consider when passing on assets to her children. Depending upon the circumstances of the family the husband should therefore consider making a transfer of at least some of his estate to his children].

(b) *Annual gifts* A person is allowed to give away £3,000 every year without it attracting any tax. Both husband and wife can do this so passing on £6,000 each year between them.

(c) *Small gifts* Any one person may give away £250 to any other person each year without attracting tax e.g. a father could give each of his children £250 per annum.
(d) *Normal expenditure out of income* A donor may give away funds out of his own income provided that there is an element of regularity in the gift and the donor is left with sufficient income to maintain his own standard of living.
(e) *Marriage gifts* When a child is getting married he or she may be given £5,000 by a parent of either partner to the marriage.

(iii) Life Assurance

Transfers (other than the exemptions) made before death are known as PETs or Potentially Exempt Transfers. There is always the chance, however, that the donor may die before the 7 years are up. The recipient is therefore responsible for paying any inheritance tax which becomes liable.

Insurance companies offer reducing term assurance on the life of the donor. These may be taken out in trust by the donor or by the recipient.

(iv) Relief for Agricultural Property

Agricultural land and buildings qualifies for a 50% relief if transferred with vacant possession and 30% if tenanted. To qualify either (i) the land must have been occupied by the transferror for the purpose of agriculture throughout the period of 2 years ending with the date of the transfer; or (ii) the land must have been owned by the transferror throughout the 7 years ending at the date of transfer and was throughout that period occupied for agricultural purposes.

(v) Business Property Relief

For IHT purposes business property relief reduces the value of property by either 50% or 30% according to circumstances. The 50% rate applies where the donor has a controlling shareholding in a farming company or has sole or partnership interests in a farming business. The 30% rate applies to land, machinery and plant used by a business company controlled by the transferror or by a partnership of which the transferror is a partner.

Normally the property must have been owned for 2 years or have replaced property owned for at least 2 out of the last 5 years.

(v) Reservations

If a gift is made in which the transferror retains an interest or benefit then the gift is treated as the donor's estate at death, or if the interest is given up before death the gift is then treated as a PET under the 7 year rule.

(vi) Summary

In view of the complex nature of Inheritance Tax farmers are strongly advised to obtain professional help. Much can be done to minimise the loss of capital to farm businesses by good planning and making full use of reliefs.

7.3.7. Influence of Tax on Decisions

The desire by farmers to minimise tax commitments can greatly influence their investment decisions as well as the method and timing of the disposal or transfer of assets to others. A classic example often quoted is that of the farmers who make very high profits from potatoes in isolated years. They are keen to take advantage not only of the tax averaging, but also of the "writing down" allowances which can be put against machinery and buildings when the "agreed profit" is calculated for tax assessment.

All farmers should consider the benefits from these allowances. Some expect the accountant to take full advantage after the financial year is finished. This is too late since tax planning is necessary beforehand if investments are to be made. It should not, however, be a matter of estimating the profit shortly before the end of the year and deciding to buy a tractor to save tax. The planning must be done seriously to ensure that the money could not be better used elsewhere in spite of the tax relief. Far too many farmers have precipitated financial problems for themselves in relation to future cash flows by investing in machinery in order to save tax.

Farmers, especially those with surplus finance, may decide to make improvements to their property to obtain not only the short term advantages from taxation on farm profits, as opposed to the higher taxes on unearned income, but also the longer term benefits associated with Inheritance Tax. Other farmers may invest in more land. In all cases it is necessary to ensure that the investment is in the best interests of the business and that working capital is not reduced below safe levels.

When making an investment the extra profit generated has to be considered in relation to the additional tax which may be payable. The added profit may put the farmer into a "higher tax bracket" so that the rate of tax on it is above that on current profits.

7.4. TYPES OF BUSINESS STRUCTURE

7.4.1. Sole Trader

The farmer operating in sole charge of his business is the simplest form of trading in agriculture. This has the disadvantage that when good profits are made he suffers the penalty of high marginal rates of tax. However, his wife and children can sometimes be employed and since each has individual personal allowances the problem can be reduced.

One of the main advantages of the sole trader is that he can take decisions with a minimum of outside interference.

7.4.2. Partnerships

(i) Formation

Partnerships may be formed for a variety of reasons. Frequently sons join their fathers in part to gain an entry to farming, but also to obtain the benefits of splitting profits so that each partner is in a lower "tax bracket" than the father would be on his own.

Alternatively two or more farmers may join together to pool their capital and other resources in order to obtain the benefits associated with scale of business or of sharing expensive items. Usually the maximum number of partners allowed by law is twenty. In spite of the fact that a verbal agreement can establish a partnership it is advisable that a formal deed is completed.

Partnership is not something to be entered into lightly because under partnership law one partner can commit his co-partners in relation to business transactions, even without their knowledge and consent. Although partners may contribute different amounts of capital to the business, they can be held to share losses and profits equally unless the partnership deed specified otherwise.

The deed must be carefully worded and should include statements defining how and when accounts are to be prepared, how profits and losses are to be shared, who signs cheques, who manages the business, who buys and sells, how much capital, each partner contributes and whether any interest will be allowed on this capital, whether any partner is allowed any salary or wages, how long the partnership should last, and what happens if one partner dies or wishes to withdraw.

The major worry is that, in spite of the deed, if the business goes into liquidation one partner is not only responsible for his share of the losses, but also for that of his partners if they cannot pay their shares or cannot be traced.

(ii) Taxation of Partnerships

The taxation of a partnership is raised in the name of the business and each member can be held responsible for all the tax if the others cannot or will not pay their share. In practice each member's share of tax is assessed on the assumption that the profit is divided to conform to the particular member's entitlement specified in the agreement for the current and not the previous year. This last point is only important if there is a change in entitlement.

Each individual is then taxed on his share of the profit after deduction of appropriate personal allowances.

In some cases where the wife is a partner there can be advantages in electing for "separate taxation" from the husband. This applies where profits are high and the wife's proportion of the profit is above an amount which on an overall basis results in less tax being paid.

There can be some benefit from the point of view of Inheritance Tax by appropriate arrangement of revenue profit sharing ratios and partnership salaries. Partnerships can arrange for a retired partner to have an annuity and the remaining partners can put these as charges on income and obtain tax relief.

Expert advice should be sought to establish the position with reference to CGT where the land is owned by the partnership rather than by one of the partners who then leases it to the partnership. The question of IHT needs considering when the partnership is established using land owned by one of the partners. It needs to be clear whether the partnership has the right to continue using the farm on the owner's death or whether occupation will be terminated and the land be treated as being subject to vacant possession.

7.4.3. Companies

(i) Formation and Operation

The main difference between a partnership and a company is that whereas the former is generally regarded as a collection of individuals the latter has a specific legal identity. As stated above, members of an insolvent partnership can be held responsible for the debts of the business, but shareholders in a limited liability company can only lose the money they invested in the business to buy the shares or which they gave as loans.

To comply with the law companies have to be registered. The first step is to produce a memorandum of association which gives the company's name, its activities and registered office, the maximum nominal capital of the business, and the liability of its members. Next the articles of association are agreed. These give the company's rules, including those which govern the directors or members. The last step is to file with the Registry of Companies a statement of the nominal capital and a declaration of compliance with the Companies Act.

A private company must have at least two shareholders and a public company a minimum of seven. For most farmers a private company is normally formed since it avoids the necessity for public meetings, the publication of a prospectus, the need for directors to retire at seventy, and has other advantages. Public companies do not have the same restrictions on the transfer of shares and are not forbidden from seeking public money, but a private company can go public later if desirable. An "exempt" private company can make loans to directors, does not have to submit a balance sheet to the Registry, and does not have to have a qualified accountant. Public companies must have at least two directors, but a private company need only have one. Directors and not shareholders run the affairs of a company although shareholders can make their presence felt at the statutory meetings.

A company has certain benefits over individuals in relation to the provision of security associated with borrowing money. Unlike the individual the farming company can offer not only land as security but also its other assets.

It is possible for a farmer and his wife to sell their business to a company which they form. They receive shares in the company and provide funds for it by providing a mortgage on a large part of the assets. If the company subsequently fails this latter security makes them primary creditors.

The name of a limited company must include the word "Limited" at the end to indicate the fact to suppliers of goods that shareholders cannot be pursued for debts. In addition the law also states that anyone carrying on a business who uses any word in its name other than the real surnames (and possibly first names or initials) of all partners, must register the name.

In view of all the legal factors involved anyone considering forming a company is advised to obtain professional advice. For example, at the time of writing it was more beneficial for the shareholders of a farm business, rather than the company, to own the land, and to occupy it under a licence or tenancy. This was because of the possibility of a double chargeable gain occurring on liquidation; first for the sale of the farm and then on any distributions to the shareholders.

(ii) Taxation of Companies

The profits, income, and chargeable gains of companies for an accounting period, less allowable deductions, are subject to corporation tax. The tax rate is based on an accounting year of 1 April to 31 March, but the accounting period used by the company is usually acceptable. If this does not end on 31 March and the rate of corporation tax changes an apportionment is necessary.

Corporation tax for small companies is less than that for large companies. Most UK farm companies are "close companies" because they have five or less people with shares in them.

Distributions of profits to shareholders are not deductable when establishing the profit subject to corporation tax. The tax is payable on all profits whether they are distributed or not.

Advance corporation tax must be paid by the company to the Inland Revenue when distributions are made to shareholders. At the time of writing this was 27/73rds. This advance corporation tax can then be used to reduce the companies total liability for corporation tax for the accounting period.

A shareholder who receives a dividend statement also receives a record of the tax credit associated with the dividend. This has to be added to the dividend to obtain the total income received by the shareholder for tax assessment purposes. If the shareholder does not pay tax at the basic rate, part or all of the tax credit may be reclaimed, but if he pays tax above the basic rate he will have additional tax to pay.

From the standpoint of taxation a company is usually better than a partnership where profits are high. However, there is the possibility of additional tax charges when profits retained in the business are extracted.

Companies can offer favourable pension arrangements for directors and the cost associated with these is allowable when establishing the company's profits. One benefit of companies compared to partnerships is that, whilst in the latter the division of profits is normally controlled by the partnership agreement, distribution of income to directors of a company is flexible which can prove advantageous in some circumstances. Directors may decide to leave some of the profit in the company so that only corporation tax is paid on this part.

7.5. INTEREST RATES AND INFLATION

7.5.1. Interest Rates

In the United Kingdom borrowing and lending rates are very much influenced by bank "base rate". This is in turn influenced to a marked degree by the supply and demand for money. If demand is high in relation to supply, interest rates tend to increase, whereas they generally fall if supply exceeds demand. For example if banks have a high demand from borrowers they will attempt to obtain more money from the "inter-bank" market, which creates a simulus for increased interest rates. This in turn can push up returns to depositors who may invest more, but costs for borrowers increase. The latter may then reduce their demand for borrowing so that a balance is created.

Although the above paragraph illustrates the general principles many other factors are involved. Governments aim to control money supply and/or the exchange rates between their own and foreign currencies by influencing interest rates. They also recognise that there can be a relationship between interest rates and inflation.

7.5.2. Inflation

Inflation is effectively a lowering in the purchasing power of money. It can have a very severe effect on the economy of a nation and thus on an individual farm business. The value of money is virtually halved if inflation is at 20% for 3 years and it only takes 7 years to halve it if the annual rate is 10%.

Identical models of machinery, buildings, and other equipment, can cost significantly more if purchase is delayed. Workers tend to press for higher wages and the fixed costs of farming increase rapidly. The value of outputs may not always keep pace with these increases and profit margins may fall.

Farmers adopt their own measures to assess the impact of inflation on their businesses. They frequently talk about the number of litres of milk or tonnes of wheat which are currently equal in value to the cost of a tractor or item of equipment on their farm. Care is necessary in interpreting such comparisons because factors other than inflation can alter these price relationships.

7.5.3. Borrowing During Inflation

Assume that a farmer borrowed £20,000 to be repaid at the end of 10 years and that the annual inflation rate during this period was 7%. The value of the £20,000 by the time it was repaid would have been halved. In other words, although the loan had to be repaid in full the impact on the business would have been less than if money had retained a constant value.

People borrowing money to purchase assets which can grow in value with inflation, such as farmland or private houses, have the added benefit that they can help to keep pace with the inflation using money which is not their own. Assume that £20,000 was used to purchase a house. Ten years later the house might be worth £40,000 or more. Admittedly if the assets were realised the owners would have to spend a similar amount to obtain a property of comparable quality, but they might justifiably consider themselves to be better off than people who decided not to borrow and remained in rented accommodation. The same applies to farms but on a larger scale.

Borrowing money to erect farm buildings might enable a farmer to obtain them at cheaper prices than if he had to wait until he had generated enough from his business to cover their cost. He would have their use earlier which might help the working capital associated with the use of the buildings to keep pace with inflation.

It would appear from this that borrowing during periods of inflation is very attractive, but it must be noted that if deflation occurs the reverse situation would be true. It is not as simple as this and many other factors have to be considered.

Much depends on the rates of interest associated with the inflation. If these are high and profit margins are declining because of inflation, borrowing for a project can have a serious impact on the spendable surplus which the farmer has, with a consequent effect on living standards. This is especially the case if it takes a few years for the new project to produce substantial returns. It would be even more significant if the borrowing was to finance something which might simply ease work load, such as a large tractor, and which might not produce an extra cash return.

Some farmers therefore face a dilemma. They have to ask if they can afford not to invest and still remain viable and keep pace with inflation. At the same time they have to question whether they can remain viable in the short term if they do invest. In many cases a large element of the investment will be from borrowed capital. Careful investment appraisal is therefore essential.

7.5.4. Interest, Inflation and Money Supply

Inflation usually increases the demand for credit because people wish to buy before prices rise. They also recognise that repayment of a loan may be easier because if they borrow a specific sum effectively the value of this sum when repaid will, because of inflation, be less. Depositors, however, become concerned that the value of the money they have invested is falling. They may look for alternative investments. This tends to reduce the money supply available in relation to borrowing demand with the result that interest rates rise. Often high inflation is therefore associated with high interest rates.

The farmer should always take note of the rate of interest on borrowed capital and the rate of inflation. Inflation can essentially reduce the real cost of interest. For example if interest is 15% and inflation 10% the cost of the money might be considered to be 5%. There are times when inflation is higher than the interest rate. However, when interest rates greatly exceed inflation business investment and spending by the public can become less attractive.

If a Government can control the money supply in circulation they can help control inflation. In the United Kingdom the Government keeps close control of what is known as Sterling M3. This essentially reflects bank deposits. By selling some of its own debt to potential investors it takes up part of the money that would have otherwise gone into banks, which in turn would have been available to borrowers. The Government can also help contain investment/spending, and hopefully inflation, by keeping interest rates high. Through the Bank of England it provides funds to the money market. If these funds are increased money supply in relation to borrowing demand is high and there is a tendency for lower interest rates, but if the funds are curtailed the tendancy is for interest rates to rise.

A country can be concerned about inflation for many reasons. These are not restricted to internal political considerations. They include the stability of the exchange rate between its own currency and those of other countries. This in turn influences international trade and the investment by people from other countries in its currency.

7.5.5. Accounting and Inflation

In times of inflation the P & L accounts prepared by methods shown in Chapter 4 can show profits which are higher than reality. One method of overcoming this is current cost accounting. Basically this means that the costs of production of goods sold from a farm are entered at their current values rather than the amount paid for them when they were actually purchased. Also instead of depreciating items on the basis of their written down values the depreciation is based on their current value to the business.

Each farmer is interested in his spendable surplus and its purchasing power. Great care has to be taken when comparing results from different years of business operation to take account of the effects of inflation so that comparison reflects the value of money in real terms.

7.5.6. Deflation and Reducing Values

When the value of assets fall considerable problems can be created especially for those who have borrowed a substantial proportion of the initial cost. This happened to many farmers in the mid-1980s who had paid high prices for their land, frequently obtained with the assistance of large mortgages. When profit margins became difficult to achieve some found that their main security for further borrowing, their land, was not worth what they had paid for it initially. They still had the mortgage to pay on the original amount borrowed and with no more loans being available to them some became insolvent, or sold up whilst they could still extract some money from their businesses.

CHAPTER 8

LABOUR AND MACHINERY

CONTENTS

8.1. LABOUR, MACHINERY, AND THE FARMING SYSTEM

8.1.1. Importance and Interrelationship

The marked influence on profitability of the quality of work performed by the staff of a farm must never be underestimated. This not only includes the workers who undertake the day-to-day operations but also those responsible for management. The variability of the factors associated with the running of a farm means that there can be few other industries where non-managerial staff have responsibility for taking so many important decisions. Basic examples at the stockman level include the need for observation and appropriate action in relation to disease, dystokia, milking, breeding, and nutrition; or in the case of crop production attention to details such as drill setting, depth of drilling, efficient operation, maintenance and care of machinery, and adaptation to ground conditions.

Labour cannot entirely be divorced from mechanisation. In countries where workers are plentiful and cheap, men and women tend to undertake many tasks which elsewhere are performed by machines. In more prosperous countries capital invested in machinery substitutes for labour. Where competition from other industries for staff is high it may not only be necessary to employ machines to take a lot of the drudgery out of farmwork in order to attract men to, and retain them, on the land, but the wages and other conditions offered to agricultural workers must also improve.

The productivity of the men remaining on highly mechanised farms should be increased as a result of the mechanisation. In addition the application of new technology in both crops and animal production may also be fostered by the application of progress in the design of machinery and equipment. Speed and timeliness of operation are fundamental to the success of many farming tasks. For example, suitable machines integrated into a well-organised system can make a very substantial difference to silage quality.

8.1.2. Balance between Men and Machines

It is essential that each farm has the correct balance between men and machines if profits are to be maximised. Many factors are relevant including the farming system, availability of capital and labour, and size of farm. Machines are very expensive and most have a short working season. The opportunity cost of the capital invested in them must be considered. Unit cost, e.g. the cost per cereal hectare of keeping a combine, should be established. It may be difficult to justify ownership of some machines on small farms where contractors or a form of sharing arrangement may be more appropriate.

When interest rates on borrowed capital are high or when inflationary factors such as increases in the world price of oil push up the cost of machinery, the correct balance between men and machines for a particular farm may change. However, apart from casual labour and overtime, men can only be employed in whole full-time units. Whilst it might be considered ideal that staff and machines should be selected to suit a particular production programme it can be equally important to modify the selection of the production programme to suit the labour and mechanisation available.

Machines do not always reduce labour costs. If the introduction of a machine does not actually reduce wages because a man cannot be dispensed with, or overtime reduced, then it must be justified in other ways, such as improved productivity of the stock or crop, or by easing the work for the staff. Some farms have seasonal peak demands for labour and certain staff kept for these periods may not be fully employed at other times of the year. In these cases changes in mechanisation may reduce the labour peak and eliminate the need for a man.

When selecting machines for a farm they must not only integrate with the labour and production system, but must also relate to each other. Farmers who make silage know how important it is to have the right size of tractor for their forage harvester, to have machines which present a swath of grass to suit the forager, and to have the trailers and clamp or tower facilities to deal with the throughput of the harvester. Staff availability will also be a major factor in deciding a suitable routine, as will the distance from the field to the store, and both will influence which machines would be suitable for the operation.

The ideal balance between men and machines is specific to each particular farm, but many of the basic principles covered in this chapter can be applied to all farms to help establish what this balance is and what size the machines should be. Any capital invested in machinery should be justified and must not prejudice the success of the business.

8.2. PLANNING LABOUR

8.2.1. The Farmer

The first task of any farmer when planning a new business or reorganising an existing one is to plan his own work. He has to remember the management load that he will carry and should ensure that he has time to undertake it effectively. This type of statement may cut little ice with the farmer who is on such a small scale that he is the sole work force. Frequently it is just such a farmer who, because of the pressures upon him, cannot devote the time to it that good management requires. The work load may prevent him from "standing back" to look at his farm objectively and seeing what improvements could be made, taking any constraints into account. In many cases his position is not an easy one.

Farmers differ enormously in their attitudes to labour. The two aspects relevant here are the amount of authority and responsibility which they are prepared to pass to their staff and the amount of manual work that they do themselves. Each farmer should identify the key management tasks for his farm and assess the time needed to undertake them. He must then assess whether, in the best interests of the business, it would be wise to employ an extra man to undertake the manual work which he would otherwise do himself. This would release him to concentrate on organisation, marketing, and attention to detail, which could possibly more than compensate for the added wage. Use of some form of secretarial or management service, or computer, to reduce his routine office work may be justified.

If the business is large a management line structure might be considered which would leave the owner to take the most important decisions, and other staff, such as the enterprise managers, to make the day-to-day decisions. It is important that the owner does not become isolated from his farm and his workers, although much will depend upon the confidence he has in his immediate subordinates.

8.2.2. Assessing Labour Requirements

Having decided his own role on the farm the owner should then progress to assess his staff requirements. Past experience on the same farm is the best guide, but this is not always available.

Standard labour data based on averages from many farms can be used, but it has very severe limitations because it does not fully consider the variable factors which can affect the labour requirement of a specific farm. Techniques using it can serve as a starting point provided that the limitations are clearly recognised.

The Standard Man Day system is one such technique. A Standard Man Day (SMD) is one man working for 8 hours. Thus if a man works 48 hours in a week, he works for 6 SMD.

Table 6 shows the approximate labour requirements for some of the most important stock and crop enterprises. It must be emphasised that some farmers who are very efficient in the use of labour and who have large machines can obtain figures which are 25–30% or more better than those illustrated in this table.

Table 6. *Standard man day requirements*

Stock	SMD per head	Crop	SMD per ha
Dairy cows: parlour	5	Potatoes (non-mechanised)	18
cowshed	8	Potatoes (mechanised)	9
Beef cows	2.5	Sugar-beet	6
Bulls	4	Wheat, barley, oats	2.1†
Barley beef: 0–12 mth	1.1	Oilseed rape	1.7
Other cattle: 0–6 mth*	2	Vining peas	2
6–12 mth	0.7	Herbage seeds	3
1–2 yr	1.5	Turnips, swedes: lifted	8
2+ yr	1.8	folded	3
18 mth beef: 6–18 mth	2.8	Kale: cut	6
Ewes, rams	0.4	grazed	2
Other sheep over 6 mth	0.2	Hay: cut	2.5
Sows	3	2 cuts	4.5
Boars	1.5	Silage: 1 cut	2.5
Other pigs over 2 mth	0.5	2 cuts	4.5
Laying hens	0.04	Grazing only	1

*Artificially reared; multiple suckled 0–6 months: 1 SMD
†If straw burnt 1.5

Once the total labour requirement has been calculated from this data it is usual practice to add 15% to cover the time spent on general overheads such as farm maintenance.

The data can be employed to calculate the theoretical requirements for Church Farm based on its current enterprises.

Stock	SMD	Crop	SMD
Dairy cows:	100 × 5 = 500	Cereals:	90 ha × 2.1 = 189
Other cattle:		Potatoes:	10 ha × 18 = 180
0–6 mth:	38 × 2 = 76	Grazing:	49 ha × 1 = 49
6–12 mth:	18 × 0.7 = 13	Hay:	6 ha × 2.5 = 15
1–2 yr:	35 × 1.5 = 53	Silage:	
2+ yr:	17 × 1.8 = 31	1 cut:	26 ha × 2.5 = 65
18 mth beef:		2 cuts:	19 ha × 4.5 = 85
6–12 mth:	46 × 2.8 = 129		
Ewes and rams:	208 × 0.4 = 83		
	Total = 885		Total = 583

Summary	Based on data in Table 6	Assuming 25% higher efficiency
Total stock and crop	1,468	1,101
Add 15% for overheads	220	165
Total SMD	1,688	1,266

Assume that the total SMD worked per year, with overtime, is 300 by stockmen and 250 by other staff. On the basis of the first set of data (1,688 SMD) for Church Farm three stock workers and three other staff would be required. With 25% higher efficiency (1,266 SMD) three stock workers and two staff would be more than sufficient. This shows the limitations of the SMD technique because it is difficult for a farmer who is new to a farm to know how efficient he will be. With an existing business the efficiency of labour use is easier to ascertain and problems would only arise with a major change in policy.

It is essential not to employ more men than are needed not only because of their cost, but also because of the Employment Protection Act and dismissal procedures. When a new farm is taken over previous staff levels on the farm can be a guide.

Unit size must be considered since maximum efficiency in the use of men is fostered if the size of each enterprise is related to the working capacities of men so that they are not underemployed and do not waste too much time swopping from one job to another.

A major consequence of not employing enough men is that enterprise performance may suffer because of lack of attention to detail. Each farmer has to estimate if the cost of an extra man will be justified by the extra, or marginal, income his labour may generate, although such factors as the easing of work loads for others may be considered.

Church Farm actually employs two stockmen, two general workers and a student from April to September, in addition to the farmer. Casual labour also assists with the potatoes.

8.2.3. Seasonal Labour Demands

The above calculations neglect the seasonal demands for labour. Data in Table 7 shows the seasonal work requirements of various enterprises as percentages of the total SMD needed for the year. This information needs very careful amendment to suit different areas of the country and such factors as lambing dates. The seasonal figures for Church Farm are shown below. The top set of data assume that it achieves average performance and the bottom set 25% above average, in labour use.

Efficiency level		J	F	M	A	M	J	J	A	S	O	N	D
Average	Stock+crop	91	103	183	139	125	124	102	128	110	160	110	93
	15%	28	30	4	17	12	12	20	16	18	10	30	23
	Total	119	133	187	156	137	136	122	144	128	170	140	116

Efficiency level		J	F	M	A	M	J	J	A	S	O	N	D
25% Above Average	Stock+crop	68	77	137	104	94	93	77	96	83	120	83	69
	15%	21	23	3	13	9	9	15	12	14	7	22	17
	Total	89	100	140	117	103	102	92	108	97	127	105	86

The peak demands occur in March and October. The 15% overhead allowance has been allocated to months when miscellaneous tasks are most likely to be undertaken. Casual labour is also employed in October for potato lifting and two men help out on the planter in April.

In a situation where overtime and casual labour will not allow a farmer to cover peak seasonal demands the other steps he could take to alleviate the problem include:

(i) use of an agricultural contractor;
(ii) purchase of alternative machinery to speed work;
(iii) adjust balance of enterprises, amend such aspects as lambing dates, change cereal varieties to obtain spread of harvest, or amend other production systems;
(iv) consider greater degree of specialisation;
(v) study work techniques, handling methods and systems, building layout, roads and other methods of improving labour efficiency;
(vi) consider sharing labour and machinery with other farmers.

8.2.4. Field Work Days

When planning labour it is essential to realise that the number of days suitable for field work varies with the area of the country, soil type, and other factors including climatic differences from year to year.

Sometimes farmers must be opportunists. For example, drilling spring barley early can be satisfactory in some circumstances. One year at Auchincruive barley was drilled on 20 February and the next spell of suitable weather did not arrive until the end of March. The yield was over 6 tonnes per hectare, but the soil conditions were right and the soil was not one which would cap with later rains. In our area in the West of Scotland we feel that we can be reasonably selective in waiting for the right soil conditions for drilling barley until about the end of the first week in April. If it is not drilled by this time we cannot afford to be so choosy.

Table 8 shows the possible work days available based on an average season.

Table 8. *Field work days available*

Western Britain				Eastern Britain			
Jan.	12	July	23	Jan.	12	July	25
Feb.	11	Aug.	22	Feb.	12	Aug.	24
Mar.	16	Sept.	21	Mar.	18	Sept.	23
Apr.	20	Oct.	19	Apr.	22	Oct.	21
May	22	Nov.	12	May	24	Nov.	13
June	23	Dec.	11	June	25	Dec.	12

8.2.5. Gang Work Days

When farmers are planning their labour they must recognise that a team or gang of men is necessary if certain jobs are to be undertaken efficiently. The gang work day planning technique is particularly suitable for large scale cropping farms at peak times of the year, and can be applied to silage and hay making.

A gang work day lasts 8 hours, but the number of men in the gang must be specified. Assume that a gang of four men make silage from 5 ha in one 8 hour day. The gang work days per hectare would be

$$\frac{\text{one 8 hr day}}{\text{ha per day}} = \frac{1}{5} = 0.2 \text{ gang work days for 4 men}$$

Gang work day charts such as that for Church Farm for September and October (Fig. 67) can be constructed. The number of field work days in each month is considered together with the gang size for each job and the time that the gang requires to undertake each particular task.

Table 7. *Seasonal labour requirements as percentage of total annual SMD demand*

	Jan.	Feb.	Mar.	Apr.	May	June	July	Aug.	Sept.	Oct.	Nov.	Dec.	Total
Winter cereals	—	—	4	8	4	—	—	22	30	25	7	—	100
Spring cereals	—	4	22	8	4	—	—	37	14	4	5	2	100
Potatoes—main crop	1	1	7	20	2	4	4	1	10	38	11	1	100
Potatoes—early	—	16	19	4	3	25	29	—	—	—	3	1	100
Turnips, swedes	5	5	5	10	7	20	8	—	—	—	20	20	100
Sugar-beet	1	—	9	5	16	17	2	—	2	19	22	7	100
Oilseed rape—Spring	—	4	18	16	5	3	—	—	19	25	5	5	100
Kale—grazed	10	5	5	25	15	10	—	—	—	10	10	10	100
Kale—cut	10	5	10	20	10	10	—	—	—	10	15	10	100
Vining peas	—	12	8	5	4	2	30	30	—	3	6	—	100
Hay + grazings	—	—	3	10	3	55	25	4	—	—	—	—	100
Silage—1 cut + grazings	—	—	15	5	35	37	4	4	—	—	—	—	100
Silage—2 cuts + grazings	—	—	11	4	25	28	28	4	—	—	—	—	100
Grazing only	—	3	20	10	15	17	15	15	5	—	—	—	100
Dairy cows	10	10	10	9	7	6	7	7	7	8	9	10	100
Other cattle	12	12	11	7	7	5	5	5	5	8	11	12	100
Sheep	4	8	30	10	6	10	8	4	4	8	4	4	100
Pigs	9	9	8	8	8	8	8	8	8	8	9	9	100

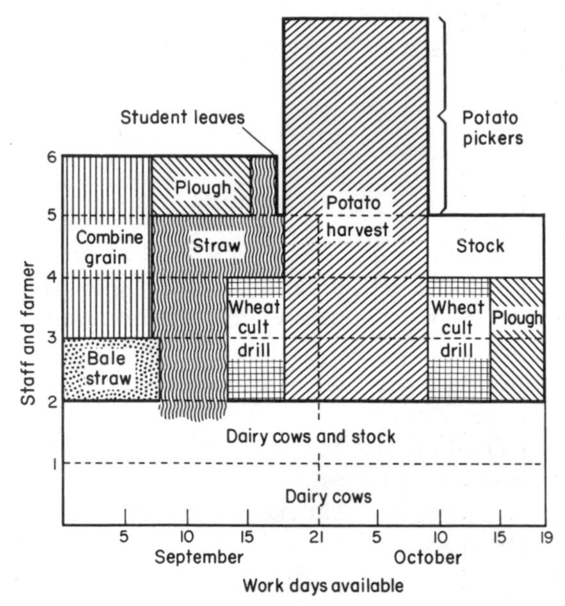

FIG. 67. Gang work day chart.

8.3. RECRUITING AND SELECTING STAFF

8.3.1. Job and Person Description

One of the first steps in recruiting an additional or replacement member of staff is to prepare a job description. The type of person necessary to fill the post and the way in which the new recruit will fit into the farm's labour force can then be considered. If it is a promoted post someone in the business may be suitable.

The job description should clearly state to whom the new employee will be responsible, the type of work including the key tasks, the degree of authority and responsibility associated with the post, pay, hours, holidays, housing, superannuation, and any other relevant details.

The person description states the experience, skills, qualifications, character, family, any particular interests or abilities, age, health, together with other points specific to the post. If training is to be given this should be included.

Both the job and the person description should be written down so that they can be referred to later and a brief description of the farm and locality added so that the information can be sent to applicants.

8.3.2. Advertising

Every farmer wants the best staff available and his advert should be good enough to attract the interest of such people. The cost of the advert can be more than repaid if a first class worker is recruited from it. Appropriate publications should be selected and an eye catching advert presented. Whilst is should not contain too many words it should derive impact by stating the attributes of the job such as wages, if such a statement will not upset existing staff, working conditions, housing, and favourable aspects of the locality. The method of application should be clearly stated. Reverse telephone call facilities are frequently used.

8.3.3. Selecting

Every effort must be made to find out as much as possible about the candidates, especially those that are short listed. References rather than open testimonials should be obtained. Telephone calls to a candidate's previous employers, or perhaps a personal visit, can be most revealing. Care has to be taken to establish the present employer's attitude to losing his worker. A check should be made on each applicant's family since they will be joining the farm's community. The way in which the interview is conducted should match the job in question and a candidate for a general farm worker's post should not be expected to have the eloquence of a potential Prime Minister. The first task is to ensure that each interviewee is fully aware of all aspects associated with the post to avoid the possibility of any resentment developing after appointment when adverse factors are discovered. The wife should be invited to see the house, but in the author's experience should not be present at the interview.

The points to look for in references and at interview include ability to undertake the job; suitability for training if necessary; attitude and care with stock and machines; aspects of character, including integrity, sobriety, reliability, and honesty; initiative if this is called for; attitude to life, work, and working hours; reaction under stress; health, timekeeping; appearance; hobbies and spare time activities.

Ample time should be given to each candidate to talk about himself and to ask questions. This can frequently indicate a great deal about the person's character. Above all the successful candidate must not only fit the job description but must fit into the work force. It is usually worth sacrificing a good contender if he does not fit this latter requirement.

When a person appointed joins the staff it is important to try to make him feel at home. Careful guidance should be provided in relation to his job and an understanding view taken to give him time to fit into the system.

8.3.4. Contract of Employment

Under the terms of the 1972 Contract of Employment Act an employer must give an employee a written statement of the terms of employment within 13 weeks of appointment.

Details included are name of both employer and employee; date when employment commenced; title of job; details of remuneration; hours of work; holidays; action in event of sickness or injury; pension scheme; length of notice of termination by both employer and employee; any disciplinary rules and the appeal procedure for the employee who is disciplined. The employee should be asked to sign a copy of the document stating that he has received the statement.

8.4. STAFF MANAGEMENT

8.4.1. Effective Man Management

It is assumed that the farmer has already selected the best people available to work on his farm. The next step is to ensure that the staff develop and maintain the correct attitude to their job. Good leadership and conditions of employment can do much to foster the motivation which all workers should possess.

People must be expected to work as a team when necessary; working for the good of the business and for each other. However, it is essential to realise that each person has individuality, with different objectives and attitudes. The good leader must get to know his staff in detail and manage them accordingly.

8.4.2. Staff Objectives

Wages, housing conditions, and job security are the basic considerations of most staff. Social needs such as proximity to schools, shops, and sources of entertainment are factors which many will have considered before they accepted their posts.

Once the basic needs have been met many individuals look for personal prestige and self-improvement. Status on the farm can be important to some since it can be related to respect in the community.

8.4.3. Leadership

The good leader will attempt to reconcile his own objectives for his staff with those of his workers. He has to be both a psychologist and a sociologist.

To prevent friction arising, and to foster sound management, it is important to make sure that all staff know the chain of command and that each individual's authority and responsibility is clearly defined. This is especially the case on large farms.

Many people enjoy responsibility, or having a degree of charge for something, and are motivated by it. Each person must be told the limits of his responsibility. Where jobs overlap a clear definition of who does what should be made. If responsibilities are clearly established there is a better chance that a high proportion of correct decisions will be made, with the attendant benefit that superiors will have to make less counter-decisions.

Pride in work can be fostered through responsibility. Each individual should be made aware of how his actions can influence the success of the business and be informed of his progress. Pay levels should reflect responsibility to avoid the worker feeling that he is being cheated.

Some farmers believe in giving head stockmen a basic office where they can develop a greater degree of importance and enthusiasm when keeping their records. There is always some concern that too great a delegation of responsibility can result in loss of control by the farmer. To avoid this the farmer should establish at the beginning that he requires frequent reports. Psychology appropriate to the individual must be used in this context. One method is to encourage the worker in the presentation of his reports by discussing their impact on day-to-day management decisions.

It is very important to ensure that there is a two-way passage of information between staff members and the farmer. In some cases results should be discussed in relation to long term policy so that the worker feels that his boss has confidence in him. Clearly this will depend upon the individual and the nature of his post.

Authority differs from responsibility since it confers with it a right to control others. Very great care must be taken to ensure that the right people are given authority. They should be of high character, with personality, integrity, and good powers of judgement. The limits of the authority must be clearly defined and overlap situations considered.

Few things are worse than having too many bosses, especially if they give conflicting orders. The line of authority must be clearly established and preferably restricted to one source for each man.

Supervision of others needs experience and decisions which are carefully arrived at if good team spirit and staff happiness are to be maintained. It is important that workers are treated fairly in every sense and they will then respect discipline if it is necessary. Quick tempers can often result in all parties later regretting what has been said. Discipline given in a calm manner can frequently be very effective. There should not be too great a delay between an event and appropriate action. In most cases there are two sides to each story and the worker must be heard. A judgement must be reached with understanding of the individual and the effect it will have upon him, his work, and relationships in the business.

If the leader has the respect of his staff they will be more likely to accept his decisions. Respect can be obtained by setting a good example in both work and life. Fairness, impartiality, being approachable, the keeping of promises, and prompt decisions are very important. Assisting staff when they have problems and helping them to fit into the community can greatly help staff relations.

To summarise, every farmer's aim should be to motivate his staff so that they have a desire to work to do a good job, for which they are rewarded, and derive satisfaction and pride in their results. This is achieved by understanding and sound leadership.

8.4.4. Day-to-day Staff Management

All the qualities of leadership are necessary to achieve a good standard of day-to-day staff management. It is important to establish priorities for each day, to consider fair times necessary to undertake tasks, and through the appropriate chain of command to give assignments and instructions clearly and concisely.

Good communications are fundamental to success. Many unforeseen events can occur, and it is essential that the right person knows about them so that appropriate action can be taken.

Grievances and dissent may arise and must be dealt with. If possible potential sources of grievance should be spotted. Some may appear small, but can be important to the individual. For example, the head tractor driver may be upset if one of the other workers is given a task which he feels he should do.

Conflicts can occur between staff. If the farmer gets involved, emotional statements must be matched to the person who made them and true facts derived.

Performances by staff may be inadequate or of superior standard. Appropriate action, including praise in the latter case, must be considered.

8.4.5. Training

Farmers benefit if they have a well-trained work force and it is their duty to identify training needs and to encourage appropriate staff to undertake suitable training. Attitudes and ambitions of individuals differ. Some need encouragement, but the benefits to the worker and the farm must always be pointed out. If higher wages are justified they should be paid.

Training must be appropriate to age and experience. Training Board courses have proved beneficial to many, including some who considered that they had little to learn. Day release classes might be appropriate to others.

Good training can be an important source of staff motivation. Admittedly there is always the danger that if a person improves his abilities he may move to another farm. Some farmers are not philanthropic enough to be satisfied that the training will benefit the industry, even if their staff leave them, and make this an excuse for not arranging appropriate training.

If a worker does go on a training course he should be given a chance on his return to put the training into action, and for him to discuss what he has learnt. Whilst on the course he may have benefited from contact with workers from other farms and have exchanged ideas with them.

Farmers and managers must also identify their own needs for training.

If the farmer and some of his staff have been well trained they may be able to train others. The quality of training for students on farms frequently bears no relationship to farm size or type, but depends upon the time devoted to explaining policies and practices to them.

8.5. PAY AND CONDITIONS

8.5.1. Levels of Pay

In Britain the minimum levels of pay and related conditions are fixed by the Agricultural Wages Boards. Different boards exist for Scotland and for England and Wales and consist of both farmer and employee representatives, in equal numbers, together with independent members. Copies of the wages orders are available to both farmers and their staff, and the Wages Inspectorate have the responsibility to see that they are complied with.

A high percentage of workers receive more than the minimum levels, particularly those with responsibilities for stock.

8.5.2. Incentive Schemes

Farmers differ in their opinions and experience of incentive schemes. One view is that each person should be given a good rate for the particular job, without additions, to act as encouragement, and to avoid the concern, together with possible resentment, if bonuses or other incentives fall.

Numerous incentive schemes exist, however, ranging from simple bonuses to profit sharing. Profit sharing is frequently not satisfactory because employers are usually unwilling to show staff audited profits, and profit depends upon many factors outwith the worker's control.

Piecework is common on cropping farms. This produces fluctuating weekly incomes, involves careful record keeping, and requires supervision for quality control. Weather can influence incomes, care has to be taken in fixing rates for the job, and workers must have no cause for concern about levels paid to other staff.

Bonuses on crop yield are an alternative, but yields fluctuate with factors outwith the worker's control. Sale values also have similar limitations.

For machine work, the area completed can determine the bonus. The quality of work is important and the care of the machine. Specific fields under certain weather conditions may prove difficult and be a source of discontent.

Stockworkers receive a wide variety of bonuses. In dairying a bonus per litre of milk produced is common, but must be related to concentrate usage. For this reason it might be based on the margin of the milk value over concentrate costs. Staff are usually happy when their bonuses are rising, but the reverse is true when they fall, and discouragement may set in. At least it gives the workers some insight into the farmer's problems.

Bonuses for a good calving index, or the number of pigs reared per sow per year, have their merits since these are two factors which significantly influence profitability.

Many other forms of bonus exist, but most staff prefer a good weekly wage in recognition of their conscientious efforts, loyalty and ability.

8.5.3. Time Sheets

Time sheets can be used not only as a basis for the calculation of gross wages, but if completed conscientiously can serve as an aid to the keeping of other records. Much depends on the staff.

Many time sheets are submitted on a weekly basis. At the West of Scotland Agricultural College farm we find no difficulty in getting our staff to complete the daily sheet shown in Fig. 68. Stockworkers complete different records. If the records are submitted daily there is less chance that items will be forgotten at the end of a week.

8.5.4. Calculating an Employee's Tax

When a new employee is engaged he should be asked for form P45. The details on this form include his National Insurance (NI) number, through which his tax records can be traced, and his tax code when he left his last post. If the employee does not have a P45, the form P46, which gives his NI number and other details, such as an indication that he is a school leaver, should be sent to the Tax Office.

The code issued to the employee, and notified to the employer, should be used in conjunction with tables A and tables B to assess the amount of tax to deduct. These tables are issued by the Tax Office in booklet form so that there are pages for each week of the year. Tables A are known as the "free pay tables".

Assume that the employee has a code of 116L and that his weekly pay is £100. At the start of the new tax year, i.e. 6 April, look at the week 1 sheet in table A and find the code 116 which will give the "total free pay to date" alongside it. Assume this figure is £22.50. Deduct this free pay from the £100 to obtain the taxable pay, which is £77.50. Now examine the week 1 table in tables B to find the total taxable pay to date on £77. Adjacent to this is the total tax due to date. In the tables used by the author this was £24.25 but it must be noted that these tables can be amended by changes in levels of tax.

<table>
<tr><td colspan="2">COLLEGE FARM DAILY TIME SHEET</td><td colspan="2">Date ...... Initials ....</td></tr>
<tr><td rowspan="2">Description of work done: Give field name, crop worked for, and job done. Give class of livestock worked for and job done. If carting, give carting .... from ...... to ......</td><td colspan="2">Hours worked</td><td rowspan="2">Tractor hours</td></tr>
<tr><td>Ordinary</td><td>Overtime</td></tr>
<tr><td></td><td></td><td></td><td></td></tr>
<tr><td>Total hours per day</td><td></td><td></td><td></td></tr>
<tr><td colspan="3">Diesel into tank today</td><td>l</td></tr>
<tr><td>Tractor No.</td><td colspan="2">Lub. oil into sump today</td><td>l</td></tr>
</table>

FIG. 68. Daily time sheet.

Tax Week 1

			£
		Weekly pay	100.0
(Table A)	*Less*	Tax free pay	22.50
		Taxable pay	77.50
(Table B)		Tax	£24.25

From this point on the total pay is treated on a cumulative basis. The second week's pay is £100 and the previous total pay this year was £100. The cumulative total pay to date is therefore £200. Examination of tables A, week 2, shows that against code 116 the total free pay to date is £45.00. This is deducted from the total pay to date and the net figure, £155, is found in tables B, week 2. Alongside this figure is the tax due to date, which is £48.80. Since to date £24.25 has been paid the tax to deduct is £48.80−£24.25=£24.55.

The reader is advised to try to obtain a set of tax tables from a farmer or the local tax office in order to practise these calculations.

The tax deducted must be remitted to the Tax Office by the 19th of each month but if the total for all the employees is small it can be remitted quarterly by 19 June, 19 September, 19 December, and 19 March.

Tax Week 2

		£
	2nd week's pay	100.00
	Previous pay this year	100.00
	Cumulative weekly pay	200.00
(Table A)	*Less* Tax free pay	45.00
		155.00
(Table B)	Tax due to date	48.80
	Less Tax paid to date	24.25
	Tax due week 2	24.55

Tax Week 3

		£
	3rd week's pay	100.00
	Previous pay this year	200.00
	Cumulative weekly pay	300.00
(Table A)	*Less* Tax free pay	67.50
		232.50
(Table B)	Tax due to date	73.05
	Less Tax paid to date	48.80
	Tax due Week 3	£24.25

8.5.5. True Cost of an Employee

The true cost of an employee is much greater than many people think. In addition to the basic wage there may be wage premiums, perquisites, incentive payments, bonuses, and overtime. The Employer's National Insurance Contribution, Employer's Liability Insurance, and any contributions to a staff superannuation scheme must be added.

8.5.6. Dismissal and Redundancy

Various Acts of Parliament have given employees considerable protection and rights. People who have been employed for more than four weeks are entitled to at least a week's notice although the terms of the contract of employment apply if this stipulates a longer period.

People who reach retirement age can usually be dismissed without being able to claim unfair dismissal, but other full-time workers, who have been in post for 52 weeks, may have the right not to be unfairly dismissed. Irrespective of these points, dismissal for joining, or proposing to join, a trade union is not permitted. An employee who feels that he has been unfairly dismissed can make a complaint to an Industrial Tribunal.

Dismissal takes place not only at the instigation of the employer, but also when "constructive" dismissal occurs. There are many ways in which it might be held that the latter has taken place, but essentially they include cases where certain forms of pressure by the employer have caused the employee to terminate his contract.

It might be held that dismissal was justified in certain cases of misconduct, such as theft and drunkenness, although it is advised that farm staff be given a warning first. Dismissal for inability to do the job, for incompetence or ill health, could be held to be fair if specific conditions, including the right of the employee to present his case, had been complied with, but this should only take place as a last resort. There are other "substantial reasons" which might be accepted, e.g. where an employee refused to work changed hours necessary to the business, but many cases have been lost by employers who felt that they had acted in an appropriate manner.

Employees who have been employed for 52 weeks can demand a written explanation for their dismissal. Where unfair dismissal has taken place the tribunal may order re-instatment. Alternatively compensation, which can be substantial, may be awarded against the employer.

An employee who is made redundant after two years' service is eligible for compensation even if he obtains another job provided that the 65th birthday (60th woman) has not been reached. The amount paid varies with the years of service, age, and wage at redundancy. An employer can claim 41% of the payment from the National Redundancy Fund if he employs nine workers or less.

Casual labour employed for 12 weeks or less is exempt from dismissal and redundancy regulations.

8.5.7. Health and Safety at Work Act

Many regulations have been introduced to promote safety but the Health and Safety at Work Act 1974 has probably been the most important. All employers must be familiar with the Act, but there is only space here to mention one or two of its main clauses.

The Act is such that the employer must not only be concerned with the health and safety of employees but also that of others who may be affected by the activities of the farm. The farmer must, so far as is practicable, ensure that his farm is a safe place to work, without health risks. The staff should be informed of safety measures and trained as necessary. Particular efforts must be made to ensure safety with machinery, equipment, appliances, and chemicals. Employees have certain responsibilities to themselves and to others, and can be expected to use protective equipment provided and operate within safety regulations.

Health and safety inspectors have the right to visit premises, can stop a particular activity continuing if they consider it to be dangerous, and may prosecute the employer or even employee for certain offences.

8.6. WORK STUDY AND RELATED ASPECTS

8.6.1. Introduction

Work study is considered here with labour and machinery because it is under these headings that it finds most frequent application. The discipline of thought, attention to detail, and decision-making processes which it involves could, however, be equally well applied to almost every aspect of farm management.

A prime objective of management is to make maximum economic use of resources available, namely men, money, machines, materials, time, space, and energy, to achieve an end product. Work study shares this objective. To be undertaken effectively it requires someone to "stand back" from the business and to look critically, systematically, but objectively, at it. This enables a clear analysis to be undertaken of such factors as "what is being achieved", "what should be achieved", and "how best could it be achieved"? Too many farmers are so busy working that they fail to undertake this type of exercise.

On many farms labour and machinery represent in the region of 25–40% of total costs and 60–75% of fixed costs. If only a 5–10% reduction can be made in these it could be argued that the exercise to achieve such a result is worthwhile, although it must be remembered that men and machines are usually only employable in whole units.

Work study, therefore, principally aims to increase the effectiveness of work or, in a wider sense, it aims to increase the productivity per unit of input of a resource. It has two main components—method study and work measurement. Method study aims to find improved routines, methods, or tools which may reduce the requirement for time, materials, or energy to undertake tasks. Work measurement establishes the time required for a worker to complete a task by a particular method.

As a result of the application of work study, work may be made easier, and any time saved could permit more productive work to be done. In addition to improved work routines, ideas may be promoted for such things as new building layouts or designs, and better machinery and equipment.

8.6.2. Method Study

When examining the method by which a particular job is done it is usual to follow a set procedure. The steps generally suggested are to select, record, examine, develop, install, maintain.

(i) Select

The first step is to identify the problem to be studied. This may be highlighted by analysis of the farm's physical and financial results or could be indicated by such factors as a high seasonal demand for labour.

A word of caution is necessary at this point. There is the danger that method study by effecting an improvement might bolster-up a situation or an enterprise which is not really viable or is basically economically unsound, e.g. an investigation of the work with a dairy herd might reveal the possible saving of, say, one or two man hours per day, but the real answer to the problem might be more cows, or fewer cows, or perhaps no cows at all. In other words the first essential is to put the management right, and then carry out the method study.

(ii) Record

The existing method of doing the job in question must be recorded in detail. Various charting techniques and diagrams can be employed.

When constructing charts, standard symbols are used to identify specific activities. They can record a man, a material, or a piece of equipment, which ever is most appropriate. If the material type is selected then one unit, such as one bag of fertiliser, is followed at a time.

The activities and their symbols, with examples from silage making are:

○ Operation (produces, accomplishes, e.g. trailer filled).
□ Inspection (quality or quantity checked, e.g. if trailer full).
⇨ Transport (e.g. trailer from field to clamp).
∇ Storage (holds, keeps, e.g. grass stored in silo).
D Delay (e.g. forage harvester waiting for trailer).

Operations ○ and inspections □ add to the value of the product, whereas the others do not necessarily do so, but they add to the costs. Thus transport of silage adds nothing to the value of the product.

Outline process charts, which record the main operations and inspections, are prepared as a preliminary to further recording. These should be of the material type.

Flow process charts, in addition to operations and inspections, record transports, storages, and delays. They can record the activities of man, material, or equipment. An example is shown in Fig. 69.

Multiple activity charts can be constructed when men and/or machines have to be recorded so that the activities of one can be related to the others. A common time scale is adopted for all.

Figure 70 shows a chart for making silage. When examining multiple activity charts the main task is to reduce man idle time or the time an expensive piece of machinery is out of commission, e.g. it will be impossible to maximise the output of a forage harvester if it is idle waiting for the next trailer to arrive.

If plans of buildings are drawn to scale, distances involved in working in and around them can be measured using string diagrams. String held at strategic points by pins is used to represent movement, and shows areas of congestion. The length of the string, adjusted for scale, gives the distance travelled.

Flow diagrams, in plan, elevation, or three-dimensional form, can also be used to record paths of movement.

Childrens' toy farm implements, or matchboxes, can be used to trace routes on architects plans or on sketch plans. These show up potential problems for access, turning, and movements of stock and materials. It is important to note the tendency for tractors and lorries to get taller and bigger.

(iii) Examine Critically

This is the most important aspect of method study and must be undertaken very

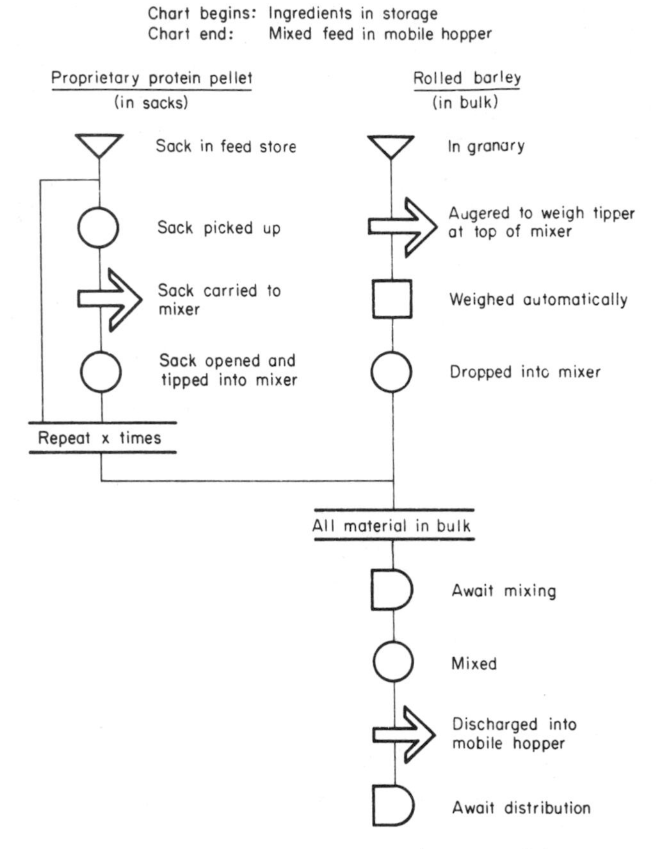

FIG. 69. Flow process chart—material type (mixing a beef ration).

carefully. Most jobs on a farm can be divided into "make ready", "do", and "put away" activities, e.g. when pail-feeding calves, milk substitute is got ready, *the calves are fed*, and the buckets are washed and put away.

The first essential is to identify the "key activity", i.e. *the calves are fed*. This is the most important one, or the "do". Where there is more than one "key activity" it is necessary to establish which is the most vital, and to examine it first. Usually it will be an operation ○, but it could be an inspection □. There is little point in proceeding with critical examination of "make ready" or "put away" activities if the "key activity" can be dispensed with or radically altered, e.g. if pail feeding of calves can be replaced with an automatic feeder. This is why it is the "key activities" which are subjected to critical examination.

An example of a critical examination sheet is shown in Fig. 71. This process appears to be extremely detailed, but if such attention to detail was paid to all factors associated with management, and not just to work study, many farmers would benefit enormously.

The first part of the examination involves an analysis of the present situation. The second is concerned with establishing alternatives and should reflect both short and long term factors. Its target is to dispense with the requirement for an activity and

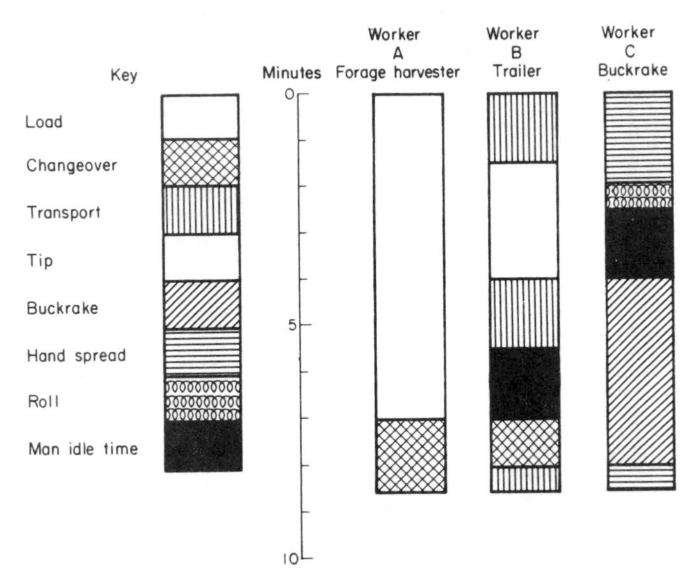

FIG. 70. Multiple activity chart—silage making.
Worker A: drives tractor and forage harvester directly coupled to trailer.
Worker B: drives tractor and trailer, exchanges empty for full trailer in the field, drives to pit and returns.
Worker C: is at the pit; fills pit by buckrake, hand spreads and rolls

where this is not possible to suggest other alternatives which produce some advantage over the present situation. The final part is to select from the alternatives remembering all the factors relevant to the particular situation. They must be practical and economic.

(iv) Develop

The alternatives selected will provide pointers to the new method. First an outline process chart should be constructed based on the operations and inspections suggested by the critical examination. This should then be developed into a flow process chart using as few supporting activities as possible which would enable the "key operations" to be undertaken.

Such aspects as the total amount of work involved and the allocation of this work to particular staff can then be considered. The savings over the previous method can also be checked.

If the new method has been developed on behalf of a farmer it will be necessary to submit a report on it to him for approval. This report must clearly state the recommendations, the reasons for them, and the results and benefits expected from them, e.g. increased safety, easier work, and better quality.

(v) Install

An important aspect is to prepare the work force for the new method before it is installed, to plan ahead, and to arrange for everything necessary so that the installation

Description of element: . . . .   Study No: . . . .

The present facts		Alternatives		
Statements	Reasons	Alternatives	Assessment	Alternative selected
PURPOSE *What is achieved?*	*Is it necessary?* Why?	*What else could be achieved?*	*Evaluate alternatives*	*What should be achieved?*
SEQUENCE *When is the purpose achieved?* After: . . . . Before:	*Why then?*	*When else could it be achieved?*	*Evaluate alternatives*	*When should it be achieved?*
PLACE *Where is the purpose achieved?*	*Why there?*	*Where else could it be achieved?*	*Evaluate alternatives*	*Where should it be achieved?*
PERSON *Who achieves the purpose?*	*Why that person?*	*Who else could achieve it?*	*Evaluate alternatives*	*Who should achieve it?*
MEANS *How is the purpose achieved?* 1. Materials 2. Equipment 3. Operators method 4. Working conditions	*Why that way?*	*How else could it be achieved?*	*Evaluate alternatives*	*How should it be achieved?*

Fig. 71. Critical examination sheet.

can be undertaken smoothly. If possible the new method should be rehearsed. Careful supervision is necessary when the new method first operates, and checks should be made to see that it is proceeding correctly.

(vi) Maintain

From time to time the new method should be reviewed to check if any factors have changed which require amendments. It is particularly important to ensure that the defects of the old method are not revived by habit in the new method. Some changes may have been made as a result of experience which are beneficial. All changes should be evaluated and accepted into the method where justified.

The application of method study to the various sectors of work on a farm can produce substantial cumulative benefits. Where it is applied appropriately, staff who previously followed a set routine can be motivated by the interest being shown in their work, especially if they can see some aspects being made easier or some of the drudgery being reduced or eliminated.

8.6.3. Work Measurement

Work measurement involves ascertaining the time required by a skilled and experienced worker to carry out a given task, using a defined method, making due allowance for rest.

The results can be used to allocate tasks to staff realistically and fairly; to help in labour planning generally; to assist in method study by evaluating alternative methods; and as a sound basis for incentive schemes.

8.6.4. Ergonomics

The word ergonomics is derived from *ergos*, work, and *nomos*, natural laws. It is the study of the capabilities of people and of how work, including their conditions of work, affects them. Such factors as stress, fatigue, heat, noise, vibration, ventilation, lighting, and the influence of rest pauses are studied. Anatomical, physiological, and psychological studies are involved.

Ergonomics, in common with work study, aims to adapt work, and conditions of work, to suit the person doing the job so that more effective results can be obtained with consistency. Machinery manufacturers use ergonomic principles to design their products. Tractors have been improved considerably from the driver's point of view. The seat and controls are built and positioned to reduce fatigue; comfort is fostered by the introduction of quiet cabs which reduce vibration; and heating and ventilation can be adjusted. The design of milking parlours incorporates both work study and ergonomic principles. The depth of the pit is fixed to reduce bending for the operator of average height, and duckboards can be added for a person of lower stature. Controls are positioned to reduce movement, effort, and fatigue.

8.6.5. Network Analysis

Network analysis is a diagrammatic method used to illustrate the interrelationship of events involved in a project which consists of several tasks. An example is the construction of a farm building. It can be used to determine the critical path, which is the sequence of events determining the minimum length of time to complete the project. Delays in this sequence will postpone completion dates.

When constructed the network can be used to check on progress and to help organise the resources required for each task.

Arrows are drawn to represent each "activity". Although these are not drawn to scale the estimated times can be written on them. "Events", which do not consume time, indicate the beginning and end of activities (e.g. when foundations are complete), and are represented by circles.

For each activity it is necessary to ask: (i) What activity precedes it? (ii) What activities run concurrently? (iii) What activity follows it? (iv) What controls the start? (v) What controls the finish? The diagram is then constructed.

Activities A and B run concurrently,

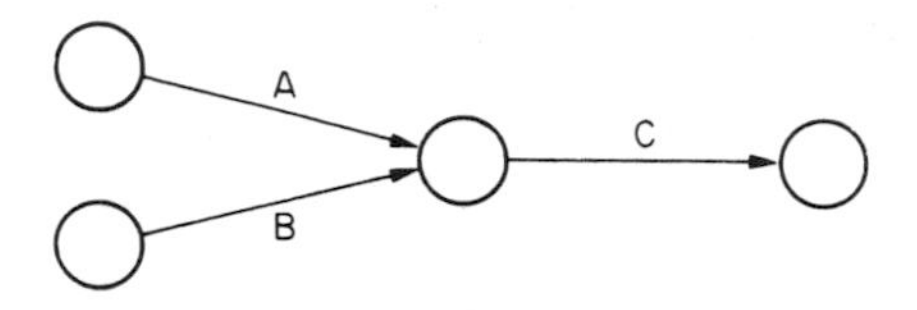

and must be carried out before C can take place; e.g. A, assemble roof trusses; B, build wall; C, put on roof.

Dummy arrows, which are dotted, are used to indicate that an event cannot occur before another event, although no specific activity occurs between the two. For example, when a puncture occurs on a landrover the activities are A remove wheel, B get spare, C fit spare, D mend puncture. C and D must follow A; C must follow B.

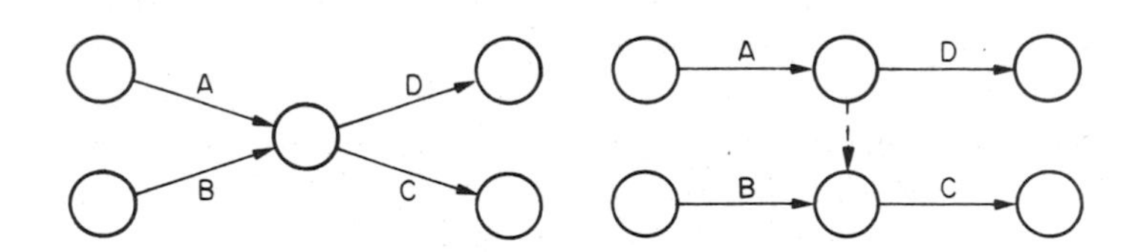

The left hand diagram is wrong because it suggests that D could never take place without B, which is not true. The diagram on the right is correct.

By convention time is represented as moving from left to right on the diagrams. The important point is that no activity can take place before essential activities in the chain beforehand have been completed. Thus it is not possible to erect the wall of a building before the foundations have been prepared. Equally the foundations could not be completed before arrangements had been made to dig them and appropriate materials bought.

The aim is to represent each activity and event associated with a project so that materials, machines, and staff are on hand when required so that there is no hold-up with the project.

Many other industries have saved large sums of money and time by the application of network techniques. Frequently computers are used to plan them. This does not mean that they are not suitable for agriculture where the networks would probably be drawn by hand. The logic and ordered thought processes involved provide a complement to the other management techniques and could even be applied to advantage to some of the basic agricultural tasks without perhaps completing the diagrams.

However, in agriculture there are occasions, such as the erection of a new building, when a network diagram could help. If the network was produced it would reveal how long the project would take and the activities which were on the "critical path", together with the job priorities essential to ensure the completion of the project on time. These are vital factors for someone constructing a new silage clamp for the next crop, or a new milking parlour before the bulk of the cows calve. It may even reveal a shorter way of doing the job. The critical path is used to consruct a timing schedule which monitors the progress of the project.

8.6.6. Materials Handling

Materials handling is the study of picking up, moving, and putting down materials, manually or mechanically. Its aim is to increase the efficiency of handling, or even eliminate it where it is not necessary. Safety of operators and damage to materials must be considered. Together with method study it can help to improve the effectiveness of work and lead to economies of time, money, and effort.

A high percentage of agricultural work is materials handling in one form or another. Grain, potatoes, hay, silage, dung, and fertiliser are all moved to a greater or lesser extent. Consider the annual input of work with fertiliser on some farms. It may be manhandled to the back of the lorry on arrival, lifted off, carried to the store, lifted back out of the store to the trailer, manhandled on the trailer, lifted off the trailer, carried to the spreader, and finally tipped into the spreader.

The first step is to eliminate unnecessary handling. The farmer should ask What is being achieved by moving a particular item? Is it being put in the right position in the first place, or could it be used where it is, or nearer to where it is? For example, concentrates could be put directly above a milking parlour when they arrive on the farm instead of being put into the feed store.

The second step is to minimise any handling that is left, and the third step is to mechanise anything else as far as is practicable and economic. If there is insufficient room to store the concentrates above the parlour an auger might be used to move them from the feed store, and the feed store might be moved nearer to the parlour.

In the case of the fertiliser a pallet system or some other bulk-handling system might be adopted.

Some of the most important rules are: (i) avoid double handling if possible; (ii) use gravity where possible; (iii) endeavour to reduce the bulk of material to be handled; (iv) keep materials at the height at which they are to be worked on; (v) standardise handling equipment; (vi) reduce vehicle turn-round to a minimum; (vii) move material in unit loads; (viii) keep stores and working areas as close as possible to reduce handling and transport.

There are many ways in which farmers can reduce work on their farms.

Manufacturers are also assisting—at a price. Electronic aids are now being employed. One example is the out of parlour feeder. This enables dairy cow concentrates to be fed without men having to lift them. The concentrates arrive in a bulk delivery lorry, are blown into a hopper above the feed point, and are automatically released—under control from a programmed unit—to the cows.

The advances in bale-handling machinery, including the production of big bales, are typical of manufacturers' efforts to produce machines which will have a good market because of the saving in labour which they can produce.

Work study and the related aspects which have been discussed all aim at increased efficiency at minimal cost.

8.7. MACHINERY PLANNING

8.7.1. Introduction

There is an optimum level of mechanisation for each farm and ideally this level should be selected as an integral part of the overall farm plan. Many machinery planning techniques have been developed, but most are outwith the scope of this book. In practice a high proportion of machines are selected on the basis of personal judgements based on previous experience, related to such factors as the labour and work days available, together with the enterprises on the farm.

Each farmer has to decide how much machinery to have, including the size and capacities of individual machines. He also has to establish whether he should buy them or obtain them by some other means such as leasing.

8.7.2. Machine Capacity

Culpin in his book *Profitable Farm Mechanization* (published by Crosby Lockwood Staples) gives the effective capacities of many machines. This data, or similar information from other sources, can be used as a guide to the machine capacity required for a particular farm. Another possibility for tractors is to study the kWh per hectare needed for the various operations on a farm. The total requirement, and particularly the requirement at times of the year when tractor demands are highest, is used as a guide to selection. It must be remembered, however, that for every 100 kW of tractor power only about four-fifths, and in many cases less, is effective by the time it reaches the back wheels because of transmission losses. Wheel slip and rolling resistance have to be added to these losses before the power for effective work can be estimated.

Most farmers do not employ such a sophisticated method. They identify the peak periods of demand for tractors and then consider the whole machinery system in relation to these. For example, in the case of silage making the number of men, and the distance from the fields to the clamp, initiate many of the decisions with reference to the number of tractors and trailers. When deciding on the size of the forage harvester they consider throughput, the amount of silage to be made, and the need for speed. In selecting the size of the tractor for the forage harvester they may consider the possibilities of fully utilising it at other times of the year. Many interrelated factors are therefore involved in these decisions, but the capacities of the most important machines, in this case the harvester and the tractor used to pull it, play a big part when the final selection is made.

Timeliness of crop operation is a major aspect in relation to the use of machines because of its effect on yields. The farmer has to decide whether to base his machine capacity on the requirements of an average year or to have extra capacity, and if so, how much, as an insurance against seasons with adverse weather conditions. The farmer's attitude to the risk involved is therefore relevant.

Weather and its effect on different soil types can be a major reason why standard data for machine capacity must be treated with caution. Drawbar pull from tractors varies with soil type and soil conditions. The skill with which a machine is used, the state of repair and maintenance, and the shape and size of fields, are other factors which influence results.

8.7.3. Machine Selection

The need to select machines so that they relate to each other for use in a particular system was illustrated above with the case of silage making. Many other factors determine choice of each particular machine. The availability of a local agent who can provide reliable service can be very important. The cost of obtaining a machine is a major consideration, but if spare parts are not readily available problems may arise.

The performance, and in particular the quality of work, of a machine must be kept in mind. Demonstrations and observations on other farms can give some idea of the standard achieved, but unfortunately the number of test reports is still limited.

Few farmers introduce all new machines into a particular system at the one time. When one of the machines is replaced care has to be taken that it will fit in with the others. It is equally important to look forward to establish if a change will be made to the system for which the new machine might be an integral part.

There can be some drawbacks to having an entirely new design of machine which has not been widely tested on other farms. If it proves to have limitations it may not only turn out to be a liability to the farmer, but could have a poor second-hand value.

Some manufacturers offer their equipment at very competitive prices in order to get sales penetration into a new area. Before buying their machines long term factors must be considered. The farmer may lose the goodwill of local dealers, and getting repairs done promptly could be difficult.

8.7.4. Machinery Running Costs

Depreciation, interest on capital invested in machines, repairs, spare parts, routine maintenance, and fuel costs must be considered by a farmer who has purchased machines. Depreciation, or wear and tear has already been discussed in relation to taxation. In many accounts constructed for management purposes depreciation is taken at 20% per annum on a straight line basis.

Culpin (Fig. 72) gives estimates of the useful life of different types of machinery and Baker (Fig. 73) suggests depreciation figures for various categories of machine according to the frequency of renewal. Culpin also gives estimates of the annual cost of spares and repairs as a percentage of the purchase price at various levels of use (Fig. 74).

Equipment	Annual use (hours)				
	25	50	100	200	300
Group 1: Ploughs, cultivators, toothed harrows, hoes, rolls, ridgers, simple potato planting attachments, grain cleaners.	12+	12+	12+	12	10
Group 2: Disc harrows, corn drills, binders, grain-drying machines, food grinders and mixers.	12+	12+	12	10	8
Group 3: Combine harvesters, pick-up balers, rotary cultivators, hydraulic loaders.	12+	12+	12	9	7
Group 4: Mowers, forage harvesters, swath turners, side-delivery rakes, tedders, hedge-cutting machines, semi-automatic potato planters and transplanters, unit root drills, mechanical root thinners.	12+	12	11	8	6
Group 5: Fertiliser distributors, combine drills, farmyard manure spreaders, elevator potato diggers, spraying machines, pea cutter-windrowers.	10	10	9	8	7
Miscellaneous:					
Beet harvesters	11	10	9	6	5
Potato harvesters	—	8	7	5	—
Milking machinery	—	—	—	12	10

	Annual Use (hours)					
	500	750	1000	1500	2000	2500
Tractors	12+	12	10	7	6	5
Electric motors	12+	12+	12+	12+	12	12

* The above figures are the author's estimates and are suggested as a general guide only.

FIG. 72. Estimated useful life (years) of power-operated machinery in relation to annual use. (*Source:* Culpin, *Profitable Farm Mechanization*, 3rd edn., Crosby Lockwood Staples, 1975.)

Good routine maintenance, care in the use of machines, and suitable storage can all help to reduce running costs in the long term as well as the short. It is false economy to try to save on maintenance. If, for example, the forage harvester or combine breaks down, then the work of the rest of the machines and men in the harvesting operation will be halted.

8.7.5. Replacement

Many techniques have been developed to predict the optimum time to replace farm machinery. The problem is complicated by the large number of factors involved.

Frequency of renewal (years)	Complex, high depreciation rate, e.g. potato harvesters mobile pea viners, etc. (%)	Established machines with many moving parts, e.g. tractors, combines, balers, forage harvesters (%)	Simple equipment with few moving parts e.g. ploughs, trailers (%)
1	34	26	19
2	24½	19½	14½
3	20*	16½*	12½
4	17½†	14½	11½
5	15‡	13†	10½*
6	13½	12	9½
7	12	11	9
8	11	10‡	8½†
9	(10)	9½	8
10	(9½)	8½	7½‡

* Typical frequency of renewal with heavy use.
† Typical frequency of renewal with average use.
‡ Typical frequency of renewal with light use.
(*Source:* V. Baker, Bristol University.)

FIG. 73. Depreciation: average annual fall in value. Percentage of new price.

Repair costs tend to increase as machines get older and there comes a point when replacement is preferable to holding onto a machine. Much will depend upon the type of equipment involved and the way in which care and maintenance can reduce the need for repairs. The extent to which a machine is used can also be important. Reliability and risk of breakdown must feature in any replacement policy.

As they age, machines may become obsolescent, the cost of which can be difficult to evaluate. Tax considerations play a big part in most farmers' replacement decisions. Many look at their possible profits shortly before the financial year end to assess the tax saving which would be derived from buying machinery. They must look at their cash flow to see that the business can afford the purchase and must establish that alternative investments would not be more worthwhile.

Discounts, or trade-in values, on old machines will influence some farmers' decisions and these can vary from dealer to dealer. Each farmer should consider the implications of retaining rather than replacing a particular machine in the current year. During times of inflation prices rise substantially, although second-hand values should also rise. He must determine if the cash will be available next year to buy the machine and assess the trend in interest rates if capital has to be borrowed. This is in addition to factors such as increasing breakdowns and repair costs which may occur if purchase is delayed.

Finally, farmers have to consider the motivational influence of new machines on their staff. It can be very important for the workers' morale to have machines which are at least equal to those on neighbouring farms.

8.7.6. Dis-economies of Scale

Elsewhere in the book it has been suggested that farmers with large farms can have advantages of economy of scale compared to those with small farms because they can use machines over greater areas and so spread fixed costs further. This is true, but on some large farms an attempt is made to speed up work by the purchase of bigger machinery.

Consider the example of silage making. The reader is encouraged to calculate the current cost for a mechanised system capable of completing 0.6 ha per hour compared to one capable of 1.2 ha per hour. Almost certainly the latter system will cost more than twice that of the first. It will require a bigger harvester, more tractor power, and bigger trailers.

Such an example illustrates dis-economy of scale. Examples could be quoted for many other machinery systems, including those for potato harvesting. When devising a machinery system the farmer should ponder long enough to consider the necessity for large machines and the economics of the benefits to be derived from them.

8.7.7. Computer Based Techniques for Work Rate Assessment

Many of the techniques employed in labour and machinery planning before the application of computers to agriculture had limitations because they failed to relate to circumstances on the individual holding such as travelling distances, field shape and size and even crop yield. ADAS have developed a method which aims to provide a more accurate assessment of work rates for a specific farm. It takes note of the type of work being undertaken, for example drilling, spraying or combining and its dependence upon weather conditions.

Initially a spot work rate is established. This is the area covered by a machine of a specified width travelling at a constant speed i.e.

$$\text{Spot Rate (ha per hour)} = \frac{\text{Machine width (m)} \times \text{working speed}}{\text{(km per hour)} \div 10}$$

This spot rate is then reduced to allow for relevant factors e.g.:

(i) travelling distance from building to field;
(ii) area and shape of field;
(iii) number of workers and their specific duties;
(iv) yield of crops;
(v) size of trailers.

The system also employs standard data from a data bank for such aspects as:

(i) effect of headlands on speed reduction for the particular machine;
(ii) time to add for working corners of fields;
(iii) time to travel from buildings to fields;
(iv) time to change trailers;
(v) time for machinery preparation.

Superimposed on this is an allowance for the days on which work can take place according to the dependence of the particular field operation on the weather.

	Approximate annual use (hours)				Additional use per 100 hours ADD (%)
	500 (%)	750 (%)	1000 (%)	1500 (%)	
Tractors	5	6.7	8.0	10.5	0.5

	Approximate annual use (hours)				Additional use per 100 hours ADD (%)
	50 (%)	100 (%)	150 (%)	200 (%)	
Harvesting machinery					
Combine harvesters, self-propelled and engine driven	1.5	2.5	3.5	4.5	2.0
Combine harvesters, p.t.o. driven, metered-chop forage harvesters, pick-up balers, potato harvesters, sugar-beet harvesters	3.0	5.0	6.0	7.0	2.0
Other implements and machines					
Group 1: Ploughs, cultivators, toothed harrows, hoes, elevator potato diggers } normal soils	4.5	8.0	11.0	14.0	6.0
Group 2: Rotary cultivators, mowers, binders, pea-cutter windrowers	4.0	7.0	9.5	12.0	5.0
Group 3: Disc harrows, fertiliser distributors, farmyard manure spreaders, combine drills, potato planters with fertiliser attachment, sprayers, hedge-cutting machines	3.0	5.5	7.5	9.5	4.0
Group 4: Swath turners, tedders, side-delivery rakes, unit drills, flail forage harvesters, semi-automatic potato planters and transplanters, down-the-row thinners	2.5	4.5	6.5	8.5	4.0
Group 5: Corn drills, milking machines, hydraulic loaders, simple potato planting attachments	2.0	4.0	5.5	7.0	3.0
Group 6: Grain driers, grain cleaners, rolls, hammer mills, feed mixers, threshers	1.5	2.0	2.5	3.0	0.5

FIG. 74. Estimated annual cost of spares and repairs as a percentage of purchase price* at various levels of use. (*Source:* Culpin, *Profitable Farm Mechanization*, 3rd edn., Crosby Lockwood Staples; 1975).

* When it is known that a high purchase price is due to high quality and durability, or a low price corresponds to a high rate of wear and tear, adjustments to the figures should be made.

A computer program is finally employed to use all the information to give a profile of labour and machinery use on the farm. It produces a series of tables which show, on a half monthly basis, labour availability and requirement; labour availability in relation to its requirement in each workability class; work carried out in each workability class; work carried out for each crop; tractor hours needed for each task and crop; work rate (i.e. man hours per ha) for each task.

It is possible to identify bottlenecks from the tables and establish which machine is causing the problem. The computer can then be instructed to consider an alternative machine, or combination of machines, to establish if there is an improvement. This is a particularly useful facility when the farmer is planning the purchase of new machines, but can also be employed to calculate the cost benefit of selling existing machines and replacing them with others.

Other programs, many not unlike the one outlined above, have been developed by several organisations which can also be employed to assist in labour and machinery planning.

8.8. METHODS OF ACQUIRING THE USE OF MACHINES

8.8.1. Possible Methods

The following possibilities exist:

(i) Ownership by purchase using own capital or bank loan.
(ii) Hire purchase.
(iii) Contracting (in various forms).
(iv) Leasing.
(v) Syndication or sharing.

Before deciding which method to adopt the following points should be considered:

(i) Capital—how much is available, what alternative investments exist for it, how much will it cost if borrowed.
(ii) Taxation—which possibly would save most tax.
(iii) Size and performance—can a machine of adequate size with good performance be obtained.
(iv) Reliability and availability of machines obtained by respective methods.
(v) Labour—impact on periods with peak labour demands, quality of work.
(vi) Responsibility for repairs and maintenance.
(vii) What the unit costs will be.
(viii) Other specific benefits.

8.8.2. Taxation

It has already been indicated that the taxation allowances can very much reduce the cost of machine ownership, (see p. 158), but it must be realised that the initial cost of the machine has to be found first except in the case of a machine acquired by hire purchase. In addition it is essential to note that the level at which the farmer pays tax, i.e. his highest rate, can determine how much is saved.

Interest on money borrowed to purchase machines is allowed by the Inland Revenue as a cost, which then reduces tax. Although a farmer does not technically own a machine bought by hire purchase until the agreement is completed, he can claim wear and tear allowance from the outset, as well as being allowed to put the interest, but not the capital, part of the instalments as a cost.

If the farmer does not own the machine any premiums charged for its use, and any interest charge on capital borrowed to obtain the use, can help reduce taxable profit.

8.8.3. Ownership by Purchase

Ownership confers certain advantages. It creates a feeling of independence and machines should always be available, subject to breakdowns. With careful maintenance reliability should be good. The farmer has better control over the quality of work than if a contractor is employed. He may even consider contracting his machines himself. Unit costs may be less than for other sources, but this will largely depend upon the area on which each machine is employed in relation to its capacity.

The disadvantages include the high capital cost, which might be at the expense of profitable, alternative investments, and the high cost of borrowed money. The farmer is responsible for maintenance and repairs, and has to find his own labour to operate the machines in contrast to the use of a contractor who usually supplies operators. It is possible that the farmer may not be able to afford the size of machine which is desirable so that speed and work performance is limited.

The purchase of secondhand machines should be considered, especially where capital is limited, but possible repair costs and breakdowns must be kept in mind.

8.8.4. Hire Purchase

The full cost of the machine does not have to be found at the time of purchase, but a deposit of up to one-third is usually required. The total cost is extremely high because interest is charged on the initial principal throughout the term of the agreement. This agreement usually lasts for 1–3 years, and the true interest rates are well above that for a bank overdraft.

This method confers many of the advantages of ownership, but there is great danger of becoming over-committed to hire purchase.

8.8.5. Contracting and Contract Hire

The possibilities under this heading are:

(i) Short term hire over a period of days or weeks with associated labour.
(ii) Short term hire but without labour.
(iii) Contract hire over a period of 2–3 years.

(i) Contractor Supplying Machine and Labour

Many contractors offer very good service and a wide range of machines. Use of a contractor reduces the farmer's need for capital. It may allow more investment in productive areas of the farm and can save interest charges. In some cases total farm machinery costs are reduced by using a contractor compared to ownership. The contractor is responsible for repairs and maintenance. In addition the labour he supplies, quite apart from adding to the farm staff complement, may be very experienced and produce a high standard of work. He may provide a larger machine than the farmer could afford. This might be important where marketing is critical or in adverse weather conditions. The size of the machine supplied may allow the farmer to grow a larger area of a particular crop, and should he change his policy the services of the contractor can easily be dispensed with.

Contractors are not always available when required and losses may result, especially at harvest. A better service may be provided if a large area is offered for contract, especially if cultivation, seeding, and harvest are involved. With some contractors the quality of work is poor.

(ii) Short Term Hire Without Labour

Many machinery dealers offer machines for hire for a short period. This is particularly useful if the machine required would be used for a limited period of the year and if purchased would remain idle for long periods.

(iii) Contract Hire over 2–3 Years

A limited number of firms offer this facility, especially for tractors. Profit to the hire firm may be limited, but they often consider that their customers will buy other machines from them. Also, because of extra throughput, they recognise that their service facilities will be better utilised. This is because in addition to licensing appropriate machines, the hire company agrees to repair and maintain the hired machines, and to replace them in the event of breakdowns. Availability and reliability are both good. The hire charge is tax deductable and the farmer does not have to find the high initial capital for purchase. It may be possible for him to obtain a bigger machine than he could afford himself.

A farmer who owns one or more tractors may hire another and get the maximum use from it by using it for well over 1,000 hours per year.

The cost can be very high and the saving through tax will depend upon the farmer's tax liability. At the end of the agreement the machine must be returned and there is no allowance for its value.

8.8.6. Leasing

Machines are generally leased for 2 or 3 years. Usually the farmer is responsible for any licensing, repairs (except those under warranty), maintenance, insurance, and spare parts. In the event of a breakdown a new machine is not supplied.

The leasing company retains ownership, but at the end of the contract the farmer may be allowed to keep the equipment at a nominal annual charge. Alternatively, it is returned to the dealer for sale and the farmer obtains a high percentage of the proceeds which could be put towards the lease of another machine. Provided that the transaction is done through a third party he may be permitted to buy the machine.

In practice many different types of leasing agreement exist. Leasing charges are based on the fall in value of the machine over the period of the lease plus an interest charge for the capital invested by the leasing company. Frequently an initial payment, equivalent to 4 months' rent, is required which is offset by a free period at the end of the lease.

All lease and interest charges are tax deductable. The amount of tax saved significantly affects the merits of leasing.

Leasing confers the advantage that high initial capital is not required which can be particularly beneficial to a farm which is expanding. Machines are replaced regularly and have the availability and reliability usually associated with ownership. It assists forward budgeting since future payments are known. A farmer benefits if he takes care of the equipment and if, after completing one agreement, he receives a rebate on the rent of the second machine, from the sale of the first, there is effectively no tax on the rebate.

The leasing costs are higher than employing capital from a bank, although tax advantages must be considered. The machine is not the property of the farmer during the term of the agreement.

Leasing charges include VAT and, although they can be claimed back, they do enter the cash flow.

8.8.7. Machinery Syndicates

Machinery syndicates consist of groups of farmers who join together for the purchase and use of machinery and equipment. Their main aims are to share the cost of purchase and to ensure that machines are used to as near full capacity as possible. This helps to spread running costs. The syndicate may operate for one or more machines, but could include grain driers and storage plants.

Quite apart from sharing the initial buying costs a machinery syndicate may have a better chance of obtaining credit than the individual farmer with the result that each member has even more of his own capital for other purposes. Frequently the syndicate has bigger machines than the members could individually justify or afford, with the attendant advantages of faster work rates, fewer crop losses, and possibly increased production or better marketing.

Generally the system is more reliable than the use of a contractor and often promotes labour sharing. Machines are usually well maintained and because of good trade-in values it may be possible to change them frequently.

The loss of independence can be a disadvantage if two people require a machine at the same time. Machines such as hedge cutters and ditchers should not present the same problem that can exist with forage harvesters or other equipment involved in activities where timeliness is important. If the rules of each syndicate are carefully worked out and goodwill prevails, problems of the latter type can be overcome.

There is always the worry that maintenance will be poor, but again this concern can be reduced by making one member responsible for repairs and the supervision of maintenance. To avoid the possibility of one farmer causing more damage to a machine than others, some syndicates arrange for one person to operate the machine on all farms. A qualified engineer may be asked to inspect the machine periodically.

With relatively inexpensive machines the initial costs may be divided into simple fractions and running costs, split according to each member's use in terms of hours, hectares, or tonnes.

With expensive machines a more accurate division of initial costs is usual on the basis of each member's expected use. The running costs are then charged in two ways. First any farmer using the machine at above his original estimate is charged on the excess at commercial rates. The remaining costs, less these excess charges, are then divided in proportion to each member's share of the cost of the machine.

In the case of very expensive equipment, such as grain-drying and storage plants, the initial cost is again shared in proportion to each member's estimated use. All work is charged at commercial rates, and any surplus is returned to the members in proportion to their share in the syndicate.

Syndicates depend upon goodwill of their members for success, but their success can also be fostered by good organisation.

8.8.8. Machinery Budgets

It must be realised that once a machine has been bought it incurs depreciation costs, interest on any capital invested, insurance costs, and where applicable road tax. These can be regarded as the fixed costs of ownership. If the machine is used on a small area, or to process a limited number of tonnes of produce, the machine costs per hectare or per tonne will be very high. As the area or number of tonnes increases the unit costs of the machine decrease.

Each machine has running costs which vary with the use of the machine. These include fuel and repairs. When completing machinery budgets they are treated as variable costs. Care must be taken not to be confused by this classification, which differs from that in gross margin calculations. In the latter case fuel and repairs are treated as fixed costs because of the difficulty involved in their allocation to enterprises.

The farmer may undertake a budget to establish the costs involved if he buys a particular machine, or he may budget to see which would be the best machine from alternatives. Budgets can vary in their accuracy in so far that the most simple may produce crude answers, which might be sufficient to give a basic appreciation, whereas the most accurate budgets can be detailed and time consuming.

Take the example of a farmer considering the purchase of a £30,000 combine. The first step in calculating the cost of its use is to calculate the depreciation. Culpin's data shown earlier indicates that a combine would last about 12 years. In the budget below, this figure is taken.

The interest on the capital could be based on the initial capital, but this might be considered unfair. The depreciation charged in the P & L account is purely a notional item and no money is actually spent on it. This depreciation figure can therefore be regarded as reducing the bank overdraft each year and hence the interest payable on the capital still invested in the machine. To obtain a fully accurate position for each year it would be necessary to undertake annual budgets. Many people overcome this by taking the interest on the average capital invested in the machine. Thus for the £30,000 combine it would be (£30,000+£0)÷2=£15,000 (Fig. 75).

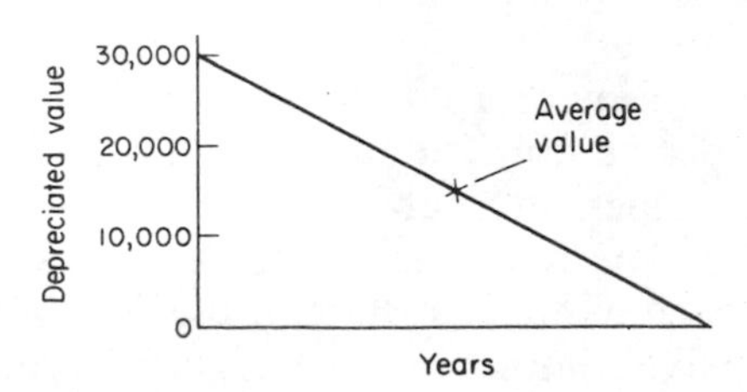

FIG. 75. Average depreciation value of a machine.

Since many machines are sold before they are fully depreciated, some farmers modify this calculation and fix the capital on which interest is to be charged on the (initial capital + trade in value) ÷ 2.

This type of "average position future" calculation is sufficient for many farmers, but others require greater accuracy. Discounted cash flow techniques, described in Chapter 11, permit calculations to be undertaken more accurately on a year-to-year basis. They also allow tax relief on interest actually paid to be taken into account.

Where a farmer does not need to borrow to finance the purchase it is still common practice to charge an opportunity cost on the capital invested in the machine.

Returning to the combine the fixed costs per annum would be:

	£
Depreciation over 12 years	2500
Interest on £30,000 ÷ 2@, say, 10%	1500
Insurance	60
	£4060

Assume that the variable costs of operating the machine are £4 per hectare. If the machine was used on the following areas the cost per hectare would be:

Area (ha)	50	100	150
Total annual fixed cost (£)	4060	4060	4060
Fixed cost per hectare (£)	81.20	40.60	27.07
Variable cost per hectare (£)	4.00	4.00	4.00
Total cost per hectare (£)	85.20	44.60	31.07

The farmer must carefully evaluate whether he should buy a combine of this size or if he should rely on a contractor. This point is covered under break-even budgets in Chapter 10.

CHAPTER 9

FACTORS AFFECTING THE PROFITABILITY OF THE MAIN FARM ENTERPRISES

CONTENTS

9.1. INTRODUCTION TO ENTERPRISE STUDIES

9.1.1. Background Points

The profit from any enterprise can basically be derived as follows:

(Average value of each unit of its product − Average cost of producing each unit) × (Number of units of the product)

At the enterprise level the farmer's task is to ensure that an optimum balance is maintained between output and costs, with relevant consideration of the scale of production. However, individual enterprises cannot be treated in isolation, and when the whole farm is considered the best interests of the business may be served by accepting less than optimum situations for some of the enterprises. Such cases can arise when there is competition for scarce resources which should be allocated in a way that gives maximum benefit to the whole farm and not just an individual enterprise. For example, capital may be limited. An injection of money into a given enterprise may allow it to become much more profitable, but the same funds put into another enterprise might generate more profit, i.e. the opportunity cost associated with the latter enterprise would be higher.

This chapter will concentrate on individual enterprises, and a later section of the book will cover the need for integration to form a complete business.

Usually several production possibilities exist for each enterprise. It is unwise to divorce the selection of the particular production system from the rest of the business because of the competition for available resources outlined above. Once the system has been selected every effort must be made to make it as efficient as possible, either by maximising the output to the resources allocated to the enterprise or by a reduction in the resources it requires.

9.1.2. Standards of Husbandry and Management

Throughout this chapter it will be seen that some farmers achieve far better results than others, even under similar conditions and using the same, or possibly less, of the relevant resources. This is a reflection of the standard of husbandry and management. Attention to those points of detail which can affect output is very important. Analysis of the cost/benefit of any particular action is also essential.

It will be seen that weaner pig production provides a classic example of how marginal additions to output, in this case additional weaners, can significantly improve profits. This is because many of the costs of weaner production are incurred irrespective of the number reared. The more weaners produced, the more thinly these costs can be spread.

Although yield is a major factor determining profit, it must be considered in relation to input levels. Where inputs are low, less than average yields may produce reasonable profits. The situation to avoid is where high costs are associated with moderate or low yields. It is possible for farmers with national average yields to be making a loss.

Careful control of inputs coupled with regular monitoring of performance can help to ensure the efficiency of an enterprise and diagnose trouble or favourable trends at an early stage. Such practices can separate the successful from the unsuccessful farmer.

9.1.3. Weak Links in the Production Chain

Lay ye down the golden chain
From Heaven, and pull at its inferior links.

This quotation from Homer can be adapted to emphasise a point which is fundamental to the success of any enterprise. If there is a weak link in the production chain there is frequently little point in improving the other links before doing something about the weak link. For example, if the nutrition of dairy cows is basically at fault the improvement of a herd's genetic potential will probably bring little or no extra returns.

Once the weak link has been strengthened then further improvement might best be achieved by developing all links in the chain so that one does not again become a limiting factor to progress.

9.1.4. Scale of Production

The advantages of large scale production have been referred to before. It was pointed out that it can justify specialist labour and machinery of a size and type which assists timeliness and produces improved results. Better buildings and other equipment may be warranted. It might also be easier to apply new techniques in such cases.

A large scale enterprise merits the particular attention of management not only in production, but also for marketing. The price received for products is a major factor determining profits. Selling large quantities can sometimes produce higher prices. Discounts may also be obtained for buying in bulk. This is particularly the case with fertiliser.

Large scale farming is not without its problems. Some of these are referred to in Section 9.3.12.

9.1.5. Quality

It is frequently said that it is important to produce goods of the correct type and quality in relation to demand, at the right time and place. Quality can significantly affect price, but the extra returns must justify any additional trouble and costs. The merits of producing goods of a given quality must be evaluated. An example will be referred to under milk production, where yield and quality may be in conflict.

9.2. GRASS

9.2.1. Good Grassland Management

A special book would be required to give this crop the attention it deserves because of its importance to British Agriculture, but it is only possible to make a brief reference to it here. In spite of the amount of information that is available about grassland management, on many farms production is less than optimum and utilisation is inefficient.

9.2.2. Grazing

(i) Fertiliser

Optimum fertiliser use depends upon several factors. The potential of a site to produce grass is particularly important. The following site classification based on summer rainfall and soil water holding capacity is often employed. In this 1 indicates the best and 5 the worst, conditions for growing grass.

	Average rainfall April–September		
Soil	Above 400 mm	300–400 mm	Less than 300 mm
Clay loams, heavy soils	1	2	3
Loams, soils of medium texture and deep soils over chalk	2	3	4
Shallow soils over chalk or rock, gravelly and coarse sandy soils	3	4	5

[N.B. In northern areas of Britain or on land above 300 m add 1]

Water is essential for grass growth. However, as a result of winter and spring rains the differences in grass growing conditions attributable to "site class"/"soil moisture" are usually not as large up to the end of May as they are for the rest of the season.

Although phosphorus and potassium are important one of the main determinants of grass yield is the amount of nitrogen used and the frequency of its application. At applications up to about 300 kg N/ha per annum the response is virtually linear with about 20 kg of dry matter being produced for each extra kg of nitrogen. There is a diminishing return to applications above this. The upper limit of the "economic yield" in response to nitrogen is reached when the return falls to 10 kg of DM/kg nitrogen. This occurs at about 90% of maximum yield at which point only 60% of the nitrogen required to achieve maximum yield has to be applied. Use of nitrogen above this level may not be justified economically.

The annual nitrogen application rate needed to achieve a 10:1 response, and hence optimum yield, varies with the site class. It is reached first with site class 5 at about 300 kg N/ha, with about 8.5 t DM/ha and last with site class 1 at about 450 kg/ha, with about 12.5 t DM/ha. These rates do, however, require modification according to stocking rates and other factors.

Ideally nitrogen should be applied just before grass growth starts. On low-ground farms this can be from the last week in February up to mid-March. Nitrogen fertiliser can be applied at this time provided the weather outlook is good.

Guides to "optimal" timing are usually published in the farming press. For example for T-sum 200 the daily air temperatures above 0°C after January 1 are added until the cumulative figure reaches 200 at which point nitrogen application is advised. With the T-value 80 technique soil temperature above 0°C at a depth of 10 cm are added up from February 1 and when the total reaches 80 application is recommended. In both cases application should be subject to weather at the time. However, work at the Scottish Colleges showed that these systems can only serve as a rough guide. The conclusion was reached that there is a period of about three weeks, centred around the "optimal" date during which *N* application can give 90% of maximum yield response.

Some farmers split the first dressing and apply say 45 kg N per ha early in anticipation of growth and 45 kg when it has started.

Phosphorus and potash applications should be based on soil analysis, applications of farmyard manure or slurry, and the method of grass utilisation. It must be realised that the potash status of a sandy soil can change quickly. Potash is not usually applied in the spring to grazing land because of the risk of precipitating hypomagnesaemia.

(ii) Grazing Systems

The aim of any grazing system is to provide highly digestible, clean material, which allows stock to achieve the desired performance.

Set stocking, which is essentially where stock continuously graze the same area of land for large sectors of the grazing season, is practised on many farms. It requires a high degree of management skill to match "grass on offer" to optimum "consumption". If the grass height is below 6 to 8 cm for dairy cows high stock performance may not be achieved, but it is essential to maintain sufficient grazing pressure, especially early in the season, to ensure that the grass does not grow above about 8 cm and lead to a reduction in quality.

With a continuous grazing system nitrogen should be applied to one quarter of the land at a time, about every 8 to 10 days. Site potential will determine the amount of nitrogen applied but the level of stocking must be considered.

Other farmers practice some form of rotational grazing such as paddock grazing, or strip grazing behind an electric fence. If dairy stock on such systems are not to have their production performance restricted the post grazing height must not be less than 8 to 10 cm.

Nitrogen applications for both continuous and rotational grazings must relate to site potential. The essential point with rotational grazing is to apply the fertiliser

Recommended target grass heights for continuous grazing (set stocking)

	Dairy (cm)	Beef (cm)	Sheep (cm)
Spring	6	5	4
Summer—cows	8	—	—
—finishing	—	7	5
—store	—	6	4
Autumn	10	8	7

immediately after grazing. During spring and early summer a total of between 1.7 kg and 2.5 kg N/ha for each day that the grass will grow before next being grazed can be justified according to site class.

Although site potential will be important a rough guide is that 1 ha of grass should graze about 5–6 livestock units, or 3 to 3.5 t liveweight, until first cut silage or hay aftermaths are available. One hectare should carry 2.5 to 3 t of animal throughout the grazing season.

Where strip grazing is practiced care has to be taken to avoid poaching in wet weather if stocking is very high. With this system it is essential to observe both the animal's production and the grass they are leaving to help decide how far to move the electric fence. In the peak grazing period of May–June 0.5 ha should graze 100–120 livestock units each half day, but there will be a variation in this because of such things as weather.

In addition to appropriate fertiliser, suitable strains of grass must be selected. Ryegrasses are most popular because of their digestibility and productivity. Strains can be selected according to their earliness in maturity to produce a succession of grass at the right stage of digestibility. In May and early June the D-value should be well over 70.

Digestibility of re-growths is not usually as high as earlier in the season, and if first grazing has been incomplete the digestibility of subsequent grazing will be depressed by old material present. With appropriate defoliation and fertiliser management, grass of reasonable digestibility (D-value 61–65) can be produced in the autumn, and although low in sugars it is high in digestible protein.

When rotational grazing is practiced early in the season it may be possible to return stock to a field about 21 days after first grazing, but as the season advances the return may be nearer to 36 days. Grass from silage or hay aftermaths helps meet any deficiency in the latter case.

9.2.3. Conservation

(i) Fertiliser

Again the rate of fertiliser used should relate to site potential. A general guide is to use 120:60:90 kg of NPK per ha respectively for first cut silage. Up to 150 kg of nitrogen may be used if the grass is to be cut late in order to achieve higher quantity but lower quality of product. A common practice is to use a split dressing, the first containing N P and K, and second, applied 2 to 3 weeks later, containing the remainder of the nitrogen as a "straight". A split dressing should only be used if the second application is before mid-April otherwise the silage will contain high amounts of ammonia nitrogen. The split application encourages early growth, helps to minimise nitrogen losses through leaching by rain, and reduces the cost of the nitrogen because straight nitrogen is cheaper than that in compounds.

Immediately after cutting a further 80 to 100 kg of nitrogen may be justified for second cut silage or 40 to 60 kg if intended for grazing depending upon the site class.

Every 5 t of dry matter in grass removed by cutting takes the equivalent of 40 to 50 kg of P_2O_5 and 120 kg of K_2O per hectare. Slurry can provide a proportion of the potassium and, subject to the time of application, some of the nitrogen. Applications of potassium usually have to be higher on sandy than on clay soils.

A very general guide for hay crops is to use 75:50:50 kg of NPK per ha for first cut and 60:30:45 for second cut.

(ii) Quality

The quality of the conserved product can be no better than the grass from which it is made. The conflict between total yield per hectare and D-value is a matter for debate. Farmers who expect high production from their stock may be protagonists for 70 D silage, whereas others who expect high production per hectare may be happier with a lower D-value and more bulk.

Ear emergence is used as a guide to cutting and digestibility, but it must be realised that grasses differ in their digestibility at a given stage of ear emergence. Timothy falls off in D-value well before ear emergence whereas early strains of perennial ryegrass have fallen comparatively little by this stage. Later strains of perennial ryegrass are somewhere in between. Ear emergence is taken to be when 50% of the ears have emerged, (i.e. first visible). For each grass this is measured as the number of days after May 1 when it takes place and gives the grass its REE or relative ear emergence.

Strains of grass can be selected according to heading date in order to produce a succession of grass for cutting at the right stage of digestibility. Intermediate and late perennial ryegrasses are ideal for silage because of the total yield they produce in a season. Some of the new tetraploid ryegrasses are rich in sugars which helps silage fermentation. They are also highly digestible and palatable.

The main factors determining success in silage making include cutting at the right stage, speedy filling of clamps, in some circumstances the use of an additive, consolidation, and sealing the clamp each night. First cut silage is usually better than later cuts if made correctly. Autumn grass does not generally make the best silage because fermentation is restricted by sugar content.

Good fermentation is essential if both quality and stock intake are to be high. The ammonia nitrogen content of the silage has a positive relationship with feed intake and should preferably be 8% or less of the total nitrogen.

Hay making is very dependent upon the weather, although barn hay drying can help. Good hay crops will be about 7.5 t/ha, with 10 t possible and 5 t being moderate. Silage yields are equally, or more, variable with 19 t/ha material being typically ensiled from first cut, although 30 t is possible. Second cuts usually yield less, say in the region of 12–16 t/ha.

Yield is very much a reflection of site potential and nitrogen application, plus time of cutting.

9.3. MILK PRODUCTION

9.3.1. Profit Factors

The main factors influencing the profitability of milk production are shown in Fig. 76. The appropriate share of the farm's fixed costs (regular labour, rent, machinery, and general overheads) must also be considered.

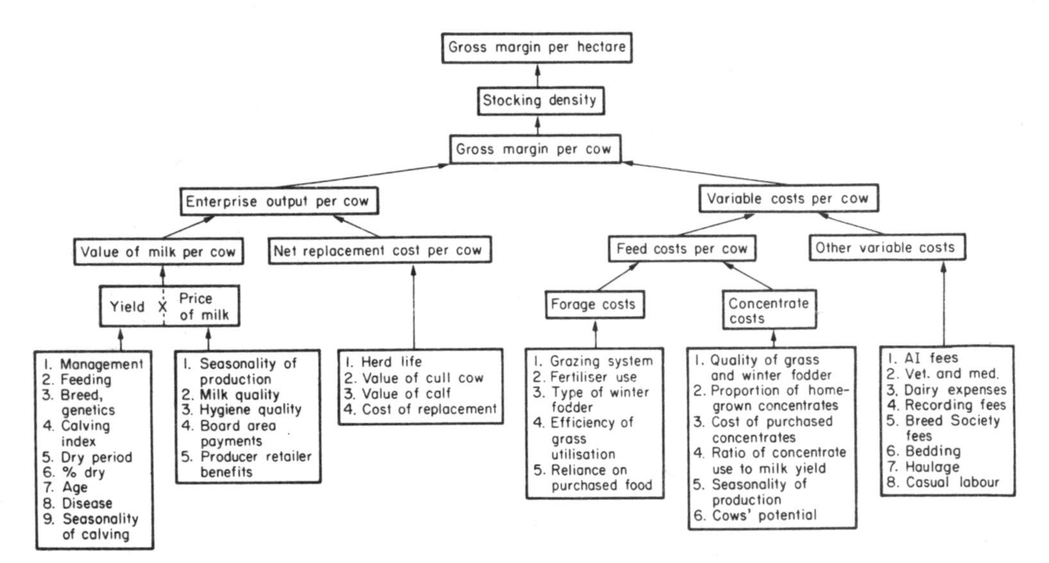

FIG. 76. Factors affecting the gross margin per cow and per hectare from dairying.

Milk quotas, introduced by the EEC in the mid-1980s to help control surplus production, gave rise to a further vital profit factor. When there is a reduction in quota each farmer must budget to minimise its effects. If quota can be leased or purchased the merits of these procedures have to be evaluated. When production has to be cut back the main options are to reduce the number of cows and to use the land released for alternative enterprises, or to cut back on feed and accept lower yields per cow.

Fixed costs associated with milk production have to be carefully examined because unless they are reduced the fixed costs per litre of milk produced must increase if total production falls. It is difficult to reduce fixed costs because it is not possible to sell part of a milking parlour, bulk tank, building, or forage harvester, and the mortgage or rent will not fall. Staff may have to accept less overtime but it is frequently impossible to reduce the labour force by a full man.

The variability in production between years because of such things as grass and silage quality makes it difficult for the farmer to be exactly on his quota target. He also has to judge whether the national quota will be exceeded and if a levy will be applied.

The level of any penalty is important because if no fixed costs are saved then the levy must exceed the variable costs of production before it becomes unprofitable to exceed quota. The variable costs of production and hence the rate of levy which would create a break-even situation differs from farm to farm, principally because of differences in feed costs. It must be remembered that concentrates tend to substitute for forage when a cow's intake is at maximum appetite. The last litre of milk produced in some cases requires 0.7 kg or more of concentrates instead of about 0.45 kg which might be calculated from energy tables. Therefore the higher the level of concentrates per litre required for the cows to exceed quota the lower the levy must be for the break-even situation to occur.

Reductions in quotas have forced some farmers out of milk production, and even out of farming, because they could not reduce fixed costs to compensate for decreased income. In some cases the problem was compounded because with less cows they had fewer bull or cross calves for beef production. Other farmers adapted to quotas and some found that with an increase in the efficiency forced upon them, particularly in relation to feeding, the impact was not as substantial as they feared.

In spite of quotas the other profit factors retain most, if not all, the importance they had in pre-quota days.

9.3.2. Yield

Total lactation yields, especially if over extended periods, are of less interest to management than average annual yields per cow. The latter can be obtained by dividing the total volume of milk produced on the farm by the average number of cows in milk and dry in the herd. Cow numbers can be derived from the average of twelve monthly livestock counts. Care must be taken when interpreting data from herds which have recently increased in size by the introduction of newly calved cows. This tends to raise the average yield because some of the animals may not have had a dry period during the year in question. The retention of cull cows at the end of lactation to await an increase in market price or to put on flesh will reduce the average yield per cow.

Genetic potential, feeding, and management are the main factors affecting yield. The cows' genetic potential can be improved by the use of nominated, proven sires, by attention to cow families, and perhaps in the future by embryo transplants. Probably 20% of the difference in performance between herds is due to genetic differences and about 80% to feeding and management. Management is concerned with almost all the factors affecting profitability. Nutrition is the most important, but this will be discussed later. The maintenance of herd health and regular calving are two of the other major aspects which determine success.

Estimates of the reduction in national average milk yields because of mastitis vary, but 10% is frequently quoted. Many cases of mastitis can be prevented by good milking routines and efficient, well-maintained milking equipment. Dry cow therapy is another important means of controlling mastitis. Attention to cows' feet will help reduce foot and leg problems, and good housing, together with appropriate diets, can help maintain herd health.

Milking cows three times daily can increase yields by up to 10% or more compared to twice daily, especially where high yields are being obtained, but labour problems tend to mitigate against this on most farms. The introduction of milk quotas also made this practice less attractive to many.

Selection of calving dates can be important, but must suit the particular farm. Autumn calvers on average produce higher yields than spring or summer calvers, with a difference of 12% in some cases. In part this may be because autumn calvers receive a boost from spring grass in late lactation, whereas many spring calvers receive poorer winter diets when their lactations are advanced.

Average data suggests that cows increase in yield up to about their sixth lactation, after which they decline. This must be interpreted carefully because many low yielders are culled after early lactations and do not feature in averages for older cows. Nevertheless, when looking at the average yield of a herd the number of heifers and the ages of the cows should be observed.

9.3.3. Calving Index

Calving index, or the average interval between calvings, is regarded by many as a major factor influencing annual average yields per cow. Targets of 365 days are frequently set. Various figures are put on the cost of the delay caused by a missed heat period. The basis of such calculations is shown in Table 9, which takes the example of two cows with total lactation yields of 5,000 l.

Table 9. *Calculating the influence of a missed heat*

	Calving interval 365 days	Calving interval 386 days
Annual average yield	5000 l	$5000 \times \frac{365}{386} = 4728$ l
Annual average calf births	1	$1 \times \frac{365}{386} = 0.95$

The missed heat in this case costs 272 l of milk and 0.05 of a calf at current values. This is partially offset by the difference in annual concentrate intakes. Assuming that an average of 0.3 kg/l is fed, the cow in the example shown with the 365-day index will consume 82 kg (272×0.3) more food in the year than the cow with the 386-day index. The illustration accepts that the second cow has no benefit in lactation yield from the longer calving interval which may not be true in all cases.

A survey of about 10,000 herds undertaken by the MMB (Report of the Breeding and Production Organisation, No. 28) indicated that with Friesians and Ayrshires intervals below 375 days were associated with reduced annual average milk yields. This gave support to the view held by some that, whilst it is important, too much emphasis can be put on a calving index.

On average cows can be empty for about 84 days or 94 days to obtain 365- and 375-day calving intervals. To delay service until 80 or 90 days after calving would be most unwise in spite of the fact that from this stage the energy balance of most cows becomes more favourable to conception. This is because disease, physiological problems, difficulties in heat observation, and other factors would prevent all cows conceiving to this service. The solution is to serve cows on the first heat after the forty second day following calving subject to consideration of each cow's condition, milk yield, and whether she is in her heifer lactation. Sound feeding and management is essential to minimise energy balance and other problems at this time. Surveys have suggested there is probably no benefit in milk yield from most cows in delaying breeding further. However, they have shown a positive relationship between the dry period and yield in the next lactation.

In spite of this an excessive number of dry cows indicates poor management. Some herds have an annual average of 30% of cows dry, whereas 16% is generally accepted as the target. Too low a figure may indicate that cows are not having sufficient time to repair alveolar tissue between lactations.

Care must be taken in the interpretation of the calving index for a particular herd. Where many cows have been culled after unsuccessful attempts at service the calving index can give a false impression of the breeding standard although a "weighting figure" is often given to these animals in the calculation of calving index.

9.3.4. Yield in Relation to Profits

In general the highest gross margins are obtained from high yielding herds, but high yields do not produce high profits if they are associated with excessive costs. Good gross margins can be obtained from comparatively moderate yields if costs are contained sufficiently.

Each farmer has the difficult task of deciding the yield level to aim for. The problem has many facets, but some general pointers can be given to factors which should be considered.

It must be recognised that each cow takes up a place in the building and that its share of the fixed costs such as rent, labour, capital, etc. will be similar irrespective of its yield. In addition there will be little or, in some cases, no variation in the cost of the maintenance ration, the vet bills, miscellanous dairy expenses, and replacement costs with changes in the yield.

For convenience these can be called the "quasi fixed costs". Food costs for milk production will, however, vary with yield, and in conformity with the law of diminishing returns not only will they increase as yield increases, but this increase will get proportionately greater as the yield gets larger. If this was not the case it would always pay to feed concentrates to maximum intake.

Figure 77 helps to illustrate one of the factors which can influence a farmer's yield target. Assume that *OM* represents the total of the fixed costs plus the quasi fixed costs for Farmer A and *ON* those for Farmer B. The dotted curve represents the cost of the concentrates added to these fixed costs. If both farmers use identical concentrate inputs at each output level the slope of the curve will be identical in each case, and it will be at the same height above the fixed cost line for both farms.

The amount of output per cow required by Farmer A to cover his total cost is *OW*, much less than the output *OY* required by Farmer B. Output is to a great extent a reflection of yield.

Farmers who have committed themselves to high fixed cost regimes therefore tend

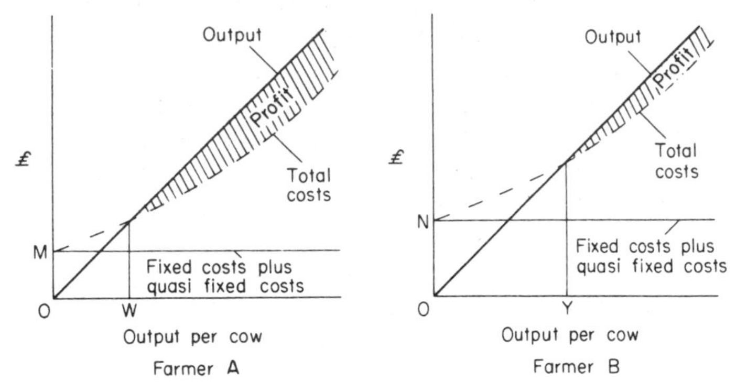

FIG. 77. Significance of fixed cost to yield.

to have greater pressure on them to aim for higher yields than those with low fixed costs. The latter have greater flexibility of choice.

In many cases, however, fixed costs are increased as a result of the provision of facilities which enable staff to handle more cows with greater attention to detail. Such things as out of parlour feeders are installed, bigger machinery is bought to improve fodder conservation, and better stock may be purchased. Many of these factors contributing to the higher fixed costs can help determine the maximum yield obtainable on a farm.

Farmer B can attempt to make the same profits as Farmer A without having the yield differential indicated by Fig. 77. If he cannot reduce his fixed costs it might be done by greater efficiency in his variable costs, e.g. making better silage and promoting better response to concentrates at any given yield level than Farmer A, plus ensuring that he has appropriate rations, but at least cost. He could also attempt to improve the genetic potential of his cows and, by better management, disease prevention, and greater attention to detail, improve the efficiency of production. His other possibility is to endeavour to keep more cows to reduce the fixed costs per cow.

Satisfactory profits can be made at all fixed cost levels provided that output is sufficient. Low profits can also be made at low fixed cost levels, especially if facilities on the farm are a major constraint.

An important point to note is that once fixed costs have been covered by output, any additional yield has only to cover the extra variable costs involved in its production. Thus, in spite of the operation of the law of diminishing returns, marginal increases in yield can be highly profitable. The growth in the zones marked "Profit" in Fig. 77 illustrate this.

The point to stop increasing yield is when diminishing returns has operated to such an extent that the marginal value of additional milk equals the marginal cost of producing it, assuming that cow intake and related problems do not occur first.

Profit per hectare is very important especially on small farms. Survey data shows that on average profit per hectare increases with yield per cow, although there are many exceptions. Extensive use of purchased concentrates can reduce the area required per cow and increase yields, but careful appraisal of all factors is necessary before a farmer becomes heavily dependent upon bought feed.

9.3.5. Price of Milk

In Britain the price of milk is controlled through the Milk Marketing Boards and the individual farmer's opportunity to increase the money that he receives is restricted. There are bonuses and deductions for compositional quality, and penalties for poor hygienic standards or for selling milk containing antibiotics. In addition there are seasonal differences in price, with the highest rates being paid in winter to encourage milk production at a time when costs of production are greatest.

Many surveys show that winter producers on average obtain more profit than summer producers, but that individual summer producers do as well, if not better, than the best winter producers. Summer production is popular in areas which have a long grazing season, but requires a farmer who is highly capable of growing and utilising grass. Farmers with a short grazing season, who purchase considerable amounts of food to supplement their own land area, or who make use of arable by-products or draff, may favour winter production.

Whilst endeavouring to maintain milk quality, farmers may aim for higher yields, which means that in some cases they have to sacrifice part of the quality premiums. Several factors influence the economics of this issue including breed, breeding, feeding, and the differential currently being paid for milk of high compositional quality.

Producer retailers have a price advantage over their colleagues who sell to the Boards, but they have extra costs and also have to pay levies to the Milk Board for their area.

9.3.6. Net Replacement Cost

Net replacement costs can significantly affect dairy cow output. The value of Friesian calves and cast cows has given this breed an advantage over Ayrshires, and it is interesting to calculate the current value of calves relative to the profit of a herd. The introduction of the Holstein has, however, reduced the price advantage of black and whites.

An ability to breed good wearing cows in terms of udders and feet can reduce the culling rate. In addition to foot problems and dropped udders, culling occurs because of failure to breed, mastitis, other diseases, and poor milk yield.

It is surprising how many farmers still do not record milk production to help identify cows to be culled on yield, as well as to give a guide to rations and to form a basis for selection of replacements.

One important principle is to always keep cow numbers up to the normal herd level if possible, because even though a cow may be a moderate yielder she will be contributing to output and helping to reduce the fixed costs per animal in the herd. This is until she can be replaced by a potentially better yielding beast, or unless for disease or other reasons she should be sold.

Market prices for cast cows can vary significantly, even within a year. When local schools or factories are on holiday they are usually down, and they may be affected when there are ample supplies of other beef animals. The condition of a particular cow is important and the merits of keeping her a little longer to improve her value should be examined.

In some herds 25–30% or more of the cows are culled each year, but in others the figure is 20% or less.

Calf prices vary with the breed, or cross, and with the current demand for beef. Subject to calving difficulties it may be worthwhile to use beef bulls on the bottom third of a herd, in terms of milk yield, since it is unlikely that these cows' calves would be retained for breeding.

Where possible replacement heifers should be genetically superior for milk production to the cows they replace. Use of a proven AI sire has definite merit, and although the price of his semen may appear to be expensive, this must be related to any increase in milking potential of his heifers over and above present herd levels, and against the unknown breeding merits of non-proven sires. Some purchased pedigree stock may not always live up to their records because of the differences in management between their original and new herds, and in particular because their sires may not be widely proven.

9.3.7. Feed Costs

Feed accounts for about 56–60% of the costs of milk production compared to labour at about 20% and other costs, including replacements, at 20–25%.

Several efficiency factors are used to examine feed use. The difference between the value of milk sold per cow and the cost of the concentrates fed, known as the margin over concentrates, provides a measure which incorporates four of the main factors influencing profits, viz. cost and use of concentrates, yield and price of milk. Beef-pulp is usually included as a concentrate, but not brewers grains.

Satisfactory figures for margin over concentrates can be obtained at a range of yield levels. It is important to relate the figure obtained to the factors which may have precipitated the particular level. For example, if a large number of heifers calving at 2 years of age are introduced into a herd, in a given year, concentrate use may be more generous because these cattle are still growing. The acceptability of the figure can also depend upon such things as the amount of land available and dependence upon purchased feed to increase the number of stock carried.

The quantity of concentrates fed per litre can also be calculated, but it is the cost of these concentrates per litre of milk which is more important. Realistic costs must be put on home-produced feed, and suitable rations which can be obtained at "least cost" should be fed. Purchase of ingredients "out of season" for home mixes, or bulk delivery of compounder's feeds, rather than in sacks, may reduce prices.

It is extremely difficult to state the optimum concentrate usage for a given herd because of the number of factors involved. These include the seasonality of milk production, price of milk relative to cost of concentrates, yield and genetic potential of each cow, stocking density, pressure for land, and the quality of bulk fodders.

The quality of bulk fodders is a particularly important factor determining concentrate use and the economics of milk production.

The potential for the production of high profits using low concentrate inputs and with the emphasis on grass and grass products has been demonstrated at several research centres, in particular Hillsborough in Northern Ireland, as well as on commercial farms. This contrasts with farmers who, in some cases using out of parlour feeders, attempt to obtain high profits by feeding over 2 tonne concentrates per cow per year. Some of these farmers still practice "lead" or "challenge" feeding and claim that with their high standard of management they achieve better results than those who adopt flat rate feeding. Certainly the latter system simplifies management and has been demonstrated to produce good profits at The West of Scotland College's Crichton Royal Farm.

Whether aiming for high yields with relatively high costs, or moderate yields with lower costs, the farmer must monitor both yields and inputs regularly to check that performance is satisfactory. Problems occur when farmers have insufficient control over their enterprises, resulting in performances at variance with their predicted levels, leading to lower profits.

Survey data frequently presents apparent anomalies. Farmers with the most profitable herds, whilst generally feeding more concentrates per cow, are often shown to feed less concentrates per litre and usually have higher milk yields than their colleagues with less profitable herds. This must not be taken as contradicting the law of diminishing returns, which still applies in any given herd, but it indicates that some farmers are more efficient in the use of concentrates than others. Where a herd with a very high concentrate use is in the lowest profit group, and there appears to be no benefit in terms of improved stocking density, then gross inefficiency is usually indicated. Thus, whilst there does appear to be a correlation between concentrate use and milk yield, there is not always the same correlation between total concentrates fed per cow and profitability. However, there is a stronger correlation between concentrates fed per litre and profitability.

Whilst survey data is interesting the farmer must concern himself with his own individual situation and set himself targets which reflect his resources. The ability to produce and utilise good bulk fodder and grass will be one important factor.

High yields are particularly important to the farmers selling pedigree cattle. Additional concentrates can therefore be justified on this basis, but the purchaser should endeavour to establish the nutritional regime practised in the breeding herd.

9.3.8. Quality of Bulk Fodders

Conserved fodder should always be analysed as a basis for compiling rations, but in the case of silage sampling errors and variations within the clamp may occur. It is frequently said that the animals provide the best analysis by their response to their diet.

The digestibility of bulk fodders affects their intake by cows. Certain farmers, especially those with limited land, may aim for quantity of silage per hectare and so sacrifice quality. This produces a conflict in cases where cows are expected to consume high amounts of concentrates. The importance of the quality of bulk fodders to high-yielding herds cannot be over emphasised. If the farmer aims to make silage with a D-value of 65 in order to get bulk he could fail and obtain low D-value, but if he aims for 70 D silage and misses the target he can still have a silage with fairly high digestibility. For this reason farmers whose target is high milk yields usually aim for high quality bulk fodder, although they should always consider productivity per hectare, including total milk production per hectare. Silage fermentation, and successful storage generally of bulk fodders, may also affect cow intake.

The method of feeding can be important. Self-feed, easy-feed, complete diet, silage blocks, and other methods are practised. Each must be related to the buildings, capital, and other prevailing factors on the particular farm. All have their merits in certain circumstances and can have their own influence on intake. Where high-yielding cows are expected to consume large quantities of feed, rations should preferably be made available so that consumption can be in comparatively small quantities at intervals, rather than twice daily.

Many farmers start ration formulation from the standpoint of the quality of their bulk fodders, and consider both quality and total supplies available for the winter. The whole process of rationing has to be an integrated one, however, since yield targets, stages of lactation, cow condition, milk quality, and other factors have to be considered in relation to all the constituents of the diet.

Butterfat problems may arise in association with high concentrate use and the length and total intake of fibre can be important. If bulk fodders are not of good quality, intake of fibre may be reduced.

The importance of variety in the diet of dairy cows is a very controversial topic. Very satisfactory results are being obtained from silage/concentrate diets, although many farmers argue that the introduction of feedingstuffs such as sugar-beet pulp, brewers grains, and roots act as stimulants to both appetite and milk production. There is no clear evidence that shows either school of thought to be right and plenty of individual cases to lend support to each.

9.3.9. Stocking Density

Analysis of survey data shows that the most profitable herds, in addition to having the highest yields per cow and highest concentrates per cow with reasonable levels of concentrate per litre, also have the best stocking density and highest use of nitrogen per hectare. Whilst a high use of concentrate per cow might be a factor in helping to increase stock carry, in general the most profitable farmers are also the best at producing and utilising grass.

In many cases gross margin per hectare can be more important than gross margin per cow, especially where land is limited. However, this must always be viewed against the particular fixed cost level. Extra cows per hectare usually means extra capital in stock, buildings, etc., but at the same time spreads the cost of the land over more animals. For each situation budgets can help to determine if it would be better to keep a high number of cows, each producing a comparatively moderate gross margin, or a smaller number each with a higher gross margin. Naturally, the possibility of keeping a high number of cows, each with a high gross margin, must also be examined.

Standards for stocking density have always to be related to land type, climate and particular problems of the farm in question. The current price of nitrogen fertiliser is also a factor. In spite of grass being one of the cheapest feeds for stock, many farmers fail to make effective use of it. Most of these would find examination of the level and timing of their application of both nitrogen and slurry worth while, and could improve their grazing management.

Many farmers set targets of 2.5 or more cows per hectare, although the majority achieve slightly lower levels. The best farmers in a survey may have more than double the gross margin per forage hectare than the poorest. In most cases low gross margins are not offset by low fixed costs.

9.3.10. Utilisable Metabolisable Energy

The essential point is that grass or its products are efficiently utilised. One technique employed to measure the efficiency of use of energy from grassland is the utilisable metabolisable energy, UME, system. It involves calculating the energy requirements of stock for maintenance, growth, milk and pregnancy. Take a Friesian herd, for example, with a stocking rate of 2 cows per hectare, averaging 5,600 l per cow per annum, and receiving 1.7 t of concentrate.

Per Cow	MJ/annum
Requirement for maintenance, growth, pregnancy	25,000
Requirement for 5,600 l milk at 5.3 MJ/l	29,680
	54,680
Less energy from 1.7 concentrate at 11 MJ/kg	18,700
UME required from forage per cow	35,980
UME attributable to forage (2 cows/ha)	71,960

The energy supplied by the concentrate has been deducted from the theoretical energy requirement to establish the energy derived from forage. The 71,960 MJ equals 71.96 GJ/ha. It has been established that there is a positive relationship between gross margin per hectare and the UME per hectare. Although the average UME for British dairy farms is probably somewhere between 65 and 70 it is possible to achieve between 90 and 100.

The UME can be related to the potential ME output. The latter depends upon the site class and nitrogen fertiliser application. The target ME yields (GJ/ha) are as follows:

	Fertiliser rate (kg/ha)			
Site class	150	200	300	400
1	79	93	114	128
2	74	87	108	121
3	67	81	101	113
4	62	76	96	105
5	57	69	88	97

Assume that the herd was on land with classification 2 and that the farmer applied 300 kg N/ha.

The efficiency of use is:

$$\frac{\text{Calculated UME}}{\text{Potential ME output}} \times 100 = \frac{71.96}{108} \times 100 = 66.63\%$$

An alternative measure of efficiency is to calculate the percentage ME derived from forage using the following formula:

$$\% \text{ ME from forage} = \frac{M+P-C}{M+P} \times 100$$

[M=energy for maintenance;
P=energy for production;
C=energy derived from concentrates]

At the West of Scotland's Crichton Royal Farm it was found that better utilisation was achieved on silage fields than on grazing fields. One system which proved of interest was to graze cows on set-stock basis during the day and house them overnight on ad-lib silage. This technique improved stocking rates, dry matter intakes and milk yields, especially in the second half of the grazing season.

Buffer feeding, or giving hay or silage to cows still receiving fresh grass, has proved useful on many farms by helping to maintain energy intakes at a consistently high level in spite of changes in the weather and in grass growth.

9.3.11. Other Variable Costs

Apart from feed, fertiliser is often the second highest variable cost. AI fees can appear to be expensive where nominated sires are used, but these are usually justified by eventual returns. The cost of recording is rewarded by its value to management. Breed society fees, whilst not always producing benefits from pedigree sales, can provide interest for both farmer and staff. Haulage, bedding, and miscellaneous dairy expenses are comparatively small, but should be kept to a minimum whilst facilitating satisfactory production of clean milk.

Veterinary fees can be kept to reasonable proportions if suitable disease preventative measures are adopted on veterinary advice and treatment of stock is promptly undertaken. Pregnancy diagnosis by a veterinary surgeon can also be justified.

9.3.12. Herd Size

Many surveys show that the highest yields, highest milk production per forage hectare, highest stocking rates, highest profits, and highest returns on tenant's capital are obtained by relatively large herds, although not necessarily the biggest. Some of the latter have moderate performance levels. It is tempting to suggest that the farmers with large herds have more cows because in the past, as well as now, they were the most efficient and therefore had more profits to finance expansion. This would, however, neglect the fact that large herds do have some benefits over smaller herds, although if they become too large they can have additional problems.

Large herds may have an advantage over smaller ones in housing, feeding, and milking arrangements. Cows can be grouped more easily; heifers separated out; feeding can be in yield groups; bulling animals may be in one section, thus facilitating greater observation; mechanised feeding practices can be more easily justified together with better machines, which may help improve silage; specialist management attention is warranted, as well as other benefits. "Time off" and holidays may be more easily managed with three or more men than with one man.

Although there are many exceptions, herds of about 120–130 cows milked by one man can be just as efficient in labour use as bigger herds, but it is always important to examine the amount of supporting labour used to feed, muck-out, move cows, and carry out routine tasks. If these are taken into account, 70 cows per man is common.

Slurry is rapidly becoming a problem for large herds. Cow movement, grazing arrangements, shift milking, and reduced personal interest because the stockman shares a group of cows with others, calf rearing and other problems have limited the success of some very large herds, and returns on capital have been unsatisfactory.

9.3.13. Fixed Costs

The quality of the labour is one of the most important factors influencing the performance of a dairy herd. Whilst labour costs should be kept within reason, an extra wage for the right man can be amply justified.

Capital invested in milking parlours, cluster removers, and other equipment can enable men to handle more cows. However, it is possible to over-emphasise such things as "parlour throughput", and whilst allowing men to handle higher numbers, modern methods must allow for adequate cow attention.

Investment in buildings and machinery requires a certain amount of courage especially at times when money values are changing or milk is in an over-supply situation. Careful budgets are necessary, with sensitivity analysis to changes in interest rates, and to the milk price/concentrate cost ratio. Efforts must be made to maintain future viability.

9.3.14. Management

Throughout this section reference has been made to the importance of management. This has many facets relating both to the husbandry of production and to economic aspects of the enterprise.

Participation in some form of group recording scheme can greatly help control. Monthly data relating to the main profit factors, such as margin over concentrates, can highlight times when performances are poor, although annual average results may be satisfactory. The competition arising from comparison with results obtained by other farmers can be a useful stimulant. Where group members actually meet the exchange of ideas is usually valuable.

A good scheme will include a forecasting element which can serve as a basis for control of the enterprise. Information such as calving dates, ages of cows, yield and length of past and current lactations, together with other information, is required from the farmer. The forecast can be presented in graphical form to illustrate the predicted pattern of milk yield, and the actual yield can be plotted on the same graph. When actual levels deviate from predictions the farmer's attention is drawn to the situation and he can endeavour to establish the reasons. In some cases it is possible to identify which group of cows is responsible for the discrepancy, newly calved, heifers, mid or late lactation animals.

Good records of milk yields, feeding, health, and breeding, are an essential aid to management. A satisfactory method of cow identification is also necessary. Condition scoring, assessed by examination of the quantity of fat in the loin area and around the tailhead, can also be a valuable guide to cow management.

Regular breeding is one of the aspects which are fundamental to success in dairying.

Heat detection is a weak point in many herds. Visual aids can be employed to indicate when cows should be in season and it is then a case of observation, including last thing at night. Paint on the tail head has proved useful in some herds to indicate when cows have been ridden.

Many aspects of husbandry management in relation to feeding, milking, and breeding have been mentioned in earlier sections. The importance of milk yield, concentrate feeding, stocking density, and nitrogen use, as being key factors determining profit, has also been stressed. It is the function of management to obtain the most efficient use of concentrates, land, and nitrogen to achieve satisfactory milk yields and gross margins per hectare in relation to the fixed cost levels of the herd.

9.3.15. Dairy Replacement

Each farmer has to decide whether to rear his replacements or to purchase them. More might be known about the quality of home-reared stock, and there is less disease risk than with bought animals.

Many farmers who have fairly intensive dairy systems fail to have the same intensity with their young stock. In spite of the widespread evidence that heifers can be successfully calved at 2 years of age, a high proportion of farmers reject this policy. This is usually because the system requires a good standard of management at the calf-rearing stage and many people fail to achieve the necessary growth rates. The aim is to bring down the non-productive period of $2\frac{1}{2}$–3 years to 2–$2\frac{1}{4}$ years, thus reducing the area of land required. The period for maintenance costs is also reduced. Extra feed costs during rearing offset this in part, but provided that the animal will calve at the right time of the year for the particular herd, suitably reared calves which have reached the right weight can be mated at 15 months. A little extra feeding in the first lactation may be necessary because the beasts are still growing.

Sufficient replacements should be reared to allow for selection. However, many farmers have a deliberate policy of rearing nearly all their heifers and sell a large number. If they calculated the total cost of producing them there would be many instances when they would change their policy. Fortunately they are not all as misguided as one farmer who said: "The cows pay for all the food that my heifers get, so the heifers are all profit when they are sold."

This does not mean that surplus heifers should never be reared for sale. If they utilise resources which would not otherwise be fully utilised there can be a place for them.

9.4. BEEF PRODUCTION

9.4.1. Intensity of Systems

There are so many different methods of beef production that it is not feasible to cover them all here. They can be divided up according to the level of intensity, and it is intended to outline the main factors affecting profitability using illustrations from semi-intensive, store cattle, intensive, and low intensity systems.

Low intensity methods are mainly found in areas where fixed cost levels can be kept low. Intensive systems, such as barley beef and veal calves, do not use land directly since they do not have a grazing period. Semi-intensive enterprises, such as 18-month beef, may be established to utilise a grass break in an arable rotation, or have to be fully competitive for their place on the farm with other enterprises. Store cattle are produced under a range of intensity levels.

Within each main heading, particularly the semi-intensive and low intensity categories, there are systems which differ widely in intensity so that their allocation to a particular heading can be arbitrary. Sixty per cent or more of British beef comes from animals born to dairy cows, most of which are reared by intensive or semi-intensive methods.

9.4.2. Semi-intensive Systems (18–22 months)

Figure 78 illustrates some of the main factors which can affect the profitability of many of the systems where the aim is to produce beef for slaughter, using grazing for at least part of the process. To maintain clarity, 18–22-month-old beef will be discussed first and some of the essential differences in other systems mentioned subsequently.

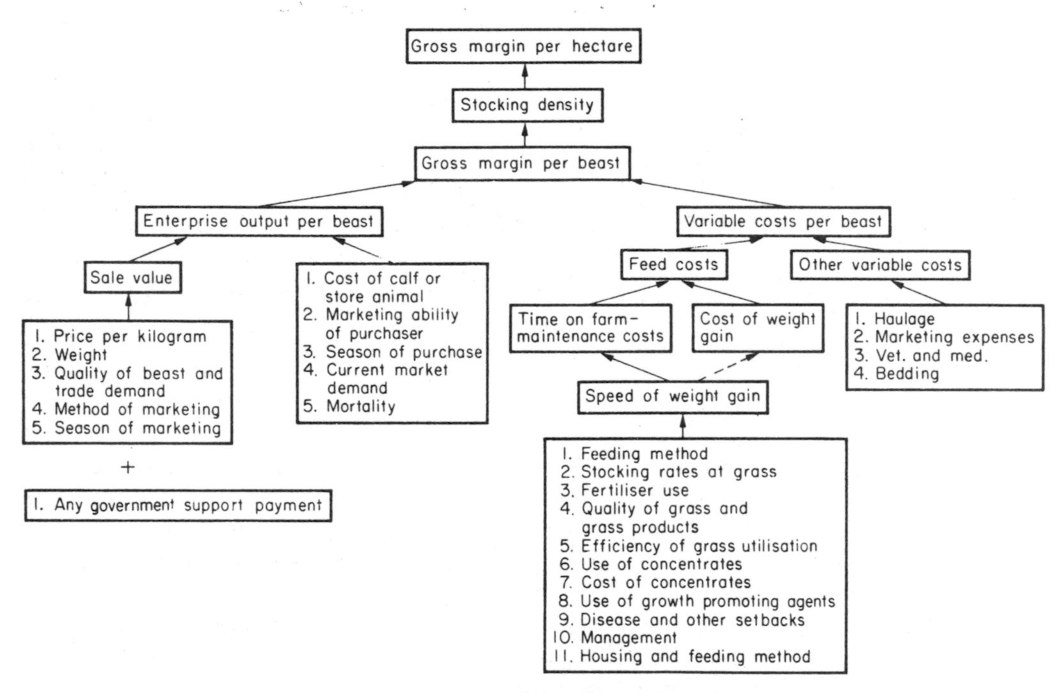

FIG. 78. Factors affecting the gross margin per hectare from finishing beef cattle which have a period at grass.

(i) Price of Calves

Many farmers produce 18-month beef from purchased autumn born animals, and competition for calves can increase prices. Demand from arable areas is so great that it encourages dealers who trade in calves from dairying areas. This movement contributes to disease problems and mortality. A contract for supply of calves with a dairy farmer may be the best solution.

Beef prices have to be forecast 18 months in advance, which is almost impossible, so that there can be little guide from this to realistic calf prices.

Many dairy farmers have 18–22-month-old beef systems as an adjunct to their main enterprise. They may have a supply problem because of the spread in calving dates which increases the difficulty of managing groups. In addition they can have pure-bred steers and cross-bred stock, which ideally require different management.

(ii) Sale Value

Initially the farmer must decide whether to market, or contract, or to sell liveweight or deadweight. Deadweight sales are influenced by killing out percentage and, in addition to saving marketing time and expenses, can, on some occasions, offer as good, or better, returns than liveweight sales. At other times the reverse is true.

At auction markets highest prices are paid for the type of animal in demand in terms of quality of finish, weight and breed. Demand varies from area to area. Any unusual, or unpopular, breed or cross can realise several pence per kilogram less than average, and unfinished animals may also be penalised.

In recent years the requirement has been for leaner carcasses. At the time of writing the "Government Support" system was implemented through a grading system involving classification on the basis of fat class and conformation. The increased use of Holstein bulls in British dairy herds meant that many of their offspring failed to grade and so did not attract the variable premium payable. The latter in many cases made the difference between profit and loss.

Prices vary significantly from season to season with the highest prices usually being in the spring when beef is in short supply, and the lowest prices in the autumn when large numbers of animals are being sold fat off grass. Substantial variations can occur from week to week and even on the same day. If cattle are in short supply, one week prices may be high, and the next week large numbers of cattle are entered with a consequent drop in price. Variations in price on a particular day can occur with the time beasts are sold, with lunch time, or the end of a big sale, producing reductions.

Opportunity and skill in marketing can make several pounds difference to the price of each animal. Factors which affect current demand must be researched and use made of statistical and report services such as those operated by the Meat and Livestock Commission.

Any government support or quality premium currently available can help the viability of a particular system.

(iii) Feed Costs

The method of feeding is influenced by many factors, but the most important are the speed of liveweight gain required, the breed or cross, and the reliance which the farmer wishes to place on grass, silage, hay, or arable by-products, compared to concentrates.

In general, Angus and Hereford cross calves will finish at lighter weights than pure Friesian or Charolais cross calves. Hereford crosses in particular tend to finish easier on grass, and grass products, than Charolais crosses which require more concentrates.

The store period must be avoided particularly with Friesians which can become leggy and finish later. Angus and Hereford cross heifers will become overfat if not managed carefully.

The longer any animal is on the farm the longer its maintenance costs have to be met. The period can justifiably be lengthened in certain systems. These include cases where the total cost of producing the beast are reduced because of cheaper diets, the sale date is delayed to a period of higher prices, or the system fits in better to the farm as a whole.

With autumn born calves intended for 18-month beef a key to success is to produce animals which are heavy enough when they go to grass in the spring. A target of 180 kg is frequently suggested and more easily obtained with September than later born calves. In spite of any compensatory growth from lighter beasts which may not have been so well fed in winter, animals over 180 kg tend to make better gains during the grazing season. Overfeeding of calves destined for summer grazing can, however, be wasteful.

A high standard of grass production and grazing management is essential if good results are to be obtained from both the stock and the land.

Stocking rates and nitrogen use must be correct. At a given stocking rate increased nitrogen use may foster higher individual gains. When stocking rates are increased at a constant level of nitrogen, the gains of the individual animals may be first impaired and then the total output per hectare reduced.

Stocking rate is one of the main factors influencing the gross margin per hectare. Some farmers use extra fertiliser but do not increase the stocking sufficiently to use the additional grass.

A paddock grazing system is frequently employed with the aim to produce a constant supply of short leafy grass. Up to 12–16 animals per ha may be grazed at the peak of the season and this is reduced to below 6–8 per ha as the animals grow, grazing deteriorates and silage aftermaths become available.

A worm control programme must be adopted if good liveweight gains are to be made. Supplementary feeding at grass will increase gains. This is usually advisable for a period before housing takes place since it keeps up performance as the grass deteriorates, and such stock receive less check in growth when they enter the yards. If possible they should weigh 340 kg at housing.

Good quality silage, or hay, is essential if low cost liveweight gains are to be made. Many farmers who make the highest profits obtain higher gains with less concentrates than those with lower margins. Some use excessive protein which may not be necessary with good silage.

When deciding the feeding programme in these final stages it is important not only to consider the type of animal, its condition and frame, the amount and quality of food available, but also the best time and weight for slaughter. An increase in concentrates will tend to increase the speed of the gain.

There have been many surveys which have suggested that a large number of farmers could feed less expensive feeds without seriously affecting profits. These same surveys have indicated the positive relationship between liveweight gain and gross margin. Regular checks on feed intake and liveweight gain are advisable because in the final stages cost of liveweight gain can be very expensive. Many cattle are marketed in an overfat condition which essentially means a waste of feed.

Many farmers modify the above system. For example calves born from late October to early March are frequently finished off grass at 20 months to 2 years of age. They probably receive concentrates whilst at grass during their first summer and are housed when weighing 225 to 275 kg. During the winter they are fed at "low-cost" with about 1.5 to 2 kg of concentrates plus silage to gain 0.4 to 0.6 kg per day. When turned out to grass they tend to make compensatory growth and many will be sold finished by midsummer.

A further alternative is to keep the cattle indoors all their lives. They receive silage and concentrates and the aim is to finish them at 14 to 17 months. Advantages claimed for this "storage feed" system include better utilisation of grassland and quicker turnover of capital.

9.4.3. Store Cattle

The range in methods of keeping these animals is enormous and some are of comparatively low intensity. Many store cattle are sold several times during their lifetime. Frequently the most important factor determining profits is the "feeder's margin", the difference between the purchase and sale prices. Both buying and selling require considerable skill. Each purchaser looks for particular types of beast which will make a profit. For example, some people who buy stores to finish look for stock with capacity for liveweight gain from compensatory growth. An old saying is: "Look for mud on their coats, not muck." Outwintered cattle, which have not been forced, grow away quicker than many inwintered beasts. The right animals will finish and produce a satisfactory margin.

(i) Finishing Spring Purchased Stores off Grass

Purchase prices are very much influenced by the season and grass growth. If winter fodder is in short supply, and grass is late in growing, store prices can be depressed by forced sale of animals at a time when buyers are reluctant. When grass grows prices can rise quickly and spring prices for stores can be higher than at other times of the year. In fact prices per kilogram liveweight can be higher than the price per kilogram for finished animals later in the summer, and profit must be derived from the increased liveweight. Buying skill and the feeder's margin are very important.

Forward stores may be sold finished in June and July before prices decline. As the summer advances prices usually fall because of increased supplies of animals. The fixed costs associated with the land and competition from other enterprises may affect the intensity with which the grazing is managed. Nitrogen use, stocking rates, and the grazing system must be judged against the value, cost, and rates of gain being achieved both per beast and per hectare. Mixed grazing with sheep may also produce benefits.

Target liveweight gains are often put at 1 kg/day early in the season from stores weighing over 390 kg, 0.9 kg/day for heavy stores later in the season, and 0.8 kg/day from overwintered suckled calves and other stores weighing less than 390 kg.

(ii) Finishing Autumn-bought Stores

A wide range of cattle is purchased for winter finishing. They may be store cattle and anything from 12 to 30 months of age, or they could be suckler calves. Even suckler calves could have been born in the summer and autumn of the previous year, or in the spring of the same year. Prices vary significantly, and part of the skill in buying is to assess what price can be paid and what type of animal to buy.

Generally buyers have a clear knowledge of their winter feed supplies at the time of the autumn sales, but the difficulty is in forecasting the prices of cattle when they are finished. If the fodder has been produced in the form of silage it may not be too saleable and stock have to be purchased to eat it, although there can still be flexibility in the choice of beast and any additional concentrate feeding.

Some farmers claim to make high profits from heifers in spite of them finishing at lighter weights. This depends on the purchase price. Generally the price per kilogram for stores is less than the sale price per kilogram in the spring, but there can be considerable variation from year to year. It may pay to buy several batches of forward stores to be finished at short intervals over the winter. Heavy stores are sometimes cheaper per kilogram to buy than lighter stores, but the lighter beasts may realise more per kilogram when sold finished. Lighter animals may also have the better feed conversion. Some farmers conclude that finishing lighter stores is more profitable especially if they are not sold until prices rise in the spring.

Speaned store calves off the hills can produce satisfactory returns provided that they are heavy enough to be finished out of yards before the grazing season. Light sucklers can also be profitable and are usually given a store period during the winter and either sold in spring as stores or sold fat off grass.

Fixed costs must be kept to a minimum, but on arable farms winter beef can use up surplus labour as well as using up arable by-products and producing dung. The source and opportunity cost of the capital investment is an important factor.

The Meat and Livestock Commission provides useful market intelligence and has produced a financial calculator which farmers can use as a guide to the price which they should pay for stores.

9.4.4. Intensive Cereal Beef

System: 10–12-week calf, weight 100 kg, gradually introduced to *ad lib* diet of rolled barley, plus vitamins, minerals, and protein supplement. Mix has 140 g per kg DM of crude protein. At 6–7 months protein reduced to 120 g per kg DM. Slaughtered at 10–12 months weighing 390–410 kg liveweight.

Barley beef was developed in the early sixties to take advantage of cheap barley and to convert it into meat through low-priced calves from dairy herds. When the system was shown to be profitable competition for calves and for barley seriously reduced profit margins. Since then profits have fluctuated with cereal and calf prices.

Cattle on this system have a shorter maintenance period than with most other beef systems, and capital is invested for less time. They do not use land directly but do require a high standard of management.

Survey data indicates that those farmers making the best gross margins have a higher proportion of bulls than those achieving poorer results. It is important to ensure that there is a good market for the stock because of prejudice in certain areas. Many are produced on contract and a premium may be obtained for tender quality and red meat.

Calves may be purchased either before or after weaning. Purchasing after weaning usually results in lower mortality, avoids the necessity to provide labour and specialist

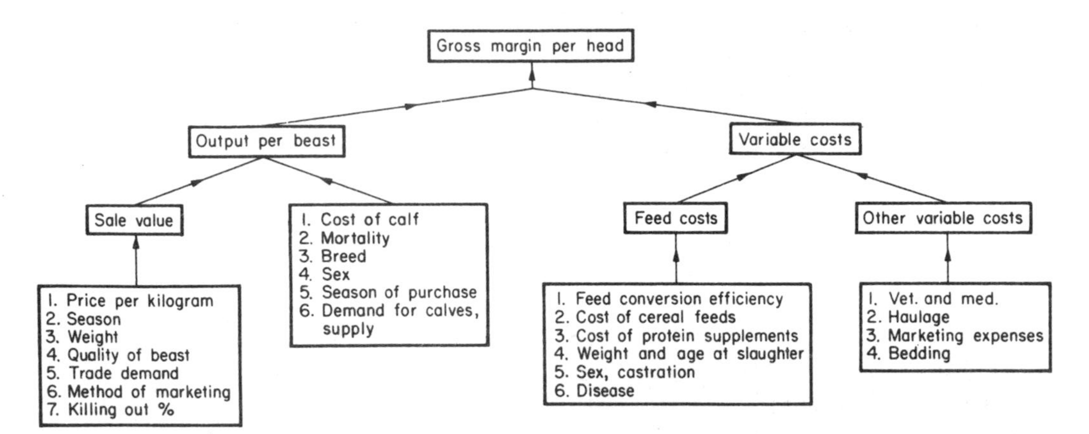

FIG. 79. Factors influencing the gross margin from barley beef.

early rearing facilities, but may produce lower gross margins. The higher price paid for the weaned calf increases the working capital requirement, but it is invested for a shorter period. In some cases more cattle can be handled than if they were bought prior to weaning.

Calf prices vary with supply and demand, quality, breed, sex and season. Demand varies according to predictions for beef prices, availability and cost of capital, and price of barley. The system is somewhat more predictable than many other beef systems which fosters greater accuracy for budgets.

Lighter cattle at slaughter may make most profit, but this must be related to breed and nutrition. Friesian bull calves do particularly well on the system. Early maturing beef crosses and heifers will become overfat at light weights.

Stock making the highest profits have lower feed costs per kilogram of liveweight gain, a higher daily rate of gain, and are slaughtered at younger ages. Care must be taken to check the quality of the barley, and protein levels should be reduced in the final stages.

Disease problems which can seriously upset performance include bloat, acidosis, condemned livers, and bad feet.

Some farmers now employ silage based diets to produce bull beef. Work at The West of Scotland College's Crichton Royal Farm has shown that Holstein-type bulls can produce satisfactory profit margins, but they must gain over 1.1 kg per day. If they do not then the delay in slaughtering increases maintenance costs and results in a significant number of animals which do not grade because of poor conformation.

9.4.5. Low Intensity Systems (Suckler Calf Production)

Suckler calves are produced on a wide range of farm types. Many are born on marginal or hill farms where the fixed cost of land is comparatively low, and the compensatory payments associated with hill suckler cows are obtainable. These payments are a major profit factor on eligible farms.

(i) The Suckled Calf

The price of the suckled calf and the number of calves sold relative to the number of cows in the herd are the major factors influencing profits. The price of the calf reflects its size, quality, and sometimes breed. Size and quality will depend on a number of factors. Date of birth and the breed and quality of sire and dam are the main ones, although creep feed and disease must be considered. Buyers at some auction marts have particular preference for certain breeds.

(ii) Sire and Dam

Sire and dam have to be selected to suit the conditions which prevail on the particular farm and the system of management adopted. In general the cow should be as big as the circumstances will permit and the bull should be selected on the basis of potential for liveweight gain. Milking ability of the dam can influence growth performance of the calf.

Surveys show that in most herds the average calving percentage is 91–94% but mortality reduces this to about 85–89%. Replacements are usually bought for those that die, but this adds to the costs.

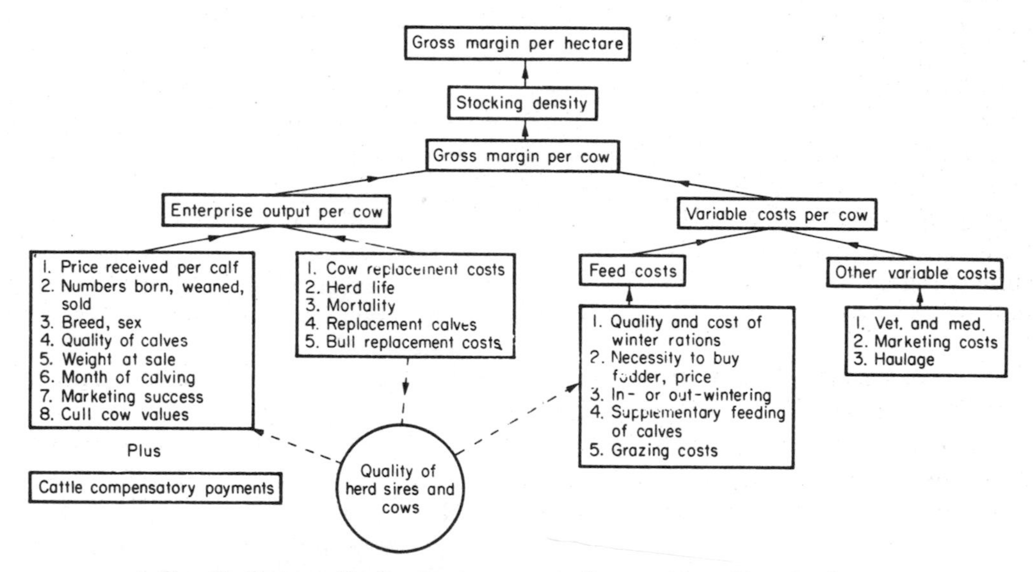

FIG. 80. Factors affecting the gross margin from suckler calf production.

(iii) Calving Date—In-wintering/Out-wintering

Time of calving may depend upon whether the herd is in- or out-wintered, but a major factor will be the quantity and quality of the winter feed which can be produced on the particular farm.

The earlier calves are born the bigger they will be by the autumn suckler sales if they are sold then. In most herds some cows "slip back" so that not all cows calve at the intended time. Irrespective of the period selected, every effort should be made to keep a tight calving pattern if even groups of calves are to be produced and profits maximised. Condition scoring of cows can assist in the management of stock and help foster improved conception.

Calving in some herds takes place in June or July. The aim with most in-wintered herds is to calve at grass in August and September, to minimise disease, before housing in November and December.

Housing avoids winter damage to grass and helps to achieve higher stocking rates. It may permit more cows to be kept, and with the same amount of labour. These benefits have to offset both the housing and additional feed costs. In practice this is not always achieved, and there are many instances of out-wintered herds producing higher margins. Much depends upon feed costs, management, and stock performance.

In-wintering is very difficult to justify if a high proportion of the winter fodder has to be purchased. Even out-wintered herds have been discontinued on many farms because of the need to buy hay.

(iv) Feeding

The feeding of the cows, whether in- or out-wintered, must relate to the work being done by them and to their condition. This is particularly important if they are milking and if they are to be put successfully to the bull.

Early summer born calves begin to rely on feed other than milk in the winter, and their dams may not need such high feed levels as autumn calvers.

Condition must be maintained on out-wintered spring calvers, but nutrition levels and costs can be much lower than for in-wintered autumn calvers, especially if good foggage is available. About 1–2.5 kg of concentrate, depending upon the quality of the basic ration, may be fed to spring calvers from 6 to 8 weeks before calving until there is sufficient growth of grass. The main suckling period coincides with grass growth. One bull to 35 cows may be used in May–June when conception is aided by nutrition levels.

There is usually a close relationship between stocking rate and gross margin per hectare. In lowland areas stocking rates range from 1 to over 2.5 cows/ha. High stocking rates are essential in such areas if the system is to be at all competitive with other enterprises. Working capital requirements increase with improved stocking rates, but return on capital should also be higher.

(v) Marketing

Marketing is just as important as calf growth rates, feed costs, and stocking rates. In addition to creep feeding of autumn born calves during winter, many producers creep feed their stock before the suckler sales. Selection of good even bunches, which are well presented, can considerably help the price at auction. The hill suckler producer is particularly vulnerable at the sales because generally he must sell. If capital is short or costly for the buyers, then trade can be poor. Prices may vary during the day, and the time when a particular group are sold can be important.

Not all sucklers are sold in the autumn. Farmers with sufficient feed sometimes finish the forward ones out of the yards, or keep the lighter ones for the spring store sales or for finishing off grass.

Some producers have contracts to supply suckled calves to farmers, especially in arable areas, who finish them. They are usually valued by an independent valuer. This system fosters improvement because purchaser and seller remain in contact and the vendor gets to know which type of animal, from which bulls and cows, finish best.

9.5. PIG PRODUCTION

9.5.1. Weaning Production (Fig. 81)

(i) Number of Weaners Reared per Sow per Year

The number of weaners reared and sold per sow per year is probably the most important factor influencing the profitability from weaner production. A high proportion of the costs associated with the system are incurred irrespective of how many there are. Numbers will make little or no difference to building costs, labour, feed for the boar and for sows during pregnancy, and interest on the capital invested in the project. A certain number of pigs have to be sold to cover these costs. Any additional weaners can, in one way, be regarded as virtually all profit because they only have to cover their creep feed, a little extra feed for the sow during lactation, a minimal addition to other variable costs, and a fractional increase in the interest on the extra working capital involved.

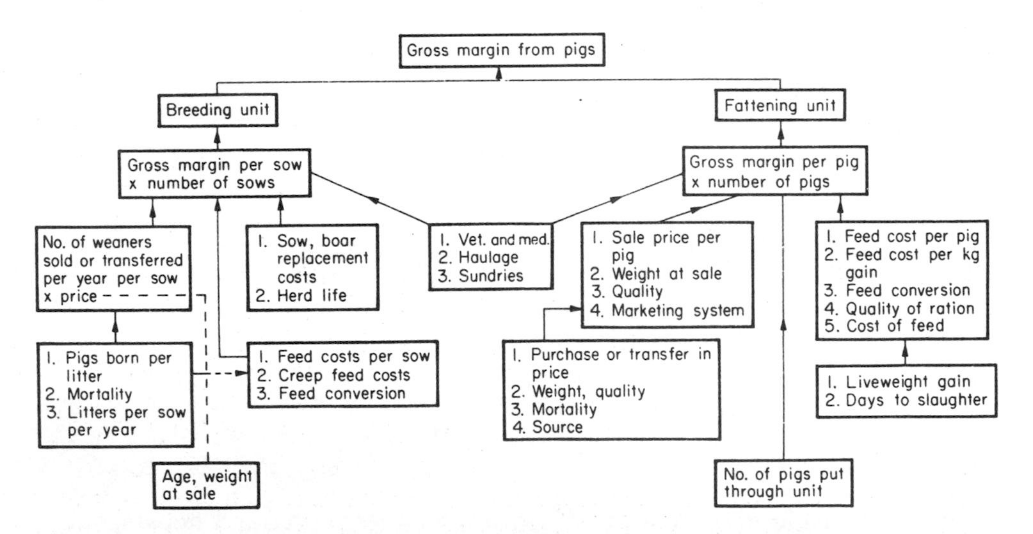

FIG. 81. Factors affecting the gross margin from pigs.

The number of weaners reared per sow per year is a product of the number of litters per sow per year and the number of pigs reared per litter. Many surveys indicate that the average number reared per sow per year is in the region of 17.5 pigs, with just over 10 pigs born alive per litter, and just over 2 litters per sow per year. The best producers rear well over 20 pigs per sow per year.

Mortality can be a serious problem and on a national basis may average 15% of the pigs between birth and weaning.

(ii) Litters per Sow per Year

The number of litters per sow per year is influenced by the age at weaning and by conception. Although many pigs are still weaned at 5–6 weeks there is an increasing trend to weaning into "flat decks" at 3–4 weeks.

The earlier the weaning the quicker the sow can be returned to the boar, but the standard of management required is higher. It is essential to get sows back into pig again as soon as possible since the advantage of early weaning would otherwise be lost both in terms of the number of litters per sow per year and also in the maintenance period whilst the sow is not carrying young.

The use of a specialist service area and the application of good husbandry are important.

(iii) Pigs Born per Litter

The use of cross-bred instead of pure-bred sows and good service management will both promote an increase in the number of pigs born. Typically the advantage of the cross-breds is about 5% in terms of pigs born alive, but because of lower mortality the benefit may be as high as 8% by the time the weaners are sold. However, without appropriate management at farrowing and the practice of good husbandry of both the sow and her litter, any benefits can be lost.

(iv) Mortality

Mortality in some herds is extremely high. Management at farrowing, the use of suitable accommodation, ensuring that the sow has milk, attention to detail, and disease prevention can all help reduce the problem.

(v) Feed Costs per Weaner

The main feed cost is incurred by the sow. The point was made above that the higher the number of pigs reared the lower the cost will be per weaner produced.

The use of appropriate rations, consistent to the condition of the sow and the stage of pregnancy or lactation, can significantly reduce the amount of feed used without adversely influencing results. Whereas some farmers use over 1.5 tonnes of concentrates per sow per year, many farmers have reduced this to 1–1.2 tonnes.

Creep feeding of young pigs is particularly worthwhile. At this stage the feed conversion is very favourable, although the cost of the feed is high. It reduces the dependence on the sow's milk yield once sufficient quantities are being consumed. With early weaning special proprietary feeds are necessary.

(vi) Price Realised per Weaner

Some farmers fatten their own weaners and in this case it is essential to put a realistic transfer value on them so that the financial contribution of the breeding and fattening herds can be accurately assessed.

Sales of weaners can be through auction marts, through weaner groups, or on a contract, either privately or with one of the commercial companies. Some of the latter finance pig enterprises through a contract to provide feed and even in-pig gilts. The weaners must be sent to the respective company and the farmer gets paid for his services.

Some weaner groups have undertaken pig improvement schemes, but not all have been successful. Disease and lack of a stable market can be problems of auction marts, and some form of contract, although restrictive at times, may have long term benefits.

The weight at sale as well as market demand or contract price will affect returns. If he has flexibility in his marketing system the farmer should calculate the optimum sale weight assuming that he has sufficient accommodation. The longer weaners are kept the higher their feed costs. A budget should show which sale weight is preferable, but an attempt must be made to predict whether prices will fall during the additional period that the pigs are on the farm.

Gross margin per sow is very sensitive to weaner prices as well as numbers reared per sow and feed costs per weaner.

(vii) Replacement Costs

Many farmers buy hybrid gilts and tested boars. The benefits to be derived from this depend not only on the price and quality of the animals they obtain, but also on the destination of the weaners. Farmers who fatten their own stock or who have a close follow up with the purchaser will benefit more than those who sell through auction marts.

Herd life is important and many herds do not average more than five litters per sow. The cost of replacement after a limited number of litters is partially offset by the possibility of quicker genetic improvement in some herds and by the fact that very old sows tend to produce smaller litters.

(viii) Records

Good records are essential. Feed records should foster the calculation of feed costs per kilogram of weaner produced and feed costs per £100 gross output.

Sow records should include service records, data to calculate the farrowing index, numbers born and reared, and deformities or other peculiarities.

"Young pigs records" should include birth, 3-week and weaning weights. Where possible follow-up records on feed conversion and grading should be included.

Financial records should include purchases and sales, food costs, and opening and closing valuations.

9.5.2. Fattening Pigs

(i) Main Profit Factors

The profitability of a pig-feeding unit is largely determined by three factors:

(i) Feed costs per unit of liveweight gain.

(ii) Growth rate.
(iii) Grading results.

Each is very important, but an attempt to improve one of the factors can result in deterioration of one if not both of the others.

Figure 82 demonstrates the relationship between energy intake and the efficiency with which it is used.

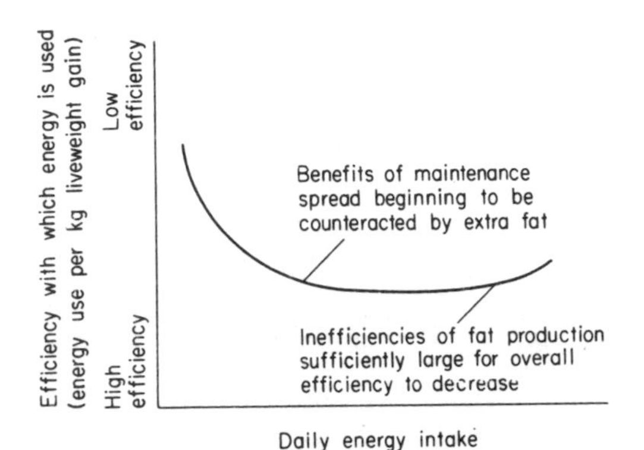

FIG. 82. Relationship between energy intake and the efficiency with which energy is used.

The more a pig eats the faster it grows. An increase in growth rate will mean lower overheads per pig, including housing, labour, and electricity, as well as a reduction in the maintenance loading of the ration. However, as a pig gets older the faster it grows and the more fat it lays down. Since fat is much more expensive to produce in terms of energy, this leads to inefficiencies in feed usage. Fat will also reduce grading premiums.

The farmer must decide the optimum balance between the three factors for his particular circumstances. For instance, bacon pigs are heavily penalised for poor grading, and this aspect must receive priority from the bacon producer. He may restrict the feed intake of his pigs to delay fat retention. A producer of manufacturing pigs, on the other hand, is not so heavily penalised on carcass quality, and his main concern would be to optimise feed costs and growth rate.

Whatever the system of feeding, the pig producer is advised to choose a strain of pig selected for its growth and lean meat potential.

The importance of feed costs is illustrated by the following figures which show feed costs as a percentage of all costs excluding rent and depreciation: weaners 83%, porkers 80%, baconers 89%, heavy pigs 82%.

(ii) Records

Feed records should enable the total quantities of feed used to be ascertained and feed conversion to be calculated for each batch of pigs.

Weights of pigs should be recorded at entry to the unit, at slaughter, and at intervals to monitor performance so as to modify feeding and management if necessary. Disease and mortality records, monthly numbers, records of sales and purchases, and opening and closing valuation numbers should be kept together with grading results.

Financial records should include opening and closing valuations, purchase and sale prices for each batch, and feed costs.

9.5.3. Labour

Labour is a major fixed cost and in common with most livestock units the performance of the enterprise can be seen to change with a change in pigman. The right man is therefore worth extra money.

Very wide ranges in numbers of pigs handled per man are found in practice and much depends upon the buildings and labour-saving devices.

In a fully automated unit, with slats, one man should be capable of looking after at least 1,000 pigs in each batch. In a Danish house, where there is less automation, this may be reduced to a batch of 600, especially if home feed mixing is undertaken. There are many farms where these figures are not reached and others where they are exceeded.

With breeding units one man should handle 90–120 sows but some fully automated units have much higher numbers.

9.6. SHEEP PRODUCTION

9.6.1. Sheep Production Systems

There are very many different systems of sheep production in Britain and a proliferation of sheep breeds and crosses. Figure 83 shows the pattern of stratification in which large numbers of sheep produced on the hills and uplands move to lower ground for finishing.

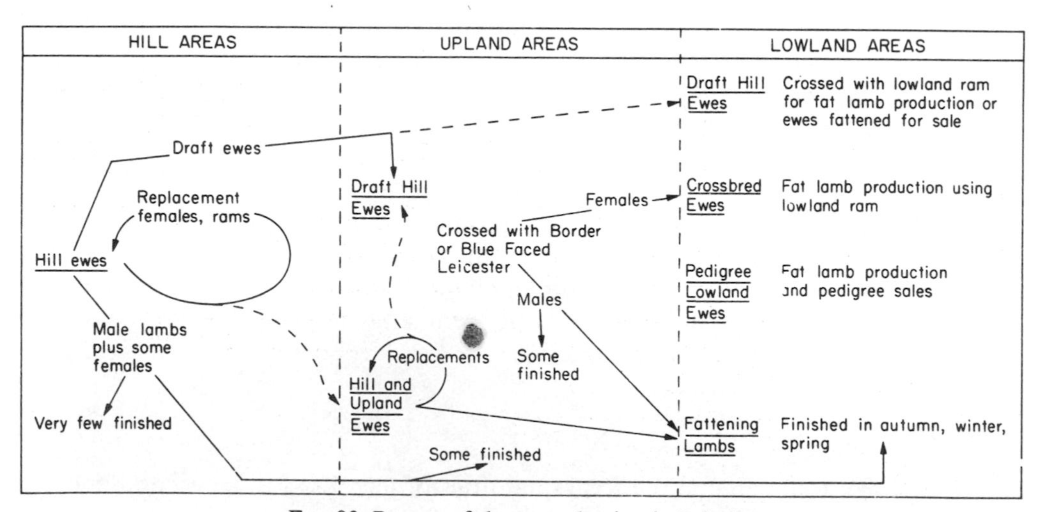

FIG. 83. Pattern of sheep production in Britain.

One of the main difficulties in describing a system of production for hill and upland areas is that the reader may not have the correct concept of topography, climate, suitability of the respective farm for the production of grass and winter fodder, and many other factors such as distances to markets. Hill farms can range from those capable of carrying several sheep to the hectare to those requiring over a hectare to the sheep. Here it is only possible to give some of the main profit factors and it must be appreciated that there can be wide variations with prevaling circumstances.

9.6.2. Hill Sheep Production

(i) Compensatory Allowances

Many farms rely heavily on the Hill Compensatory Payments which in some cases can be as high as the net farm income. This is paid on the number of "specially qualified" ewes and gimmers which must be in regular ages so that there are at least three successive age groups. It is limited to ewes or gimmers for each hectare of qualifying land available for their maintenance, and it is important to maintain appropriate numbers so that they are eligible on the qualifying day.

(ii) Number of Lambs Sold

Although wool and cull ewes make a contribution to output, the most important profit factors are the number of lambs sold and their price. The higher the number sold for a given set of resources the greater the spread of labour costs, rent and the costs of maintaining the ewe flock (Fig. 84).

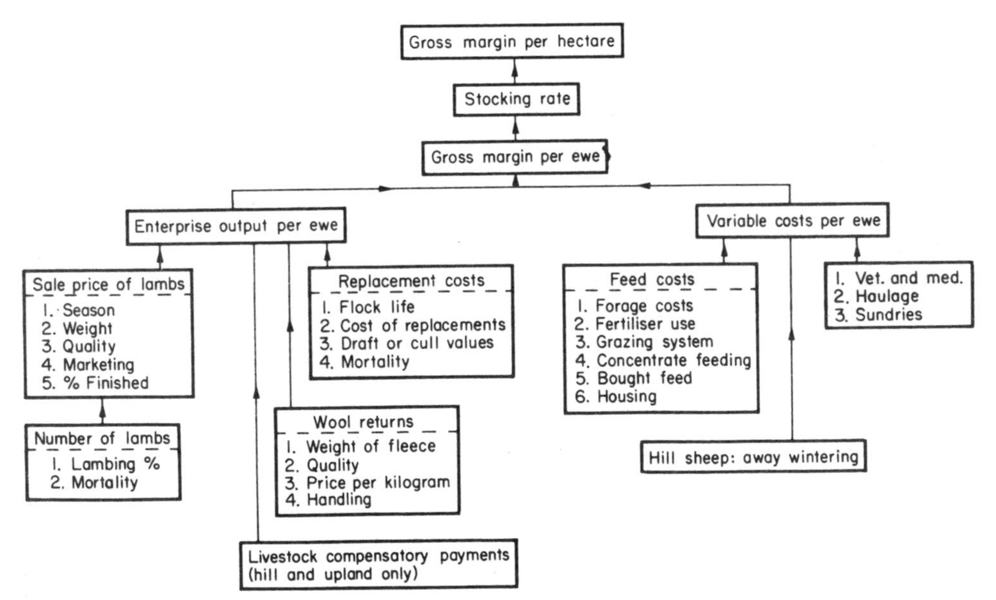

FIG. 84. Factors influencing the gross margin per hectare from lamb production.

Lambing percentage and mortality require careful attention. On the poorest hill farms 80% may be considered acceptable and with very severe winters 40–50% is possible. As the quality of land improves, 90% becomes a target and, as land and management improve further, up to 130% or more becomes attainable.

Results obtained very much depend upon nutrition and management at key times of the year, at tupping, and during lactation. The two-pasture system, in which an area of better low ground is fenced and improved for use at tupping, lambing, and for suckling ewes with twins, whilst the hill ground is used for the rest of the year, has improved productivity on many hill farms. In the worst hill areas this is not possible because there is insufficient lower ground, and the hill is not accessible except on foot or by air.

Scanning to assess foetal numbers has proved of value especially on hill farms which have some in bye land and on upland holdings. Apart from identifying barren ewes its main benefit is that it establishes which sheep are carrying more than one lamb. This has two benefits to the farmer. First, it enables him to select those ewes which require extra feeding prior to lambing. Second, because he can establish which are carrying twins, he can be more confident in providing extra flushing prior to tupping to stimulate twinning. Although scanning is particularly valuable on hill farms and upland farms with lambing percentages over 130 its merits have also been established on flocks with below 100% lambing.

The importance of lambing percentage on numbers sold is emphasised by the following example. In all cases 100 ewes producing four crops of lambs is assumed, and 27 ewe lambs are retained for replacement to allow for mortality.

Lambing (%)	Male, wethers sold	Ewe lambs retained	Ewe lambs sold	Total lambs sold
100	50	27	23	73
90	45	27	18	63
70	35	27	8	43

(iii) Price of Lambs

Most hill lambs are sold at special sales in August and September, and are bought by lowland farmers. Good presentation is essential and great care must be taken to draw out bunches of lambs of even quality. Poor lambs, or those with undershot jaws, can spoil both the looks and price of a pen. Groups should be in marketable numbers. At any market the biggest lambs appear to be two to three times as big as the lightest, and this is reflected in the price.

In some areas some farmers retain second and third draw lambs to fatten indoors, but the profits derived from this depend upon the price of store and fat lambs, the price of barley, and the housing. On the best hill farms a few singles or even twins may be sold fat off grass, and those with sufficient low ground may grow special forage crops to finish a proportion.

Lamb marketing groups have met with mixed success. The best ones have produced close links between the hill farmer and the fattener to mutual advantage.

The hill sheep farmer is very vulnerable and can be seriously affected by changes in the lamb market. Efforts have to be made to avoid the fluctuations in price which occur at auction marts by either finishing some at home or by having a contract with a lowland farmer.

(iv) Replacement Costs

The target for most hill farmers, which is not always achieved, is to produce four crops of lambs off each ewe and then sell her, sound of feet, teeth and udder, to a farmer on better land for further breeding. In severe winters serious ewe losses can occur. This means that less lambs are born and an increased number of ewe lambs have to be retained for replacements. Since there is a delay before the lambs become mothers themselves, it can take several years to recover.

Housing is generally too costly although it has a place on some farms and can result in improved productivity.

Most replacement females on hill farms are home-bred because of hefting and disease problems. Many ewe lambs are away-wintered on lowland farms from November to April, which adds considerably to costs. Housing of ewe lambs, depending on both building and feed costs, is economic for some, and appropriate budgets should be undertaken. If all feed has to be purchased and the housing is not available, or cannot be constructed with home labour and second-hand materials, it may not be feasible.

Good tups are a means of introducing new blood of a superior type. Much local prejudice exists but, for example, the Newton Stewart type of Blackface may be superior to many other local strains, and has attributes such as the suitability of its coat for inclement weather.

(v) Wool

Synthetic materials have replaced wool in many garments, but the wool cheque can produce more than 10% of the output from a flock. Fleece weights, qualities, and prices vary, but it is surprising how many fleeces are ruined by poor handling and contamination on the farm.

(vi) Feed Costs

Feed costs for hill sheep are extremely variable. Some receive no supplementary feeding at all, and the land they graze has no forage costs. Many farmers have seen the benefit of feed blocks and a little supplementary feeding, although at times when this feed is needed transport of the materials can be a problem. Where all hay and concentrates have to be bought, costs can be high.

Two sward systems and other improved methods cost money. The essential point is to ensure that the extra grass produced is effectively utilised and, through good management, results in much higher productivity.

(vii) Other Variable Costs

Haulage and veterinary and medicine costs can be high on some farms especially if ewe lambs are away-wintered and diseases present on the farm require an extensive vaccination programme.

(viii) Fixed Costs

Fixed costs must be kept to a minimum on hill sheep farms because of the low output. As suggested above, housing is difficult to justify in many cases, machinery has to be very limited, and labour used effectively. Extensive methods may be practised on some holdings which can be so extreme that sheep are only gathered twice, for clipping and for drawing lambs for sale. On other units fencing and even housing are employed, so that each man can look after more ewes and achieve higher productivity.

9.6.3. Upland Sheep Production

(i) Compensatory Allowances

Ewes and gimmers in upland areas qualify for the lower level of compensatory allowances if in a flock composed of three or more age groups and they are a normal part of the farming enterprise. It can also be available on "flying flocks" of draft ewes kept for lamb production.

(ii) Production of Cross-lambs

Many farmers keep pure bred hill and upland sheep flocks and the factors affecting their profitability are similar to those described for hill flocks. The main difference is that upland farms have better land which can make it easier to provide appropriate nutrition at critical times, with attendant benefits in lambing percentage and the possibility of finishing more lambs.

Some upland farms produce cross-bred breeding sheep for sale to lowland farmers. Those ewe lambs or gimmers suitable for breeding frequently realise more money than the wethers, which are either finished or sold to farmers on lower ground. Usually the contribution to income of the females is over 50% more than that of the males.

9.6.4. Lowland Fat Lamb Production

(i) Gross Margin per Hectare

The fixed costs associated with sheep enterprises are less than for most other farm activities. No enterprises should be selected entirely on the basis of gross margin per hectare, and sheep are no exception. They can dovetail into some businesses using up surplus resources, or land which is difficult to cultivate.

Success in sheep farming, as with any other enterprise, depends upon the standard of management. The best farmers can achieve gross margins per hectare considerably above those obtained by cereal producers. This requires not only good sheep husbandry, but also good grassland management.

The data in Table 10 indicates the range in physical results often shown by surveys for spring lambing flocks. The farms are categorised by gross margin per hectare and the information refers to each 100 ewes.

Table 10. *Physical results from fat lamb production*

	Top one-third	Average	Bottom one-third
Number of lambs born	178	160	145
Number of lambs born alive	166	150	133
Number of lambs reared	160	140	123
Ewe deaths	3	4	5
Number of empty ewes	4	5	6
Percentage of ewe lambs in breeding flock	20	20	20
Concentrates—ewes (kg)	43	45	46
—lambs (kg)	8	8	8
Summer grazing—ewes/ha	20	14	8
Overall grassland—ewes/ha	15	11	6
Grass and forage—ewes/ha	14	10	5

(ii) Lambs Sold per Ewe

Standards of husbandry, management, feeding, genetic factors, weather (or housing) and disease control are the main factors determining the number of lambs sold. Most surveys show that the superiority in gross margin per ewe of the top one-third of producers compared to average is due to higher lamb numbers sold.

Many poor results are related to a lack of attention to detail, particularly at tupping, in late pregnancy, and at and shortly after lambing. Excessive feeding, however, can bring its own problems, such as very large lambs and a number of ewes putting out their "lamb-beds".

Condition scoring of ewes with regular checks on their progress during pregnancy is strongly advised. If only a limited amount of housing is available scanning can identify ewes carrying multiples. These can then be housed and given special treatment.

Some producers can obtain 200% or more lambs born, but only the best rear more than 180%. In some flocks as many as 25% of the lambs born alive die within a short period. Many of these flocks have low numbers born.

(iii) Prices of Lambs

It is important to produce lambs of good quality, which will give carcasses at a weight which is in demand. At the time of writing lambs attracted a variable premium, provided they reached stringent grading standards, and this was a significant proportion of the income from selling lambs for slaughter.

Good lambs will kill out at about 50%, and whilst the lambs must have the right finish they are still relatively immature. Slaughter weights are usually at about one-quarter of the combined weights of the parents, depending upon sex, and taking them beyond this often leads to excessive fat. Seasonality can have some effect on the weight demanded, however, and there are major differences in price between spring lamb and autumn lamb because of supply.

Prices rise as winter progresses, and some people argue that if the lambs are costing virtually nothing to keep, because they are using grass and labour which would not otherwise be used, it pays to carry them on to a heavier weight. This needs to be examined carefully, but the carcasses must still comply with the quality demanded. Some producers will have sold most of their lambs in the summer before the autumn price fall and, subject to disease risks, may have bought in hill lambs to finish in the winter.

(iv) Stocking Density

In many surveys the major difference in gross margin per hectare between the top one-third of producers and the average is due to better stocking density. The importance of correct grassland management, including fertiliser use, grazing system, and stocking is therefore emphasised. Figure 85 shows the relationship between performance per animal and per hectare with changes in stocking rate, all of which must be related to nitrogen use. If individual performances are affected in order to attain maximum productivity per hectare the impact of this on the delay in finishing date per lamb and any change in lamb prices must be considered.

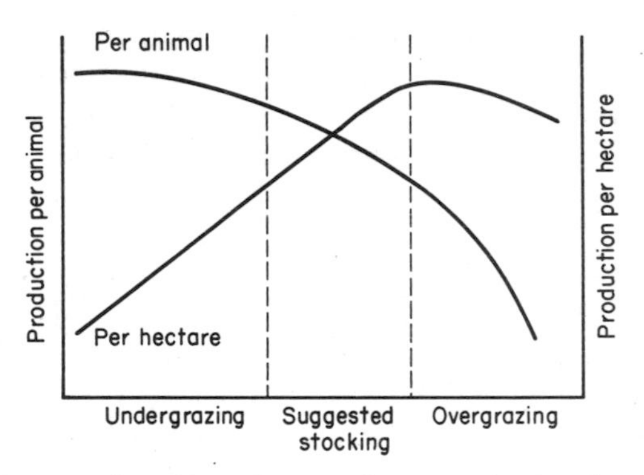

FIG. 85. Impact of stocking of production per animal and per hectare.

Mixed grazing of sheep and cattle may also alter the results so that the decline in animal performance with increased stocking per hectare is at least temporarily arrested.

(v) Replacements

Increased ewe life and breeding from ewe lambs can help increase returns especially when replacements are particularly costly. High lambing percentages should not be expected from ewe lambs.

Alternative sources of sheep are worthy of examination. On some farms purchase of draft ewes at the right price, which are tupped, and sold with lambs at foot in the spring before they become competitive with other stock for grazing, can leave satisfactory profits.

(vi) Feed Costs

A high proportion of the feed costs are derived from forage production. These can be justified if good stocking rates and high liveweight gains per hectare are obtained. Some of the best farmers can achieve higher productivity in terms of lambs sold per ewe using the same, or even less concentrates, than producers who attain average profits. This is a reflection of better grass and stock management.

With early lambing, around Christmas and the New Year, concentrate feeding has to be increased. Lambing percentages may be lower than with spring lambing flocks. However, the use of hormones coupled with sponges to synchronise oestrus not only helps improve fertility, but also concentrates lambing over a shorter period. The latter fosters earlier marketing and reduces the labour input devoted to lambing.

High levels of creep feeding for the lambs are necessary. Early lambing can only pay if the lambs are sold at peak prices, and preferably all should be finished by mid-May. Care has to be taken with slightly later lambing to ensure that if the high feed costs are incurred, the lambs are ready for sale before prices fall too far.

One benefit of early lambing is that the lambs are weaned at an early stage and the dry ewes can be tightly stocked on a restricted area, or put on poor land, so that they do not compete with summer grazing stock.

9.6.5. Finishing Store Lambs

The success of any store lamb finishing system is very dependent upon the feeder's margin. Store lambs can be finished using many different feeds and feeding practices, including grass, arable by-products, swedes, turnips, rape and in some cases with the addition of concentrates or, if indoors, primarily on concentrates.

Farmers who grow crops such as swedes specially for finishing lambs tend to be more or less committed to buying stock to eat the crop, even when prices are high. In any case it is essential to realise the full potential of the crop. With swedes 4,000 lamb days per ha can be achieved, at about 70% utilisation, whilst on dry land up to 5,000 lamb days can be obtained with about 80% utilisation.

Farmers who rely on grass for finishing have more flexibility in purchasing when prices are high, although they may want some stock to eat off the grass and prevent winter kill.

Animals must be bought at the right price, but they must also suit the feed and other factors prevailing on the specific farm. Short keep lambs are the most expensive. They may be finished off autumn grass. The cheapest, long keep stores can be on the farm well into the new year, possibly requiring concentrates to finish them, and perhaps being an embarrassment when grass growth for spring grazing of other stock commences. There is a range of stock on sale in the autumn of different sizes between these two extremes.

Prices for finished animals usually rise as winter advances to spring. If they fit into the farming system, store lambs of lighter weight, but which will finish when the prices are high, can produce acceptable carcasses without being too heavy and fat, and produce good profits.

9.7. CEREAL PRODUCTION

9.7.1. Factors Affecting Profitability

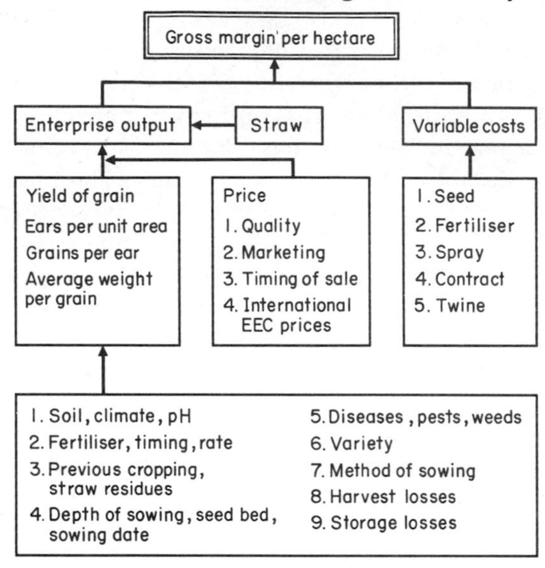

FIG. 86. Factors influencing the profitability of cereals.

9.7.2. Yield of Grain

(i) The Components of Yield

Cereal yield is a product of:

Ears per unit area × Number of grains per ear × Average weight per grain

Many of the factors which determine the components of yield can be influenced by husbandry and management. The difference between farmers who make most profits and those with average results is extra yield and better quality. These can be achieved without a proportionate increase in costs.

The ear number is determined in the early stages of growth, the number of grains per ear in the middle stages, and the grain weights, within a variety, during the ripening stage.

Ear numbers are firstly dependent on a satisfactory plant population and even distribution of plants. Good seed, good conditions for germination, and protection from seedling diseases and pests are all essential. An excessive plant population may lead to sparse tillering or to a high proportion of blind tillers due to inter-plant competition, particularly for light and water. A low plant population cannot be fully compensated for by extra tillering since many ears will be late maturing, possibly leading to harvesting problems and green grain. Higher seed rates may be used when conditions are judged to be unfavourable and seed rates for winter wheat may be increased if sowing is delayed during autumn.

Tillering can be influenced by nutrient supply, particularly of nitrogen and phosphorus. On phosphate responsive soils this element can be more efficiently utilised by placement or combine drilling into the soil so that the young seedling roots can reach it rapidly.

(ii) Sowing

Drilling seed is generally more satisfactory than broadcasting, but broadcasting is usually quicker. Broadcasting is sometimes used where timeliness of operations is important, i.e. where drilling is delayed because of adverse weather.

Where several fields have to be sown it is sound practice to concentrate resources on one field at a time and to get each field finished as quickly as possible. Many problems can arise when the weather breaks in the middle of the process of sowing a field.

Seedbeds may be prepared using conventional ploughing followed by discs or tine equipment and harrowing. Alternatively, reduced cultivation techniques may be employed using chisel ploughing and tines. Direct drilling is a further possibility.

The incorporation of straw from previous crops can be a problem and in these cases ploughing gives best results.

For optimum sowing conditions a firm seedbed is desirable. A soft seedbed leads to uneven sowing depths resulting in delayed emergence, weak plants and an uneven plant distribution. The speed with which a drill is operated can, in some situations, influence the depth of sowing.

Spring cereals require fine seedbeds whereas a degree of roughness will help protect autumn cereals from winter weather. The effectiveness of soil acting herbicides can, however, be reduced by clods of soil.

(iii) Order of Sowing

The general recommendation is to sow oil seed rape before winter cereals, and winter barley before winter wheat. It is particularly important with rape and barley that good root development has taken place before winter. Spring wheat should be sown before spring oats, and spring oats before spring barley.

Land with a stable soil structure should be sown before that which is less stable, e.g. prone to capping. In particular, crops grown after a ley seldom suffer from soil structure problems and can be sown early in the spring with reasonable safety, subject to the soil being dry enough to work.

Patience must be exercised before spring sowing to ensure satisfactory soil conditions. The soil must be sufficiently dry to the full depth of ploughing so that it is not sticky. If sowing is delayed by bad weather some fields may have to be sown in less than ideal conditions, but this should not be risked until it is getting late for the particular crop in the specific locality.

Seedbeds for autumn sowing, prepared in late August or early September may dry out excessively if over cultivated. Consideration should be given to reduced cultivation or perhaps rolling after each cultivation to reduce moisture loss.

(iv) Grain Numbers

Nearly all cereals are capable of producing more grains per ear than they actually do. Ear primordia are formed at about the time when tillering is complete, and potential grains are thought to atrophy up to the time of ear emergence and fertilization due to shortage of carbohydrate. Effective photosynthesis is therefore important. This in turn depends on healthy leaves growing on tillers which produce ears. Quick establishment of a canopy of leaves is affected by nitrogen supply. Effective photosynthesis will be reduced by competition for light from weeds and blind tillers. It will also be reduced by leaf damage caused by foliar diseases and damage, or scorching, from sprays which have been incorrectly used.

Water stress will slow down photosynthesis, but severe water shortage does not normally affect cereals in the United Kingdom at this stage unless soils are shallow or they have been allowed to dry out after lack of consolidation.

Plant growth regulators which reduce apical dominance have been shown to reduce blind tiller numbers if applied during tillering.

(v) Grain Size

This varies between varieties and there may be a maximum size to which grains of a given variety will grow.

Undersized grains may be produced by a number of factors. Water stress is a common cause reducing photosynthetic activity and translocation rate. Water stress may be induced by drought; shallow panned soils; excess nitrogen use; root (take all) or stem (eyespot) disease; lodging; or aphid attack. Small grains are also produced by secondary tillers and in crops prematurely harvested.

Disease on the flag leaf in wheat or on the ears of barley will reduce photosynthetic production and hence grain filling. This problem is also enhanced by a shortage of potassium.

Another hazard at this stage is the possibility of aphid attack, but it is comparatively easy to control by spraying.

(vi) Crop Protection

The cheapest form of crop protection is disease/pest resistance. Varieties with improved resistance are continually being developed. Crop diversification with several varieties (Diversified Groups) should be part of a planned, integrated protection programme.

Rapid developments are taking place in the development of crop protection chemicals, and materials using different modes of action should be used sequentially to reduce the build up of resistance in the disease population.

With the increase in cost and complexity of crop protection chemicals many intensive arable farms now employ crop consultants to manage all aspects of the weed, pest and disease problem. It is essential to assess the cost/benefit of crop protection in each situation.

(vii) Harvesting

Shedding, head losses, laid crops, secondary growth, weeds, and other factors can reduce the grain which is actually harvested. Selection of suitable varieties, appropriate fertiliser treatment, and the use of straw length inhibitors can help reduce the chances of such losses.

Very large quantities of grain can be lost because of incorrect combine settings, and these losses may be greater with increased forward speed.

A problem can arise when several fields of grain are ready at the same time. There are varietal differences in ripening date, and knowledge of these may help to reduce the problem. Cereals grown after leys will tend to be later, because of increased organic nitrogen, than cereals after a previous grain crop.

If the crops are all ready at once it is advisable to harvest those fields first in which quality of grain for sale is of prime importance for seed, malting, or for milling wheat. Lodged crops, even if slightly unripe, varieties subject to head loss and shedding, or which are prone to laying, should also be considered. Where wild oats are present these fields should be left till last to avoid the possibility of transfer. The herbicide glyphosate can be used pre-harvest to dessicate weeds (especially grass) in badly infested crops. This aids crop drying and improves harvesting efficiency.

Maximum overtime is justified if conditions are right, but early starts in the morning will add to drying costs. The book of Genesis promises a seed time and harvest once a year. Farmers should realise that it does not promise them twice a year.

(viii) Drying, Storing

Grain loses weight on drying and all yields should be quoted in relation to their dry matter. Dried grain is usually stored at 14% moisture to reduce respiration, overheating and spoilage.

The following formula can be used to establish weight loss on drying:

$$X = \frac{W1\ (M1 - M2)}{(100 - M2)}$$

where X is the weight loss, $W1$ is the original grain weight, $M1$ is the original moisture content, and $M2$ is the final moisture content.

Dried grain can take up moisture from the air in humid weather. Precautions against storage losses should be taken, such as cleaning, fumigation, and sealing of possible rodent entry points before harvest. Artificially dried grain should always be cooled to below 15°C before storage to deter insects. Methods of moist grain storage have advantages in some situations, but are unsuitable for seed grain, malting barley, or milling wheat.

9.7.3. Price of Grain

Prices received for grain can be influenced by grain quality and marketing.

(i) Grain Quality

Good samples of grain are those which would conform to the EEC intervention requirements in terms of specific weight, cleanliness, moisture, and damage, and are suitable for the use to which they are to be put. Barley can frequently be sold for malting at a premium compared to that for feed. Varietal selection and restrictions in the use of nitrogen to ensure that the nitrogen content of the grain is such that it qualifies as a malting sample can reduce yields. There is always a risk that it will not be suitable, and on rare occasions there is no premium. All these factors must be considered in budgets.

Wheat of milling quality can also attract a premium. Seed grain realises higher prices but is subject to inspection during its growing period.

(ii) Time of Marketing

In theory grain prices are lowest when there are ample supplies being sold off the combine, and get progressively higher as the winter advances. This pattern can be upset by a considerable number of farmers delaying sale in the hope of higher prices and then selling together to upset the supply and demand situation.

Anyone contemplating a new grain drying and storage unit must budget carefully to calculate the depreciation cost, interest on capital invested in the stored grain and the capital involved in the unit, handling and drying costs, and losses in weight of grain through drying. The possibility of joining a grain group with central drying facilities should be considered, and it might be the only justifiable approach to the problem for some.

Grain can be sold by forward contract, but it may be impossible to obtain a contract for more than a limited period ahead, or if the contract was made prior to harvest there might be difficulty in complying with the quality requirements. Disputes over contracts can also arise.

An alternative is to consider the futures market. Brokers in London will act for farmers and provide legally binding contracts for up to 10 months ahead. The broker is informed of the number of tonnes for sale, but might require an initial payment and, if grain prices rise, even a subsequent payment. These are returnable, but they act as protection for the broker to prevent the farmer from defaulting.

The farmer must now follow the futures market. For example, if in October the March "futures price" is high enough for him he telephones the broker to say: "Sell a March futures contract for X tonnes at £y or better", X standing for the quantity he wishes to sell and £y the futures price, the farmer now has a guaranteed price for his grain.

In March the grain can be delivered to an authorised grain futures store, if it is of appropriate quality, by prior arrangement with the storekeeper. A charge is made for each contract and a further charge for handling. Payment is received in seven days.

Alternatively, as March approaches the farmer can endeavour to sell his grain to anyone at the best price available. If he decides to sell he must, before actually committing himself to the sale, instruct the broker to buy a futures contract for the same amount of grain.

Assume that the grain future was sold in October at £100 per tonne but that in March the futures price, and the price the farmer sold his grain for, were both £90. Having sold a contract for £100 and bought one for £90 he receives £10 from the broker without the grain having gone off the farm. In addition he gets the price which he actually sells the grain for, i.e. £90, giving a total of £100.

If the futures price for March was £110 per tonne, i.e. a rise rather than a fall, and he sold the grain for £110 locally, he would have to pay the broker an extra £10, which represents the difference between the futures he bought and the ones he sold. He actually receives £110 per tonne for his grain so that he ends up with a net figure of £100 per tonne.

In practice the broker makes a charge for services, but this is a comparatively small percentage.

9.7.4. Variable Costs

Variable cost control should always be advocated, but this must be related to output. Saving a few pounds by using inferior homegrown seed, or seed of moderate quality or variety, can cost more than it saves through loss in yield. Equally, failure to use optimum fertiliser dressings, or justifiable sprays, can be false economy.

Soil analysis to help ensure the correct use of fertilisers, buying in bulk where possible, cultivations to control weeds, and similar practices, all fall into the realm of legitimate cost control.

The essential point to remember is that efficiency in cereal production depends upon husbandry and management, and how effectively the factors of production, e.g. seed and fertiliser, are used to produce grain. Again, as was pointed out for other enterprises, many of the costs of production are incurred irrespective of yield. Marginal additions to yield once these basic costs have been covered can add significantly to profits. This is probably more true with crops than is the case with many animal enterprises. Additional weaner pigs require some extra food, but much additional yield from crops can be attained by such things as timeliness, which may not cost anything extra.

9.8. POTATOES

9.8.1. Yield

(i) Early or Maincrop

With early potatoes comparatively low yields can be acceptable if the price per tonne is high. Crops bulk quickly provided there is sufficient moisture, and prices fall rapidly as supplies increase. Suitable soils and climate are essential for early crops.

The objective of maincrop potatoes is to obtain maximum "economic" yield with good quality. Tuber growth is largely a reflection of photosynthesis and both water and nutrient availability. Yield is greatly influenced by husbandry control of these or their related factors, but the attainment of the maximum "economic" yield starts with the "seed" tubers.

(ii) Seed

Good quality, healthy seed is essential. Potatoes undergo a period of dormancy after harvest irrespective of storage temperature. After this chit or sprout development is related to temperature and starts in most varieties at about 4°C.

Sprouting effectively advances the ageing of tubers and the term "physiological age" is used to describe it. The length of the longest sprout and total sprout length per tuber have been shown to be directly related to the number of day degrees above 4°C that the tuber is subjected to after dormancy ends.

The greater the physiological age at planting the: (a) earlier plants emerge; (b) earlier tubers are initiated; (c) fewer tubers set in most varieties; (d) less the final plant size in some varieties, therefore reducing bulking rate; (e) more the susceptibility to water stress; (f) higher the optimum nitrogen use; (g) earlier senescence in most varieties.

For early lifted crops the greater the physiological age at planting the greater the yield. A strong apical sprout is required rather than a proliferation of sprouts. Sprouting brings forward "first lifting" date which is so critical to price. However, although there are varietal differences, if the physiological age of maincrops is advanced by sprouting, yields can be reduced because smaller plant size and earlier

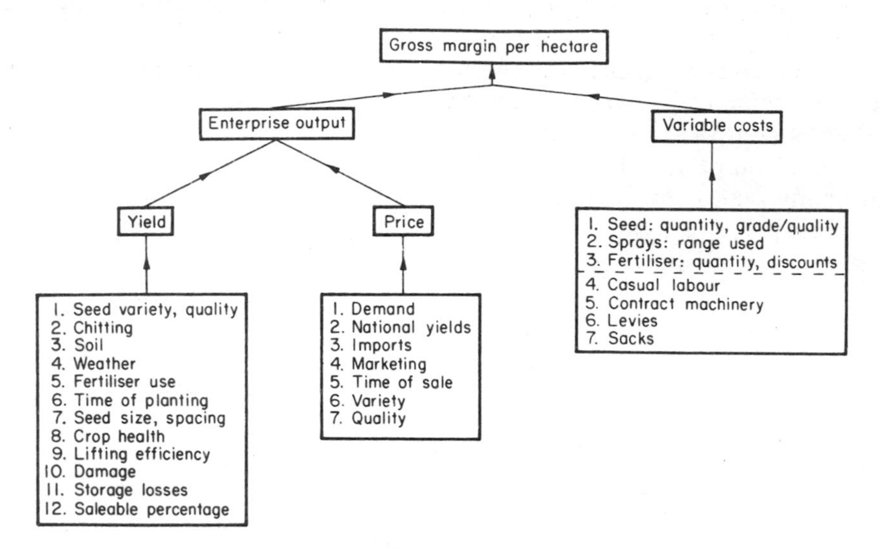

FIG. 87. Factors influencing the gross margin per hectare from potatoes.

senescence results, especially if soil moisture is limiting. A short period of sprouting to produce several mini sprouts per tuber is therefore recommended for maincrops to ensure that the sets are viable and growth starts immediately after planting. The latter point is especially important in years when planting is delayed by bad weather.

(iii) Seed Rate—Maincrops

The number of seed tubers or sets planted per hectare multiplied by their average weight determines the seed rate. There are complex interactions between set numbers, average set size, variety, numbers of sprouts per tuber and spatial arrangement which influence optimuim seed rate and both yield and size of tubers eventually produced by the crop.

The smaller the seed the higher the number of sets required per hectare, but this is not a proportional relationship. The optimum seed rate is less for small sets than for large so that, with the notable exception of one or two varieties, smaller sets are financially more attractive for ware production than large. It is recommended that because of varietal differences in the shape of sets, which influences the number of sets of the same weight going through a given riddle size, the number of tubers per 50 kg should be counted to determine seed size.

It might be considered that the optimum seed rate would be that which gives maximum yield of tubers of the desired size. However, as maximum potential yield is approached the diminishing return of ware yield to increased seed rate is such that the marginal value of the extra yield does not cover the cost of the additional seed. Therefore the cost of the seed planted must be related to the value of the ware crop it produces. Invariably the seed rate which is most profitable is below that which would give maximum saleable yield.

Tables can be obtained by farmers which give the optimum number of sets per hectare for different varieties according to seed size (number of sets per 50 kg) and the ratio of the cost of seed to the projected value of the ware crop.

Farmers producing only early or seed potatoes select larger sets and plant at higher rates per hectare than maincrop producers.

(iv) Planting

Many farmers prefer to use 90 cm row widths because this speeds up planting and harvest compared to 75 cm drills. With the wider spacing there is less damage from tractor wheels and less clod formation. It is also possible to make ridges with shallower depths of tilth.

The grower should aim to produce a comparatively fine tilth, without clods, since this helps the effectiveness of residual herbicides in addition to facilitating mechanical harvesting.

Ideal conditions do not always prevail. Every effort should be made to plant main crops before mid-April, because after this yield losses will be significant. However, planting can be too early if the soil is cold and wet. When conditions are right overtime is justified. Autumn ploughing and winter frost moulds can ease soil preparation. The date to plant earlies depends upon the district, although most are planted in February or early March.

The depth of planting for earlies is usually less, 5–8 cm, than for maincrop, 8–13 cm, since this speeds up growth. Depths for maincrops depend upon whether subsequent cultivations are intended, when they are shallow, or if weed control is to be by herbicides, when ridges are deeper. Greening of tubers must be avoided.

(v) Fertiliser

Fertiliser applications should promote both growth and quality at economic cost. Nitrogen can help encourage growth, but excessive quantities can lead to large, hollow, or cracked tubers. Phosphate in water soluble form is particularly useful since few soils release phosphate quickly enough for yield potential to be realised. This is particularly the case in relation to bulking of early crops.

The optimum quantity of fertiliser is related to a number of factors which influence the achievable yield. These include the type of soil, rainfall and moisture retention. The existing nutrient status of the soil is also important.

Recommendation (kg per hectare)

	N Soil N status			P2O5 Soil P status			K2O Soil K status		
	Low	Moderate	High	Low	Moderate	High	Low	Moderate	High
Early	200	150	100	200	150	100	200	150	90
Maincrop (expected yield 35 t/ha)	200	150	100	200	150	100	250	200	150
Sandy soil	+25	+25	+25						
Average rainfall >800 mm	−25	−25	−25						

(vi) Weather

Weather can significantly influence yields. Sunshine and moisture are both required at the right time of year. Crops on light soils suffer most in dry weather. Potatoes have a shallow root system and large leaf area for transpiration. Under dry conditions 30 mm of water can increase yields by 3 tonne per hectare. Early potatoes are often irrigated as soon as they reach 130 mm growth height. However, irrigation should be determined by the soil moisture deficit and used when it is limiting to production. This will vary not only with rainfall, but also soil texture.

It has also been established that the number of tubers can be increased by irrigation at tuber initiation. Afterwards it helps maintain leaf area and function whilst reducing irregular shaped tubers.

(vii) Weeds, pests, disease

Weeds compete for light, water and nutrients and must be controlled especially in the early stages of growth. A wide range of effective herbicides is available.

Pests Potato-root eelworm, aphids and slugs can cause problems. Nematocides, and potato varieties which are resistant to specific pathotypes, are available to control eelworm, but good rotational practice is the only long term control method. Early potatoes are less prone to this problem because they are lifted so early.

Direct damage by aphids can reduce bulking in ware crops in southern areas. The spread of potato virus disease by aphids is of particular concern to seed producers. For this reason high quality seed production is restricted to northern areas where aphid migration is later and less severe.

Slug damage can be a particular concern in some cases unless control measures are exercised.

Blight control is necessary when warm, humid conditions prevail. Warnings are issued in the press, and on the radio, when the Beaumont period occurs and spraying is advised.

Storage disease of potatoes can also be a problem and must be prevented. Farmers can face prosecution by the Potato Marketing Board if they sell potatoes damaged by slugs and disease.

(viii) Harvesting

Haulm destruction about two or three weeks before harvest of maincrops helps minimise the chances of blight getting on to tubers, allows the skins of the potatoes to set, and makes mechanical harvesting easier. Frequently chemical methods of haulm destruction are employed both for maincrop and also in seed production since they help reduce tuber infection by blight and foliar diseases. However, with early crops mechanical haulm destruction is employed because of the speed required.

Tuber damage at harvesting must be minimised, and particular care taken when dropping potatoes into the bottom of a trailer.

Subject to them being mature, crops should be lifted when soils are reasonably dry and temperatures are still mild. Less damage will be done to tubers in such conditions and fewer storage problems should result.

(ix) Storage

Storage of immature or wet potatoes and poor store management can result in losses due to rotting and other problems. Where they are stored for long periods consideration must be given to some means of controlling temperature and humidity, as well as to the use of sprouting inhibitors. Long storage can result in considerable loss of weight due to transpiration and respiration, especially if the tubers begin to sprout. In addition there can be a loss of "bloom" and softening which detracts from their appearance to the housewife.

Initially potatoes should be cured by keeping them at about 12–15°C for up to 10 days after storage, unless there is a risk of soft rot, and then cooled to about 7°C for ware crops and 10°C for crisping varieties.

9.8.2. Quality

Quality can significantly influence profits in years when supply exceeds demand, but may be of less importance if demand is greater than supply.

(i) Variety

Now that supermarket chains are putting the names of varieties of on pre-packs it is important to research probable differentials in demand. Differences exist in the preference for varieties in different areas of the country, and the processing trade can also have specific requirements.

(ii) Size

Some varieties, such as Maris Piper, tend to produce a large number of tubers per plant which are small unless the plants are widely spaced. Others, e.g. Pentland Squire, may produce comparatively few, and consequently large tubers unless the plants are closely spaced. The size must suit market requirements.

(iii) Soils

Some soils have a particular reputation for producing good quality potatoes. In general those soils which do not stick to the tubers will produce a better looking sample, and be especially suitable for supermarkets selling washed potatoes.

Greening should be avoided by earthing up and scab minimised by avoiding liming and by irrigation if necessary starting at tuber initiation.

(iv) Grading

The importance of good grading cannot be over stated especially if by chance a sample is inspected by the Potato Marketing Board. Reliable staff are essential and they require motivation and awareness of the significance of their work. If the potatoes are to be used for processing the quality must meet the demand.

9.8.3. Price/Marketing

(i) Supply/Demand

Potato prices are very vulnerable to supply/demand changes. Variations in price occur with time of year. Imports can upset the normal patterns of supply.

Storage of potatoes is costly and must be justified by price increases as winter advances. Care must be taken to observe the probable quantity harvested in the country and relate this to supplies which have already been marketed. If too many farmers have held on to their crop, hoping for higher prices, and the winter is advancing, then many will be forced to sell, probably depressing prices further.

(ii) Central Stores

Many farmers have joined co-operative and marketing groups. Advantages attained include staff who are specialised in marketing, dealing in bulk, and possibly better prices for different grades of potatoes than the individual farmer could obtain.

(iii) Potato Marketing Board (PMB)

The PMB was formed in 1934 to foster improved stability of prices for producers. Growers of more than 0.4 ha have to register with the Board and are given a basic area allocation. Each year the PMB states the percentage of this basic area which they can grow. Farmers have to make a financial contribution, or levy, to the Board for each hectare grown. The penalty for growing above this area allocation is fixed at a penal rate to deter excess production. By using the area quota system the PMB hope to match supply with demand but weather and other factors influence yields so that their objective is difficult to attain.

The levies are used to fund the PMB activities such as the forward buying or "price insurance scheme". They aim to buy by forward contract 10% of production. Contracts must be agreed prior to May. If the market price rises above the contract price the potatoes can be released back to the grower. If the market price falls the contract price is guaranteed. The PMB has the option of denaturing the potatoes for stock feed or releasing them back onto the market when prices rise.

The Board prescribes the size and quality of potatoes which can be sold to the public. The amendment of riddle size can thus adjust supply. Merchants must be licensed by the PMB, and they are not allowed to buy potatoes below the standards specified. Individual farmers can sell at the farm gate and can obtain a direct sales authority to sell to a limited number of named shops.

9.8.4. Variable Costs

The gross income from potatoes is high. Although in many cases it takes a comparatively small increase in yield to pay for the inputs necessary to produce it the factors outlined previously, such as optimum seed rate and ratio of seed price to ware price, must be considered since it is clear that maximum yield does not necessarily mean maximum profit. The variable costs of such things as blight sprays can be amply justified in situations where yield or quality would be seriously affected. In all cases, however, the cost/benefit of any operation should be assessed.

9.9. SUGAR-BEET

9.9.1. Yield

(i) Objective

Sugar-beet in Britain is grown under contract. A contract price per tonne of washed beet is paid which reflects the sugar content. High yields of clean beet with a good percentage of sugar must, therefore, be the aim. About 45 t/ha at 17% sugar is a good target.

(ii) Seed

Sugar content, disease and bolting resistance, germination, and sowing and spacing practice all feature in the selection of seeds. Pelleted monogerm seed is suitable for precision drilling which reduces subsequent work in the crop. Seed is routinely treated against pests and diseases to maintain high emergence rates.

Early sowing from mid-March is preferable, subject to bolting, if conditions are right, since this promotes higher yields and sugar content. A deep, fine, and firm seedbed is essential for satisfactory germination and growth. The retention of soil moisture is important and is fostered by speedy seedbed preparation.

A plant population of 7–9 per m**2** should be the target. Rows 55 cm apart and precision drilling at 12–18 cm within the rows can help reduce competition between the plants in the early stages, but ideal spacings can vary from situation to situation. Modern techniques relying on "drilling to a stand" require very high standards of husbandry in order to achieve a high emergence percentage.

(iii) Fertiliser

Ideal levels of fertiliser depend upon soil reserves and the application of farmyard manure. Salt at 300 kg/ha or kainit applied at 750 kg/ha during the winter has particularly beneficial effects on sugar yield and reduces the potassium requirements. The soil pH, including the subsoil, should be above 6.5. Where boron is deficient it can be applied in soluble form with the June aphicide spray. On light soils 400 kg/ha of Keiserite may be necessary to prevent magnesium deficiency. Manganese sulphate may also be required to correct manganese deficiency.

Rates of fertiliser (kg/ha)		
N	P2O5	K2O
50–125	50–80	70–200

(iv) Weeds, Pests, Diseases

Weed competition must be kept to a minimum, especially during the early stages of growth. New herbicides are developed from time to time, and the grower must find suitable treatments for his farm, the system he operates, and the time of drilling. Soil type and moisture can be particularly important to the effectiveness of herbicides. Herbicides are very expensive and application rates are quite critical. Weed beet must be controlled by rotation, and topping or herbicide treatment of all bolters.

Of all diseases, virus yellows is potentially the most troublesome, but the use of systematic insecticides to kill the aphids which transmit it can reduce the problem if applied at the right time.

Eelworm can be a problem on some farms, but the restrictions on rotations imposed by the growing contract help to minimise it.

(v) Harvesting

Beet should be harvested as clean as possible because it is pointless to deliver dirt to the factory. The weight of beet increases rapidly in late summer and autumn, and the sugar content gets higher until late October. Factories open from the end of September to January and farmers are issued with permits so that their crop is delivered at intervals during this period. If the crop is on wet land the producer will be anxious to lift the crop at a fairly early stage.

Topping must be at the correct height to avoid loss of beet or excessive top tare. Good hard surfaces for clamps, appropriate access to fields, and dry conditions all foster harvesting efficiency. Where conditions are wet, clamping for a week or so can help to reduce the dirt tare. Frost damage must be avoided both in the field and in store.

9.9.2. Price

The price is fixed and the only way for the farmer to improve his returns is to increase the yield of beet and sugar per hectare. Dirt and top tare should be kept to a minimum.

About 40 t/ha of sugar-beet tops plus beet pulp are produced as by-products.

CHAPTER 10

PLANNING AND BUDGETING

CONTENTS

10.1. BACKGROUND TO PLANNING

10.1.1. Brief History

Planning involves the organisation of resources such as land, labour and capital so that as far as possible the future operation of the business will best comply with the objectives set for it. Budgeting is a function of planning and involves the prediction of physical and financial data for the plans.

Farmers have planned since agriculture evolved and the original methods of judgement, based on experience and astute guesses using no better computer or recording mechanism than the human brain, still play their part. In fact it would be foolish to underestimate the skill and success of some farmers using these basic tools and perhaps a white-washed barn wall or the back of a cigarette packet on which to record the budgeted data.

Increased competition and business efficiency within the agricultural industry and particularly the high cost of resources relative to product prices have precipitated a need to adopt more refined approaches to planning, especially in the case of farmers who have to pay high costs to service borrowed capital. It must not be thought, however, that the need for sound judgement has been eliminated or that the so-called sophisticated planning techniques will make all who use them successful overnight.

Many techniques have been practised. Whole farm planning gave way in many cases to gross margin planning in the 1960s and early 1970s. During the late 1970s the use of computer-based techniques featured more prominently, including the preparation of complex computer models of farm businesses. The future will undoubtedly see the development of more computer-based techniques especially now that the size and cost of the hardware has been reduced. In spite of this a large number of farmers have still to receive their first introduction to any form of planning using modern methods.

Advisory services partially financed from government resources, or fostered by commercial organisations which supply goods to farmers, have been joined by specialist consultancy businesses in the provision of services which undertake planning for farmers. The need to prepare plans in order to qualify for grants or in order to obtain loans from banks has also increased the number of farmers who employ planning techniques.

10.1.2. Value and Limitations

Replanning of a whole farm business should only be undertaken when necessary. This basic statement needs emphasising because some farmers change their plans so regularly that the capital expenditure, to quote just one factor, involved in the changes reduces the long term effectiveness of their business. No hard and fast rules can be given. It has been suggested that the minimum interval between a major change in policy should be at least 4 to 6 years, but this must depend upon the circumstances prevailing for the specific business. Agriculture is, however, a dynamic industry. Minor amendments to plans using partial budgets can be undertaken at short term intervals to great advantage, and "fine tuning" should be a regular process facilitated by routine monitoring of performance.

Planning is essential at the start of a new business, when objectives for an existing business change or the existing plan continuously fails to comply with objectives. Changes in resource availability, such as land, labour and capital, and amendments to Government and EEC policy, can also precipitate a need for significant amendments.

Budgets provide a means of assessing whether a proposed plan will be viable. Frequently more than one plan is produced and the one that best meets the objectives is selected. The quality of the data used in the budgets is the most critical factor. Anyone can make money on paper but reality must be the keyword. A balance has to be struck between pessimism and optimism. Too pessimistic an approach may result in a useful activity not being selected, whereas over optimism can result in bankruptcy if the plans fail to work out. The majority of people when undertaking their first budget tend to be far too optimistic. A few also forget the natural sequence of events.

For example, it is necessary to allow time for a building to be erected before it can be used or breeding stock must have time to grow up before they can produce young.

Data from previous experience on the same farm can be the most useful if care is taken to normalise the results and amend them for any new factors which may prevail, such as technical innovations and price changes. Estimates must also be made of the impact of the change in size of the particular enterprise on both its performance and the performance of other activities on the farm.

Data from other farms and standard data from management publications has to be treated with caution. It is difficult to obtain results from identical situations, not the least important of which is the quality of the management.

If data for a budget is recorded carefully, actual results can be written beside the target figures to ease the problem of control at a later stage. In other words it will be easier to see where the actual performance deviates from the plan. Contra account items should always be written out in full.

It is almost always impossible to produce budgets which materialise exactly in practice. The idea is more to estimate the relative superiority or inferiority of one possibility versus another and at the same time to try to assess the parameters of risk involved in understanding a particular proposal. This is why in many cases it is suggested that sensitivity analysis should be carried out to see how changes in yields, prices, the cost of money and other factors affect the plan's viability. The plan which should potentially produce most profit may not always be selected because the owner of the business is not prepared to accept the risks involved or the additional management required.

The question of whether to use present day or projected prices is often asked. Frequently the answer is that present prices should be used because there is enough difficulty in budgeting without crystal ball gazing. The real answer is to use common sense. What really matters is the price of outputs relative to the price of inputs and the number of units of outputs now and in the future. If it is absolutely clear that the relationships will change, some cognisance must be taken of the fact. What should be avoided is a hybrid situation where part of the budget contains current prices and part future prices. Also data used in assessing a present plan should not differ from that of a future plan unless there is sound justification. If it does you are not being fair to the present plan, i.e. you are not comparing like with like.

10.2. PARTIAL BUDGETING

10.2.1. Use of Partial Budgets

Whenever a small change is made to a plan a partial budget can be employed to assess the likely effects. Frequently the change is so minor that no amendment to the fixed costs is necessary, but it must not be thought from this that full farm planning techniques are necessary in every case where fixed costs change. Care must be taken to ensure that only those factors which do change are included. A frequent mistake is to include an extra charge for labour in the budget when in practice the new enterprise is making more effective use of existing labour and there is no actual increase in the farm's wage bill.

The effects of a change can be far reaching and all must be included whether they are beneficial or deleterious. For examplc, a farmer may decide to increase the area devoted to potatoes at the expense of sheep. The budget might, however, have to include the impact on winter wheat which, because of the additional autumn labour for the potatoes cannot all be drilled on time with a consequent reduction in yields. Another farmer may reduce his potato area which allows him to grow more cereals so that he can justify a bigger combine, or spread the costs of the existing one over a greater area.

It must be realised that these budgets only deal with physical and financial aspects which can be quantified. Some factors such as the capability of management to meet the increased pressures require subjective judgements.

The usual layout for a partial budget is as follows:

Cost side of budget	Benefit side of budget
Income lost + New costs incurred	New income gained + Former costs saved

The two sides are then made to balance. If in order to do this a figure has to be added to the cost side of the budget there will be a net gain from the proposal, but if the figure has to be added to the income side a net loss will occur.

Partial budgets can be used to assess the merits of (a) the replacement or partial substitution of one enterprise by another; (b) the introduction of a new enterprise; (c) the expansion of an existing enterprise; (d) a change in an item or factor of production such as an increase in animal feed.

Gross margins can be used to produce a relatively quick form of partial budgeting. Great care is necessary, however, to check that fixed costs have not changed and that an increase or decrease in the size of an activity will result in a linear increase or decrease per unit in gross margin.

Before starting to produce a partial budget it is important to decide the timing of its measurement. In the first year or so after any change there will usually be a phase in period before the new system becomes fully operational. Initially a project may produce little or no return and need subsidising by other activities even although in the long term it is highly profitable. Alternatively there could initially be some favourable factors. For example when a farmer establishes a new sheep flock with gimmers the number of replacements needed at first will be less than when the activity has been operating for a few years.

It is therefore important to study the initial years of any change especially in relation to the impact on the cash flow. However, a frequent policy is to first study the performance of a new project for a typical year when it is fully operational to see its long term potential. In the case of a substitution budget it is usual to compare the results of the new activity with a typical year's performance of the activity it is replacing.

10.2.2. Substitution Partial Budget

Assume that the owner of Church Farm wishes to budget for a reduction of 8 hectares of barley in order to accommodate an extra 100 breeding ewes. He considers that apart from a change in interest on capital invested there will be no alteration to any fixed costs and there will be no interaction effects with other enterprises.

The budget (Fig. 88) suggests that, ignoring interest, there will be a gain of £1,047 from the change, but he will want to know what the situation is if interest is considered. In addition there are several questions the farmer would ask including: (a) how representative is the data used in the budget of future physical performances of sheep and barley on the farm; (b) what are the relative futures for sheep and barley in terms of sale prices and costs of production; (c) what will be the impact of the change on timing of cash flows; (d) how competitive will the extra sheep be with cows for early grazing and with silage grass; (e) will there be labour problems because of the increased number of ewes to lamb; (f) will the extra sheep help to eat off more grass in the autumn and so reduce winter kill?

Cost side			Benefit side		
Income lost	£	£	*Income gained*	£	£
Barley grain	3,200		150 lambs @ £45	6,750	
Barley straw	288	3,488	Cull ewes (20) @ £32	640	
			Wool	325	
New costs			Ewe premium	300	8,015
Concentrates	1,033				
Vet. and medicines	325		*Costs saved*		
Miscellaneous	240		Fertiliser	576	
Forage	1,552		Seed	424	
Gimmers (21)	1,386		Sprays	224	
Share of tups	200	4,736	Sundries	32	1,256
		8,224			9,271
	GAIN	1,047		LOSS	-
		9,271			9,271

FIG. 88. Substitution Partial Budget.

Students meeting this type of budget for the first time sometimes ask why the purchase price of all 100 ewes is not included in the budget. It must be realised that this budget attempts to measure an on-going situation. True in the first year there will be a need to purchase 100 gimmers plus extra tups and the cost of these will be included in the cash flow. However, this is a once and for all situation. From this time on only the replacements have to be purchased. [Remember also that in the opposite situation if the sheep were being given up they could only be sold once and in this case again only the replacements would be included in the budget because one on-going situation would be compared with an alternative on-going situation].

The change of enterprise would alter the financial requirements of the business. The farmer may borrow all the money from the bank in which case he will have interest to pay.

Alternatively he may have sufficient money of his own. However, if the sheep require more finance than the barley it has to be found from somewhere. The farmer has therefore to give up the opportunity of investing the extra capital elsewhere in his business or even of investing it outwith the farm in such as a building society or stocks and shares. To make a fair comparison between the barley and sheep he should therefore consider either the interest, if the money is borrowed, or the opportunity cost, if it is his own capital, for the extra money required to change to sheep. Technically the opportunity cost is the best return he could get from investing the money in something else. For convenience here it will be assumed that both the interest on borrowed money and the opportunity cost are 16%. In practice the farmer would have to establish an appropriate opportunity cost rate by looking at the actual investment alternatives.

It was established in the chapter on capital that there are several different ways in which interest on borrowed money is calculated. The most basic approach is to quote an interest rate on the amount borrowed per annum. Thus if a farmer borrowed £100 at 16% per annum to invest in barley seed and did not pay it back for one year he would pay £16 interest. However, if he paid the money back after 6 months the interest would be £8. The interest has remained the same, on a per annum basis, but the time period for the loan has been halved. The first task therefore is to establish the average investment (loan) on a "per annum" basis. This can be done by multiplying the amount of the investment (loan) by the time it is borrowed i.e. the part of the year. For the case of £100 borrowed for six months at 16% per annum the average investment on a per annum basis is:

$$\pounds 100 \times \frac{6}{12}\text{months} = \pounds 50$$

$$\text{Interest} = \pounds 50 \times \frac{16}{100} = \pounds 8$$

The capital invested in the extra sheep themselves will be invested permanently because as one group is culled replacements are bought. The annual average investment in 100 ewes and the share of the rams can therefore be taken as £6,600+£800=£7,400. Calculation of the average investment in the variable costs is somewhat more complicated because the outlay on them will not be for a full year and each cost item will be invested for a different length of time. The task is to establish the typical time interval between the actual payment of a particular input cost and the receipt of income for the flock. The latter can then be regarded as paying off the expenditure.

With sheep the matter is further complicated because all of the income is not received at the same time. Detailed cash flow predictions can help to overcome this problem and as more farmers obtain computers with suitable software it is probable that most will produce such cash flows. Meanwhile a more crude method is adopted by many since if common sense is used the accuracy is sufficient to assess the merits of the two investment alternatives on a scale involved in most farm businesses.

Thus the concentrates for the ewes and lambs would mainly be purchased from February to April, say on average some 9 months before most of the income is derived from the lambs. Some of the vet. and medicine costs would be incurred early in the breeding cycle and others before and after lambing. An average investment of 8 months might be acceptable. Miscellaneous costs will include such items as transport for purchased gimmers and also of lambs to market. An average of 6 months could be taken. Some of the forage costs would be incurred in producing fodder for the winter and others for grazing. An average of 10 months might be appropriate.

The average annual investment in sheep is therefore:

	£	Time (Year or Part Year)	£
Ewes	7,400	× 1 =	7,400
Feed	1,033	× $^{9}/_{12}$=	775
Vet. and medicines	325	× $^{8}/_{12}$=	217
Miscellaneous	240	× $^{6}/_{12}$=	120
Forage	1,552	× $^{10}/_{12}$=	1,293
			9,805

The next step is to calculate the average annual investment in the barley. Assume that the grain and straw are sold off the combine and the cheque is received on the 1st of October. Also assume that the seed and fertliser are paid for on the 1st April whilst the sprays and sundries are paid for at the end of April.

The average investment in the barley is therefore:

	£	Time (Year or Part Year)	£
Seed	424	× $^{6}/_{12}$=	212
Fertiliser	576	× $^{6}/_{12}$=	188
Sprays	224	× $^{5}/_{12}$=	93
Sundries	32	× $^{5}/_{12}$=	13
			506

The difference in average annual investment between the two projects is therefore:

£9,805 − 506 = £9,299

Assuming an interest rate of 16% per annum for both the borrowed money and the farmer's own money invested in the projects this means that the sheep would have £1,488 additional interest charges above those from the barley. This exceeds the benefits calculated without consideration of interest, producing a loss from the change of £441.

This is still not the end of the comparison for anyone who wishes to take taxation into consideration. Tax relief is given only on interest actually paid. The budgeting has to be more refined if this is to be calculated with any accuracy. First a projected cash flow for the whole farm which incorporates the existing barley area has to be produced and a second cash flow is needed which allows for the replacement of the barley by the sheep. These then show the actual difference in borrowing, and hence interest, between the two situations depending upon the business's overall overdraft position.

Two projected profit and loss accounts have to be produced, one before and one after the replacement, in order to establish the farmer's level of tax and hence the comparison of tax relief between the borrowing for the barley and sheep alternatives. In the case above, where the sheep were apparently giving rise to an extra £1,488 of interest, the tax relief might be £595 if the farmer's income was such that he was paying tax at a marginal rate of 40% and his overdraft was sufficiently large. This tax relief would not be realised for at least 12 months because tax is paid on the profits from the previous financial year.

Given this situation the budget swings back in favour of the sheep.

All this may appear to be complicated to the student. The example has been specifically designed to show the importance of considering interest and taxation. It is meant to highlight the need to treat the traditional partial budget shown in Fig. 88 with a degree of caution but does not necessarily detract from it as a method of establishing the merits of a proposal provided appropriate care is taken. Fortunately computer programs are now available which allow rapid comparison of cash flow and P & L Account predictions.

Before finally leaving the issue it might be worthwhile expanding upon the concept of opportunity costs. Most people easily comprehend the case for calculating interest actually paid on borrowed money because it reduces profits and hence tax.

Opportunity costs are employed as a "management procedure" to establish if the farmer is making best use of all money available. It is important to regard the farmer as an investor in his business and to take his own money into account. If he invests (ties up capital) in one project he forfeits the opportunity of investing this money in an alternative project.

Where the alternative project is capable of producing a high rate of return the capital will have a high opportunity cost.

For example a farmer who stores grain in order to obtain higher prices in the spring may have to give up the opportunity to purchase lambs for winter fattening, because his capital is tied up in the grain. When calculating the merits of storing the grain he should put a cost on the capital equivalent to the return he could get from investing an equivalent amount of money in fattening lambs i.e. the opportunity cost of the capital.

If the return from storing grain with the opportunity cost included in the calculation is positive then storage of grain is a better prospect than selling it off the combine and buying lambs. If it is negative then lamb finishing could be considered as an alternative.

10.2.3. Marginal Returns

Rather than introduce the term opportunity cost into the substitution budget some authorities suggest that the return on marginal capital invested be compared between the alternatives. Marginal simply means extra and one way of obtaining this is to calculate the average capital invested for each case as shown above.

The respective Management and Investment Income as a percentage of the marginal capital is then calculated. For sheep the MII is £8,015−£4,736=£3,279. This as a percentage of the marginal capital for the extra sheep is 33%. The comparative figure for barley is

$$\pounds3{,}488-\pounds1{,}256=\pounds2{,}232;\ \frac{2{,}232}{506}\times100=441\%.$$

Each return can then be compared with interest rates for borrowed money or even returns on other investments if the farmer has money to invest.

Two important points must be recognised from this example. The first is that if a return on marginal capital is being calculated the numerator must be the MII. In other words no interest must have been included in the costs.

The second is that in certain situations exceptionally high returns on marginal capital can be obtained. Such cases only arise, however, when more efficient use is being made of many existing costs. In the case of the barley the costs of such things as the combine, drying and storage facilities, etc. have been ignored because for this exercise they were assumed not to change. This is not always the case and the following partial budgets have been designed to show how to take account of changes in investment that give rise to changes in fixed costs other than interest.

10.2.4. Innovation Budget

Assume that the owner of Church Farm has some old pig buildings which are not currently in use and which could be refurbished for £4,000. He decides that he could buy in and finish three batches each of 100 store pigs per year, and this could be done with existing labour. Capital can be borrowed for the project at 16% interest. Since there is no income lost or costs saved, the formal layout of the partial budget can be abandoned (Fig. 89).

	£	£
Income gained		
288 pigs @ £70	20,160	
Casualties 12	-	20,160
New costs incurred		
300 store pigs @ £29	8,700	
225 kg feed per 295 pigs* @ £145 per tonne	10,908	
Miscellaneous	820	
Depreciation on building	400	
Interest on average capital @ 16%**	990	21,818
NET LOSS		1,658

*Allows food for some of the casualties.

**For the calculation of average investment, see paragraphs (i)-(vi).

FIG. 89.

Calculation of average investment:

(i) Buildings (Fig. 90)

At the start of the project the cost of the building is £4,000 and it is written off in 10 years.

The average investment is

$$\frac{£4{,}000}{2} = £2{,}000$$

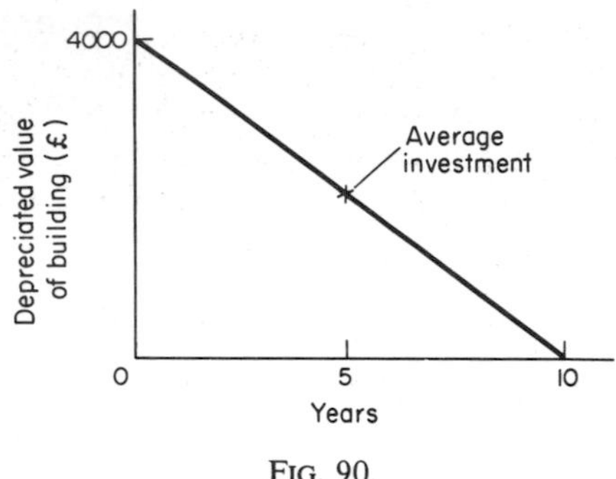

FIG. 90.

Students frequently question the above calculation. First it is necessary to realise that, when the building alterations were made and paid for, £4,000 was actually taken from the bank. When calculating profitability it is usual to spread the initial cost of a building, or for that matter a machine, over a number of years rather than penalise just the first year. This is done by including a depreciation charge in the costs, and in the example above the improvements were written off on a straight line basis over 10 years at a cost of £400 per year. Thus only £400 is charged for the building in year 1 although £4,000 was actually spent. (This is a good example of why a farm can be showing a profit and still get into financial difficulties because a farmer has overstretched his cash position. If the building had cost £40,000 and the depreciation been £4,000 it would have highlighted the point even better).

The key question is what is the position as far as the bank manager is concerned? For the first year the farmer has borrowed £4,000 and will be charged interest on that figure. At the end of the first year the depreciation charge has reduced the profit by £400, but as far as the bank manager is concerned this £400 has not been spent. The bank balance will in fact be £400 better off than the profit might suggest and it can therefore be assumed that £400 has been used to pay off the first part of the £4,000 loan. The principal on which interest will be charged for the second year can therefore be regarded as £3,600. Each successive year this figure would be reduced by £400 until the £4,000 was fully written off. Thus the amount of interest charged on this basis would vary each year. Instead of undertaking ten calculations at this stage the average position for the 10 years is taken. Hence the division of the £4,000 by 2. This is the case when calculating profitability, but from the cash flow point of view it would be important to undertake an annual calculation to see if the capital position of the business could stand the investment. In fact if the investment appeared to be reasonable from the profit point of view using average data for the future it might be wise to calculate the profits on a year-by-year basis, especially for the first year, to check the actual profit from the sternest interest charge in the first year. This aspect will be covered again later under investment appraisal.

(ii) Average Investment in Pigs (Fig. 91)

The average investment in pigs can be calculated as follows:

Total investment in 3 batches =£8,700
Investment in each batch=£2,900

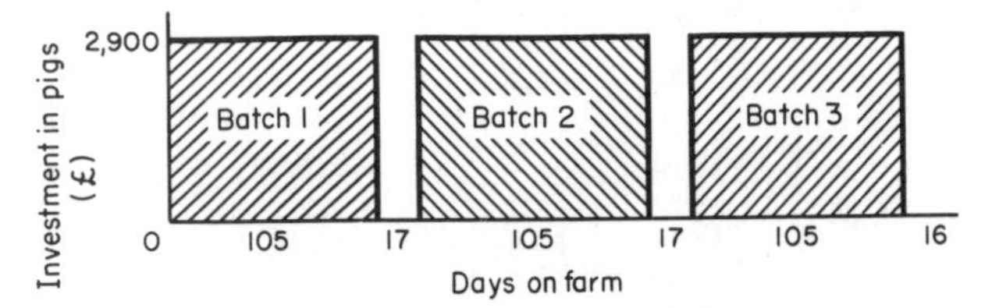

FIG. 91. Average investment in pigs.

The £2,900 is invested for 3×105=315 days, therefore the average investment per annum is

$$\frac{315}{365}\times 2{,}900=£2{,}503.$$

The maximum borrowing from the bank is £2,900 at any one time, but on average it is only invested for a total of 315 days each year.

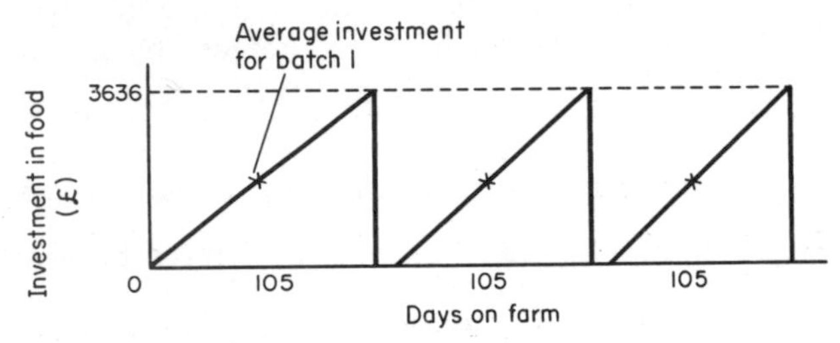

FIG. 92. Average investment in feed.

(iii) Average Investment in Feed (Fig. 92)

The maximum investment in feed is £3,636. The feed cost starts at £0 and grows to £3,636 during the 105 days that each batch of pigs is on the farm. The average of

$$£0 \text{ and } £3{,}636=\frac{3{,}636}{2}=£1{,}818.$$

When a batch of pigs is sold the sale value more than covers the feed costs and the investment in feed reverts to £0. The average investment of £1,818 is invested for 3×105 days=315 days. The annual average investment is therefore

$$\frac{315}{365}\times £1{,}818=£1{,}569.$$

(iv) Average Investment in Miscellaneous Costs

This is calculated in a similar way to that for food:

Three batches: £820 Each batch: £273.

Average investment for each batch $\frac{£273}{2}=£136.5$

Average annual investment $£136.5\times\frac{315}{365}=£118$

(v) Summary of Average Investments

	£
Buildings	2,000
Pigs	2,503
Feed	1,569
Miscellaneous costs	118
	£6,190

(vi) Interest at 16% per annum

$$£6,190 \times \frac{16}{100} = £990$$

The budget in (1) above shows that the project would have a loss of £1,658 and it is unlikely that the farmer would proceed with it. He might test the sensitivity of the plan to price change before discarding it. Thus for the plan to break even with no profit or loss the sale price of each pig would have to increase by £1,658 divided by 288 pigs=£5.76 assuming that all other factors remain the same.

10.2.5. Expansion/Contraction Partial Budget

Assume that because of milk quotas the owner of Church Farm decides that he has to reduce his dairy herd by ten cows. He cannot save anything on his buildings, machinery or labour costs but decides that he will reduce his fertiliser use by £800 per annum and use the same amount of land to keep the reduced number of cows. He assumes that interest on capital invested would be 16%.

The important thing with this type of budget (Fig. 93) is to consider only those factors which change. Thus the only change in forage cost is £800.

The saving in interest can be calculated by first calculating the average investment.

Reduced income	£	£	£
Milk 10 cows × 5800 1 @ 16.9p	9,802		
9 calves × £67	603		
2.5 cull cows £324	810	11,215	
Less 2.5 heifers @ £600		1,500	9,715
Reduced costs			
Purchased concentrates	2,400		
Homegrown barley	580		
Vet. and medicines	160		
Miscellaneous AI	320		
Reduced fertiliser	800	4,260	
Reduced interest		988	5,248
NET LOSS			4,467

FIG. 93. Expansion/Contraction Budget.

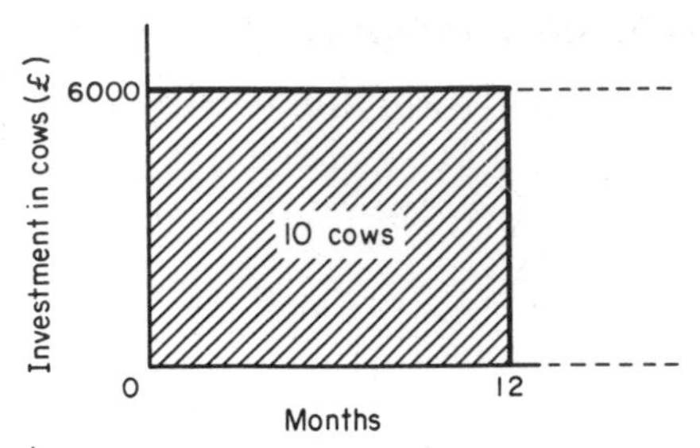

FIG. 94. Average investment in cows.

The cows are continually being replaced so that the average investment is a straight line £6,000.

Average Investment in Variable Costs (Fig. 95)

In the case of milk any investment in variable costs is usually paid off each month with the monthly milk cheque.

Although the total annual investment in variable costs is £4,260 this figure is never actually required because of the monthly return through the milk cheque. The maximum figure required is one-twelfth of £4,260 equals £355. Naturally this answer is slightly crude because there will be some variation from month to month but the degree of variation should not cause concern.

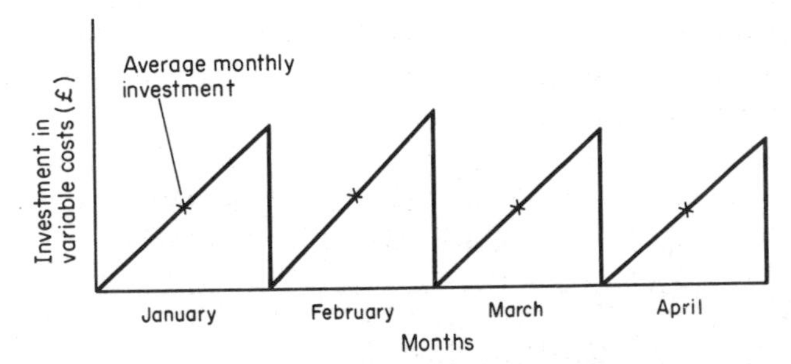

FIG. 95. Average investment in variable costs.

At the beginning of each month the investment is £0 and it builds up to a peak at the end of the month when the milk cheque arrives (in practice it is paid into the bank about the 23rd of the month but there is usually a month to pay the bills). The average investment in variable costs per month can be obtained by dividing the montly total by two.

The average investment in variable costs is therefore:

$$\frac{£4,260}{12} \text{divided by } 2 = £178$$

The total average annual capital investment is therefore:

Cows £6,000 + variable costs £178 = £6,178

$$\text{Interest at } 16\% = £6,178 \times \frac{16}{100} = £988$$

10.2.6. Change in Use of Feed

Assume that with the existing herd of 100 cows the farmer considered that if he increased his feed per litre from 0.31 kg @ £150 per tonne to 0.33 kg/l of the same mixture the yield per cow might increase by 200 l.

	£	£
Extra income		
100 cows × 200 l × 16.9 p		3,380
Extra cost		
Extra food on existing yield		
100 cows × 5,800 l × 0.02 kg × 15 p	1,740	
Extra food on increased yield		
100 cows × 200 l × 0.33 × 15 p	990	2,730
Gain		650

This budget is shown not so much to test that the reader agrees with the farmer's hypothesis but more to point out a common mistake. Note that the increase of 0.02 kg of feed is on the existing yield but that the full 0.33 kg of feed is included for the additional 200 l of milk and not just 0.02 kg.

10.2.7. Break-even Budgeting

Break-even budgets can be used in several ways, but basically they measure when one course of action equals that of another or when outputs of a system or enterprise match the costs of production. Assume that the owner of Church Farm contemplated reducing his cereals and wondered at what area it would be as cheap to employ a contractor to harvest his grain rather than keep his own machine. The following formula could be used:

$$\frac{\text{Difference between total fixed costs of alternatives}}{\text{Difference between variable costs per hectare of alternatives}} = \text{Break-even area}$$

In this case the cost of fuel would be treated as a variable cost. Such break-even budgets can be expressed graphically (Fig. 96).

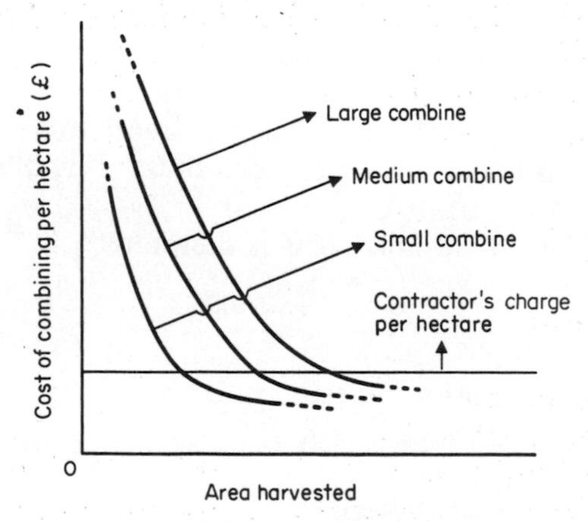

FIG. 96. Break-even budget.

The point at which the curve cuts the line depicting the contractor's charge gives the break-even point, in terms of area to be harvested, at which it would be just as cheap to have the contractor as keep the machine. This point is reached later with the larger machine because of the higher cost of depreciation and interest charges. The benefit of having a man with the combine, the availability of the contractor, and the quality of his work would be included in the factors to be considered.

Example

Assume that a farmer has recently taken on extra land and intends to grow more cereals. He wishes to know the break-even area required to justify purchase of his own combine harvester rather than rely on a contractor who charges £55 per hectare.

He decides to use the following data:

Initial cost of combine	£33,000
Out of season discount	£7,000
Trade in value after 5 years	£12,000
Insurance per year	£100
Repairs (as % gross cost)	2.5
Work rate per hour	1.2 ha
Fuel used per hour	14 litres
Cost of fuel per l	15 p
Interest	14%
Extra casual labour cost per ha (because of no combine driver)	£4

Extra Annual Fixed Costs of Ownership

(a) Depreciation

$$\frac{£26{,}000-£12{,}000}{5 \text{ years}}=£2{,}800$$

(b) Interest on capital invested in combine

$$\frac{26{,}000+£12{,}000}{2}=£19{,}000\times\frac{14}{100}=£2{,}600$$

[Note: Purchase price and trade in price are added and divided by 2 to obtain average of two figures]

(c) Repairs

$$£26{,}000\times\frac{2.5}{100}=£650$$

(d) Insurance =£100

Total extra fixed costs=£6,210

Extra Variable Costs per Hectare

Fuel £1.75 Casual labour £4

$$\text{Break-even area}=\frac{6{,}210-0}{55-5.75}=126 \text{ hectare}$$

Break-even budgets can be employed in various other situations. For example assume a farmer who purchases store cattle for finishing in the winter estimates that the sale price per head next spring will be £570 for a given pen of animals and that his costs to that point will be £100 each. He knows that the maximum price he can pay just to break-even is £470. The difference between this figure and the price that the bidding has reached gives him some idea of the profit margin he may obtain. It is then up to him to decide whether this margin is big enough remembering the risks involved in the investment, such as the death of an animal.

Break-even budgets can equally well be used in connection with cropping. The owner of Church Farm might decide that his fixed costs would be similar whether he grew barley with a gross margin of £279 per hectare or he replaced part of it with oilseed rape. He could calculate the variable costs for growing the rape, say £275 per hectare, and then calculate the yield of rape required at a given price for it to break-even, in terms of gross margin, with the barley. Thus assuming the price of oilseed rape was £270 per tonne the break-even yield (y) would be:

$$(£270 \times \text{yt/ha rape}) - \text{variable cost/ha rape} = \text{GM/ha Barley}$$

$$270\,y - £275 = £279$$

$$y = \frac{£279 + 275}{270}$$

$$= 2.05\text{ t}$$

10.3 WHOLE FARM OR COMPLETE BUDGETING

10.3.1. Introduction

Full farm budgeting can be undertaken with or without the aid of gross margins. Many refinements to the basic techniques are available most of which are outside the scope of this book. All factors contributing to costs and outputs must be considered, but partial budgets can be employed to evaluate alternatives within the overall plan.

10.3.2. Background to Gross Margin Planning

There are four main approaches to increasing profits (Fig. 97).

Method 1: Increasing Gross Margins with the Same Fixed Costs

This can be achieved in several ways: (a) by increasing existing gross margins through better husbandry to improve yields, adopting new technical innovations, producing a better product or improving marketing to increase prices, better variable cost control; (b) by expanding those enterprises which give the highest gross margins either with or without reducing those activities with lower gross margins.

Method 2: Increasing Gross Margins with Higher Fixed Costs

If the total gross margins are increased either by expanding present enterprises with high gross margins, improving existing gross margins, or introducing new high gross

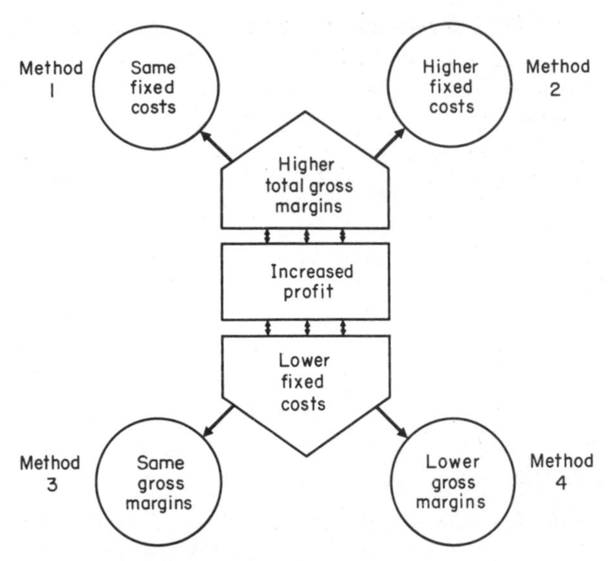

FIG. 97. Approaches to increasing profits.

margin enterprises, the profits will be higher if the total increase in gross margin exceeds the increase in fixed costs. Many dairy farmers have erected new buildings in order to increase cow numbers and at the same time intensify the management of their herds.

Method 3: Maintaining Present Gross Margins but Reducing Fixed Costs

If existing gross margins are maintained in spite of reduced fixed costs, an increase in profit must arise. In certain cases it might be possible to increase mechanisation and save more on labour.

Method 4: Lower Gross Margins with Lower Fixed Costs

If the decrease in fixed costs is greater than the decrease in gross margins, again increased profits must occur. It might be possible for a farmer to eliminate a labour intensive, highly mechanised enterprise such as potatoes and substitute an enterprise giving lower gross margins, such as cereals, which can be produced by making more effective use of remaining staff and machinery. In many cases this could increase profits.

If gross margin planning is to be fully understood it must be remembered that all these methods exist but that not all the approaches are suitable for any given farm. Method 2 has probably been the most widely adopted approach in the past whilst Method 3 might be the most difficult to achieve. All require examination for each farm.

10.3.3. Gross Margin Planning

The procedure can conveniently be divided into a series of steps:

Step 1: Clearly define objectives for the future and in the case of an existing business establish if there is a need for a full farm plan.

Step 2: Clearly establish the resources which will be available to the business in the future. These include land and buildings, labour and capital.

Step 3: Establish those enterprises which are feasible for the farm. These should reflect the farmer's interests, objectives, knowledge or willingness to learn, land, buildings, product prices, demand including forecast demand for products, labour skills and supply, and capital availability.

Step 4: Consider the limiting factors or constraints of the farm for each of the feasible enterprises. These include the farmer's wishes, his knowledge and that of his staff together with their willingness to learn, availability and quality of labour, management and marketing problems, production quotas, building requirements, machinery and equipment needs, suitability of land type, distance of some fields from the steading, the sequence of operations necessary to obtain a product, etc.

Although it is frequently advised that each of the above steps should be clearly written down to avoid anything being missed, it is not absolutely essential that this is done by a farmer producing a plan for himself.

At this stage the farmer might start to consider methods of breaking some of the constraints such as the use of casual staff to overcome a labour bottleneck.

Step 5: Calculate the normalised gross margins for each of the enterprises already on the farm, i.e. those expected from the farm under normal conditions. These can be gross margins per hectare although it will be seen later that gross margins per unit of capital or labour might be just as significant.

Examine the efficiency of each enterprise and see if improvements could justifiably be included in the budget. This might include improved grassland management which would result in higher stocking or techniques such as irrigation to reduce risk of crop failure. Adjust the normalised gross margins accordingly but avoid over-optimism.

Any adjustment of this type immediately makes comparison of the budgets for new plans with past results for the existing one unjust because such technical husbandry improvements could have been incorporated in the existing plan. This point is discussed at the end of Step 10.

Step 6: Examine each of the enterprises to establish if expansion or contraction of the area devoted to them would produce a linear response in gross margin per hectare or if the unit gross margins would change. Timeliness of cultivations, degree of individual attention to stock, incidence of disease, quality of land available, and many other factors would be considered in this context.

Interaction effects between enterprises should be checked. Amendment to the scale of existing enterprises or the introduction of new ones might precipitate far reaching consequences some of which may not always be apparent at first sight.

The sensitivity of each enterprise to quantity and price variation could also be examined to form some idea of the relative risk of the respective lines of production.

These three procedures within Step 6 require experience and judgement. They might justify amendment of the gross margins produced in Step 5 or at least a careful note being made alongside a particular enterprise. It could be that the farmer's attitude to the risk involved in an enterprise might form a constraint about its size.

Step 7: Keep in mind all the constraints listed in Step 4. By doing this the fixed costs are being considered. On most farms the rent or mortgage is fixed and only by breaking a building constraint would the cost increase. Many farmers have a surprisingly firm idea of the number of men they wish to employ and are therefore establishing labour constraints. Step 7 involves maximising the gross margins of the farm to a specific set of fixed costs, i.e. those on which the constraints are based.

The gross margins from Step 5, amended as necessary in Step 6, should be listed in a league table with the highest, per hectare, at the top.

The enterprise with the highest gross margin should be selected and the area devoted to it maximised. There will be some constraint, such as building size, preventing it from being increased further. The second enterprise should then be selected and it too should be maximised until a constraint prevents a further increase. This procedure should continue until all the land has been allocated.

Step 8: The fixed costs for the plan are then calculated and subtracted from the total of the new gross margins to establish profit margin.

Step 9: This step involves the examination of the constraints to see which could be broken and allow another plan to be formulated. It might precipitate an increase in fixed costs, such as the cost of an extra building, but could foster a greater increase in total gross margin and hence profit. The steps outlined above are then repeated for the new plan.

In essence what is being done in this case is to accept that fixed costs can be varied and justified in some cases by large scale changes in production. Remember, however, that fixed costs vary in lumpy rather than smooth patterns. Also do not forget the example given elsewhere of a case in which investment in storage facilities fostered an increase in the price of wheat and attendant gross margin. Breaking a constraint through accepting increased fixed costs does not therefore have always to be justified by a change in size of enterprises but can be warranted by improved production or marketing methods.

All approaches based on the diagram shown in Section 10.3.2 could be considered. One approach might be the elimination of some fixed costs, perhaps even producing a constraint which would prevent production from the enterprise with the highest gross margin but allowing a greater reduction in fixed costs than the gross margin lost.

Step 10: The final step is to select and implement one of the plans. In practice the farmer would produce a capital budget to see if the cash flow of the particular plan was viable. This aspect is covered elsewhere in the book.

Discussion on Steps 5 and 6

Under Step 5 the validity of comparing past results from an existing plan and budgeted results for a new plan was mentioned. There is little problem if common sense is used. The point at issue is that when making a new plan which includes amendments to performance and price levels as a result of such things as technical innovations and better husbandry, it is difficult to know how much of any improvement over the old plan is due to the superiority of the plan and how much to the amendments to performance and price levels which have probably been assessed in a subjective manner in any case. To avoid this dubiety it is often suggested that budgets should use the same yields, product prices, and costs of the various factors of production as the original plan. However, this clearly cannot hold if there is non-linear response to changes in scale, or alterations occur in the interaction between enterprises which alter performance. This leaves the aspects of technical innovations and better husbandry. If these are not reflected in the budgeted figures the latter are of little value for control purposes when comparing actual results with budgeted data. The solution therefore is that before throwing out the old plan check what its performance would have been had the technical innovations and improved husbandry been applied.

10.3.4. Planning—Church Farm

(The gross margin planning technique is employed)

Step 1: Objectives

To provide profits which will ensure a good standard of living for the farmer, a reasonable management load, and future viability for the business. In practice this should be more clearly defined since it would take the farmer to interpret "reasonable" and to define "good standard of living".

Step 2: Resources

Land: 200 ha, all ploughable, water to all fields, good fences and well drained.

Labour: Four men, student from April to September, plus farmer. Ample casual labour. Good staff cottages.

Capital: Owner occupied farm and business gearing such that banks would probably advance capital to support suitable plan.

Step 3: Enterprises

Farmer only willing to consider the enterprises already on the farm.

Step 4: Constraints

Potato quota: 10 ha with storage capacity to match. Dairy buildings for 100 cows, modern milking parlour. Ample room for housing all beef cattle and dairy youngstock. Covered area with sound roof (which was hayshed before erection of silage clamps) presently not fully utilised. Grain storage 325 tonnes.

Milk quota: 580,000 l to be reduced to 522,000 l.

Step 5: Normalised Gross Margins (per hectare) (last year's data in brackets)

	£	(£)		£	(£)
Dairy cows	1,137	(1,155)	Wheat	440	(464)
Dairy youngstock	296	(299)	Barley	282	(279)
Beef	490	(545)	Potatoes	890	(1,228)
Sheep	400	(407)			

The farmer knows that he must reduce his milk production if he is to comply with the quota. He dislikes the findings of the budget shown earlier in this chapter in which he reduced the cows by 10 and kept them on the same area of land. He therefore considers the alternatives (i) keeping the same number of cows, feeding less concentrates, and accepting a reduction in yield per cow (ii) keeping 90 cows with the same feed policy but reducing the area devoted to cows. He selects the latter and decides to reduce his dairy youngstock by 10% also.

After studying the hayshed which is not fully utilised he decides that he could keep all his ewes indoors prior to lambing. He considers that this will reduce the adverse effect of sheep on early growth of grass and when combined with an improved grassland policy, incorporating better timing in the application of fertilisers, he could improve stocking from 2.14 to 2.25 livestock units per hectare. [Remember he could have done this for his existing plan]. Allowing for some extra feeding of the sheep he calculates that this will alter the gross margins per ha for the livestock enterprises to:

	£		£
Cows	1,195	Sheep	415
Dairy youngstock	311	Beef	515

Step 6: Response to change in scale

Even with the opportunity to cull 10 of his poorest cows and with less animals to look after, he decides that he will not increase the yield per cow.

Any increase in the area of potatoes would result in both quota penalty and the need to use poorer quality land. If he eliminated his potatoes he would have the option of not replacing one of his men who is about to retire and the potato storage area could be used to house an extra 100 ewes.

He considers that an increase in the size of the ewe flock by such numbers coupled with the housing would focus increased interest plus attention on the sheep and he could improve lambing percentages by 7%. This would increase sheep gross margins per hectare to £468.

His view is that he would not wish to increase the area devoted to beef by more than that released by dairy youngstock because of housing difficulties and uncertainty about the future of the enterprise. He also thinks that any increase in wheat might lead to problems of establishment but if the potatoes were removed he could grow another 5 ha without suffering yield reduction. He could store this amount of extra grain but no more. His opinion is that the barley area could be increased without yield penalty but extra grain would have to be sold off the combine and he expects that the gross margin for any additional areas of barley would drop by £18 per ha.

Step 7: Enterprise Selection; and Step 8: Fixed Costs

Plan 1

The existing plan should be modified to incorporate the change in herd size and where practicable the other improvements outlined above. This produces a fair base with which to compare any new plans incorporating such factors.

The areas devoted to stock taking both improved stocking and the decrease in the number of cows and dairy youngstock into account are:

Dairy cows	40.00 ha	Sheep	15.32 ha
Dairy youngstock	19.43 ha	Beef	13.55 ha

Their gross margins per hectare because of improved stock carry and extra ewe feeding are:

Dairy cows	£1,195	Sheep	£415
Dairy youngstock	£311	Beef	£515

He obviously has released a total of 11.48 ha. He decides that he will use 5 ha of this for wheat, 2.27 ha to extra beef and 4.21 ha to barley.

The results for this plan are now as follows:

	Normalised and amended GM per ha (£)	Area (ha)	Total Gross margin (£)
Dairy cows	1,195	×40.00 =	47,800
Dairy youngstock	311	×19.43 =	6,043
Beef	515	×15.82 =	8,147
Sheep	415	×15.32 =	6,358
Winter wheat	440	×35.00 =	15,400
Barley	282	×60.00 =	16,920
Barley (increased)	264	× 4.21 =	1,111
Potatoes	890	×10.00 =	8,900
	Total gross margin		110,679
	Estimated fixed costs		89,889
		MII	20,790

The reduction in milk quota together with the use of normalised gross margins has more than offset the benefits of improved grassland management so that the MII is lower than that shown in Chapter 5 for the current year. It must be noted, however, that interest has not been taken into account and there will be less money invested in cows in the new situation.

Plan 2

In this plan the farmer decides to eliminate the potatoes and replace them by 7.66 ha devoted to sheep (100 ewes) and 2.34 ha of barley. He estimates he will save £7,900 in full time wages (the casual wages would be taken into account with the potato variable costs) and that the improved lambing percentage mentioned earlier would materialise. His calculations show that another £1,000 a year might be saved on specialist equipment associated with the potatoes.

The results of the new plan are:

	Normalised GM per ha (£)	Area (ha)	Total gross margin (£)
Dairy cows	1,195	×40.00 =	47,800
Dairy youngstock	311	×19.43 =	6,043
Beef	515	×15.82 =	8,147
Sheep	468	×22.98 =	10,755
Winter wheat	440	×35.00 =	15,400
Barley	282	×60.00 =	16,920
Barley (increased)	264	× 6.55 =	1,729
Total gross margin			106,794
Estimated fixed costs			80,989
		MII	25,805

This shows a benefit over Plan 1 but the farmer would wish to examine interest charges for both plans. He would have also to be absolutely clear that he could manage without replacing the man who has retired. However, he has some latitude for the employment of casual labour at peak times because of the difference in profit margins between the two plans.

At this stage the farmer might produce cash flows for both plans and establish the relative interest payments. Coupled with this he could calculate P & L accounts which would give some indication of liabilities to tax and also produce projected balance sheets.

To do this with any validity he would have to produce budgets for the phase in period through to the situation where the new plan was fully running in order to establish the true cash flow and balance sheet position. The gross margin planning technique has therefore been employed to give a comparatively quick method of evaluating plans before taking the trouble to undertake the cash flow and other calculations in detail.

Step 9: Repeat using Possible Breaks in Constraints

In common with many farmers during times when the future for farm products is uncertain the owner of Church Farm is cautious about employing large amounts of capital to break constraints. For example the erection of new drying and storage facilities to foster expansion of the area of cereals would require an element of courage given the surpluses of grain in the EEC.

Each farmer should however question if there are any ways of breaking constraints which would add to the profitability of his business. At the time the current edition of this book was written it was possible to lease milk quota at 4 p per litre for one year or to purchase it outright at 27 p per litre. [The latter had the disadvantage that once bought it could itself face a cut-back in future reductions].

The owner of Church Farm decides to budget to see if he should, at least in the short-term, lease 58,000 l at 4 p per litre and keep his 100 cows.

He produces the following league table of gross margins:

	Gross margin per hectare £
Dairy cows (90)	1,195
Potatoes	890
Dairy cows (with purchased quota)	673
Beef	515
Sheep (if 300 and no potatoes)	468
Wheat	440
Sheep (if 200)	415
Dairy youngstock	311
Barley (first 60 ha)	282
Barley (above 60 ha)	264

The number of dairy youngstock must be appropriate to the number of dairy cows if the herd is self contained. Even though the DYS may be down the league table some must be selected if the cows are to be chosen. It is therefore necessary to calculate a weighted figure for the gross margin for the dairy herd as a whole noting that the DYS occupy a smaller area than the cows.

This can be done as follows:

90 Dairy cows	40.00 ha × £1,195 =	£47,800
Dairy youngstock	19.43 ha × £311 =	£6,043
	59.43 ha	53,843

Gross margin per ha from herd = £906

"Purchased quota" section of herd

10 cows	4.44 ha × £673 =	£2,988
Share of DYS	2.16 ha × £311 =	£672
	6.60	3,660

Weighed gross margin per ha = £554

[N.B. If the original number of dairy youngstock are retained beef cannot be increased]

It will be observed that the performance of the sheep and the farm fixed costs for Plan 4 have been based on the assumptions made for Plan 2. Plans 3 and 4 show how quickly the gross margin planning technique can be used to test changes in the activities selected. The league table approach has been adopted in Plan 3 maximising the activity with highest gross margin until a constraint has been reached. The same principle has been adopted in Plan 4 except that potatoes were left out.

Plan 3 (with potatoes)				Plan 4 (without potatoes)			
ha	Activity	GM/ha £	Total GM £	ha	Activity	GM/ha £	Total GM £
59.43	Dairy	906	53,844	59.43	Dairy	906	53,844
10.00	Potatoes	890	8,900	6.60	Dairy	554	3,656
6.60	Dairy	554	3,656	13.55	Beef	515	6,978
13.55	Beef	515	6,978	22.98	Sheep	468	10,755
35.00	Wheat	440	15,400	35.00	Wheat	440	15,400
15.32	Sheep	415	6,358	60.00	Barley	282	16,920
60.00	Barley	282	16,920	2.34	Barley	264	618
199.90	Total GM		112,056	199.90			108,171
	Fixed Costs		89,889		Fixed Costs		80,989
	MII		22,167		MII		27,182

Step 10: Select and Implement Plan

The final step is to select the plan to be adopted. This would only be done after a thorough appraisal of the investment in capital terms. A careful check must be made on the validity of all data in the selected plan and it is best to write it out in greater detail than space has permitted here so that actual figures can be inserted against budgeted data. This point is covered later under budgetary control.

10.3.5. Whole Farm Planning without Gross Margins

Many farm businesses do not keep sufficient records to calculate gross margins. In such cases it might be possible to adopt standards, but applying these to a specific farm has limitations. Planning can be undertaken without gross margins and this procedure can also be divided into a series of steps.

Step 1

Define objectives for the business.

Step 2

Establish the normal picture for the farm in terms of income and expenditure and see if the objectives can be attained by better control of the existing plan rather than radical reorganisation. Identify the normal cropping programme, the average yields for crops, and the uses to which the crops are put. Obtain the normal stock carry and the normal sales of stock and stock produce off the farm.

Step 3

Consider all the limitations or constraints applicable for the future of the business. These determine results from Step 4.

Step 4: Principally Livestock Farm

(a) Decide the numbers of each class of livestock to be kept.
(b) Work out rations and formulate a cropping policy to feed the stock selected. Consider cash cropping on remainder of land if any.
(c) Consider labour and machinery requirements and alterations to buildings.
(d) Consider capital requirements and sources of capital.

(Although capital requirements can only be calculated when the final plan has been evolved, the amount of capital available must be borne in mind at each stage.)

Step 4: Principally Cropping Farm

(a) Decide the area of each crop to be grown. Consider rotations and incorporate any necessary break crops.
(b) Decide what stock should be kept and consider cropping in relation to their feed requirements.
(c) Consider labour and buildings.
(d) Consider capital requirements and sources.

(More than one plan may be produced and partial budgets may be used to evaluate alternatives. The farmer's attitude to risk and uncertainty must be considered including the question of diversification versus specialisation.)

Step 5: Budgets

Budget data should be produced for each plan and appropriate P & L accounts prepared. The basical principles with reference to selection of data, optimism, and pessimism, mentioned under gross margin planning, apply here.

Step 6: Select and Implement a Plan

The plan selected should show some distinct advantage over the present one and should be capable of lasting several years. It may not produce the largest profit but may best reflect the farmer's objectives and his attitude to such factors as risk and the level of management necessary for the plan's success.

10.4. PROGRAMMING TECHNIQUES

10.4.1. Introduction

Budgeting techniques already discussed are not the only methods of planning. Programming techniques, either with or without the aid of a computer, can also be employed. Essentially they are mathematical techniques designed to obtain the maximum, or near maximum, returns from a set of available resources.

Land is a very limiting resource for most farmers. There is a tendency for many to treat it as the most limiting factor when using the planning techniques so far described. They base their plans primarily on returns to this resource. However, labour and capital, and even labour supplies at particular times of the year, can be equally, if not more, limiting in some businesses. Procedures which could consider all these factors on their merits and deal with the mathematics involved, would therefore appear to be needed. Programming methods were designed to meet this need.

10.4.2. Net Revenues

Whilst gross margins are sometimes employed in programming, net revenues are frequently used instead. They represent the enterprise or "activity" output minus the variable costs. For most crops the net revenue is identical to the gross margin. However, there is a greater definition into "activities" than in gross margin planning. For example, grass is not simply treated as part of forage. It is an "activity" in its own right, but since it does not have a direct financial return, its net revenue is negative, i.e. by the extent of its variable costs.

Activities such as dairy cows, which use grass and other supplying activities, have net revenues which are higher than their gross margin. At this stage the variable costs of the supplying activities have not been deducted.

10.4.3. Production Possibilities

(i) Example

The principles involved in any programming technique can be explained graphically. For simplicity consider the case of a farmer with just two crops—winter barley and sugar-beet. The two are competitive for many resources.

Assume that the data shown in Table 11 is relevant.

Table 11. *Resource requirements*

	Resource requirements per hectare		Max. area (ha) which resource permits	
Resources available	Winter barley	Sugar-beet	Winter barley	Sugar-beet
Land (70 ha)	1	1	70	70
Autumn labour (800 hr)	8	20	100	40
Spring labour (390 hr)	1	13	390	30
August labour (210 hr)	3.5	0	60	—
Working capital (£12,000)	100	300	120	40
Sugar-beet contract (28 ha)	—	1	—	28

Table 12 shows the returns of each crop to the respective resources. Autumn labour can be used to explain how the figures are derived. Each hectare of winter barley requires 8 hours of autumn labour. The net revenue per hectare is £300. Therefore the net revenue per hour is £300÷8=£37.5.

Table 12. *Returns per unit of resource*

Resource	Returns per unit of resource (net revenue)	
	Winter barley (£)	Sugar-beet (£)
Land (per ha)	300.00	600.00
Autumn labour (per hr)	37.50	30.00
Spring labour (per hr)	300.00	46.15
August labour (per hr)	85.71	—
Working capital (per £)	3.00	2.00
Sugar-beet contract (per ha)	—	600.00

It will be noted that whilst sugar-beet gives the highest returns to land, winter barley gives the highest return to all the other resources for which the two crops are competitive. Programming techniques take this fact into account.

(ii) Explanation of Fig. 98

(a) The land constraint would allow the farmer to grow 70 ha of winter barley, or 70 ha of sugar-beet, or some combination of the two crops.
(b) Autumn labour restricts him to 40 ha of sugar-beet, although he could grow up to 100 ha of barley (considering labour only and not available land).
(c) When the land and labour are considered together, the feasible region of production is obtained. In practice at any point other than the enterprise combination shown by point *D* there will either be a surplus of land or of labour.
(d) This shows the ratio of the net revenues of the two crops.
(e) The net revenue ratio line has been applied to Fig. 98(c). The point where it touches the feasible region shows the optimum combination between the two crops to obtain maximum net revenue.

(iii) Explanation of Fig. 99

All the constraints have been applied to this figure to show the production possibilities from each one. Sugar-beet does not require August labour so it is supplementary to winter barley for this resource. The working capital has been left off because there are other constraints which are more limiting in this case.

The figure shows the feasible region of production, which is *A B C D E F*, with the given constraints. The task is to find which combination of enterprises, within this feasible region, gives the highest net revenue.

At point *A* 28 ha of sugar-beet could be grown which gives a net revenue of

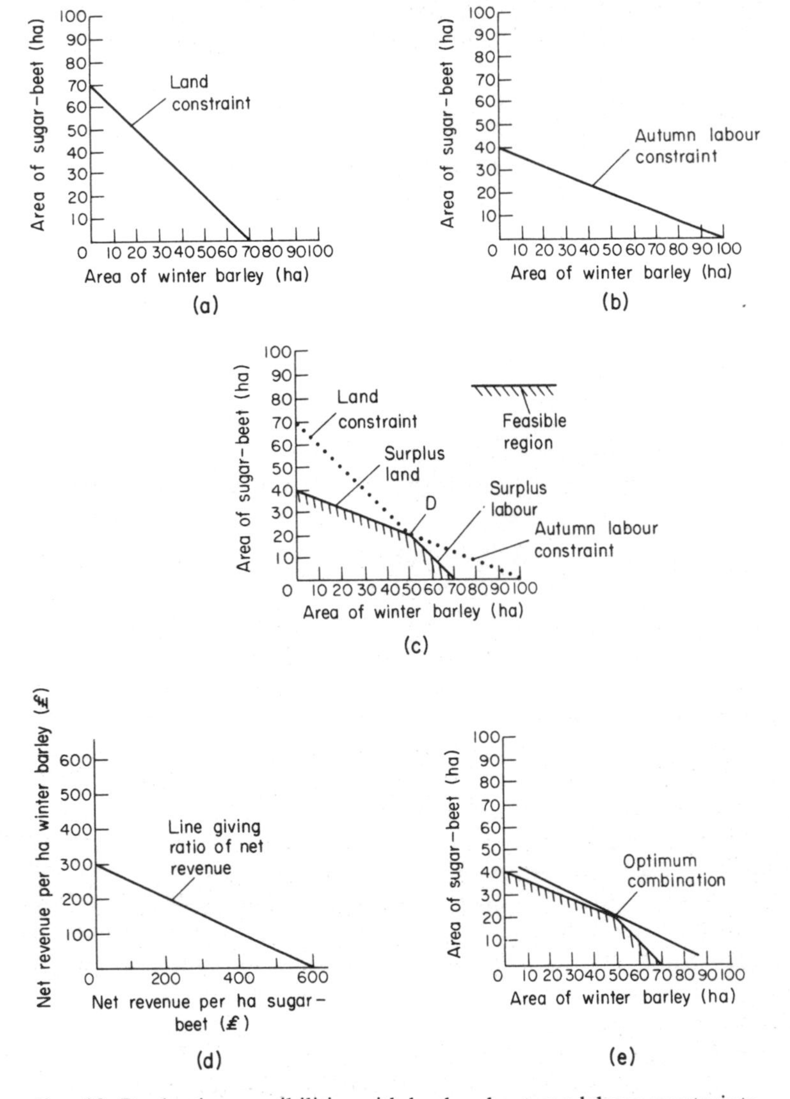

FIG. 98. Production possibilities with land and autumn labour constraints.

28×£600=£16,800. However, at this point only the sugar-beet contract constraint has been used so that we can at least move to point *B*. This shows that 28 ha of sugar-beet and 24 ha of barley can be grown, and gives a net revenue of 28×£600+24×£300=£24,000.

Between points *B* and *C*, spring labour becomes a constraint. One hectare of winter barley only requires 1 hour of spring labour, but 1 ha of sugar-beet requires 13 hours. Therefore in terms of this resource 13 ha of barley could be grown instead of 1 ha of sugar-beet.

Consider sustituting some sugar-beet by barley until point *C* is reached. Here 27.5 ha of sugar-beet could be grown and 30.5 ha of barley. This gives a net revenue of

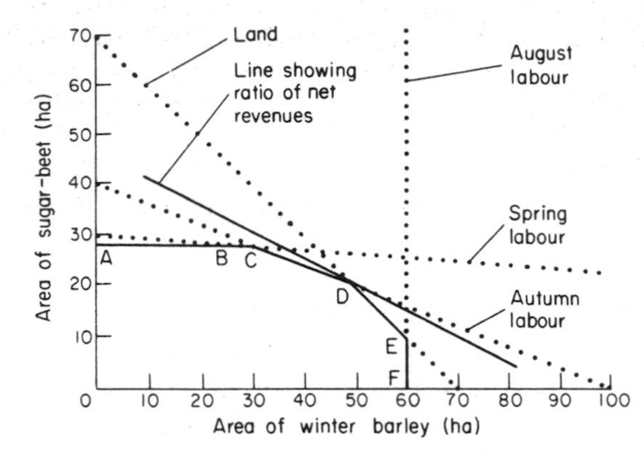

FIG. 99. Feasible region of production, *A B C D E F*, with all constraints applied.

£25,650. Note the small area of sugar-beet which has to be given up to get the rise in net revenue.

Between points *C* and *D* autumn labour becomes the limiting factor. The requirements for this resource are 20 hours for the sugar-beet per hectare and 8 for the barley. Therefore 2.5 ha of barley could be grown instead of each hectare of sugar-beet. Point *D* shows that 20 ha of sugar-beet and 50 ha of barley could be grown. This gives a net revenue of £27,000.

Between *D* and *E* land is the resource which limits production. Since the substitution between barley and sugar-beet is 1 ha for 1 ha, and the sugar-beet has twice the net revenue of the barley, there is no point in progressing further.

The point *D* therefore shows the optimum combination between the two enterprises when all the constraints have been applied. This is also shown by the application of the line showing the ratio of the net revenues between the crops.

Although it is not as simple as this when more than two enterprises are concerned, and there are notable differences in procedure, the above example illustrates many of the principles of programming techniques.

10.4.4. Programme Planning

Programme planning is one technique which can consider more than two enterprises for a plan. It can be described as a manual, or "by hand" method, and can be very time consuming. In many cases programming techniques using computers are used in preference. However, it does show more of the principles which are involved in computer-based programming techniques. The steps are as follows:

Step 1: Identify the scarcest resources and select the one resource which is most limiting. This is frequently land but it may be labour, machinery, or capital.

Step 2: Identify the enterprise which produces the highest gross margin per unit of this most limiting resource; call this resource one and maximise its size.

Step 3: Identify the enterprise producing the second highest gross margin per unit of the most limiting resource and maximise its size. This procedure is continued using enterprises with successively lower gross margins per unit of the scarcest resource until this resource has been exhausted. However, it has to stop if one of the other resources runs out earlier.

Step 4: (a) Where the scarcest resource has been fully utilised before the others; establish if any of the enterprises under consideration do not need any of this resource, e.g. if the scarcest resource is land, concrete-based enterprises such as pigs, poultry, veal calves, and barley beef can be kept without it. Select these enterprises according to their return to the second scarcest resource.

(b) Where another resource (call this resource "two") has been exhausted before the scarcest one; this second resource is now considered to be the scarcest. Each of the enterprises has to be examined to see which gives the highest return to this resource. When ascertained this enterprise is introduced into the plan (or expanded if already there) at the expense of the enterprise already in the plan which is producing the lowest return per unit of the new scarcest resource.

This substitution procedure is continued using the enterprises with successively highest returns per unit of resource "two" and always removing an enterprise still left in the plan with the lowest returns per unit of this resource. The substitution stops when another resource is exhausted. In this case it becomes the new scarcest resource and the whole procedure is repeated for this resource; or it stops if there are no further enterprises which can be introduced or expanded with higher returns to resource "two" than those already in the plan. In the latter case further enterprises not requiring resource "two" can be considered for addition to the plan.

Step 5: When the point has been reached where there are no more benefits from further substitution the total gross margin and fixed costs for the plan are calculated and profit established. The next logical step from programme planning is the use of programming techniques using computers.

10.4.5. Planning Techniques Using Computers

(i) Linear Programming

Linear programming is employed by agriculturalists for such things as least cost ration formulation and for farm planning. In the former case the analysis of the ration to be prepared is established. The analysis and cost per unit of each relevant "feed nutrient" of all the feeds available is then obtained. With this information the computer can formulate an appropriate ration at "least cost".

In planning it can be used to produce a plan which maximises the profit from a fixed set of resources and limitations. The technique requires the construction of a matrix. This gives the computer information it requires about the farm and the activities being considered for it. Extremely detailed data can be provided because the computer has the capacity to deal with a large number of factors. For example the labour availability in each month can be included and the requirement of each activity for labour each month entered into the matrix.

A program instructs the computer what to do with the data. The results for the plan are then printed out.

Improved programming techniques have removed many of the limitations originally experienced with LP techniques. For example they used to assume that labour could enter a plan continuously so that the result implied that "part of" a man could be employed. Now the programs can be instructed to consider "integer" or whole amounts.

A further problem with linear programming initially was that a single plan was produced and it was not possible to see those plans that gave almost as much profit, but which might have been less risky. Developments have resulted in the facility to produce a range of plans with information which can allow the farmer to exercise an element of subjective judgement that reflects his attitude to risk, his objectives and views on the future.

Unfortunately, programming techniques have still continued to meet with mixed success. Recognition has to be given to the fact that agriculture consists of a series of biological processes. The variability resulting from these can make it difficult to firmly establish appropriate data for input/output relationships which are always repeatable. The quality of the results obtained from any planning technique depends upon the data used. Much of the advantage of LP is lost if input data lacks reliability.

Programming techniques have tended to find more favour on large cropping farms than on grass and stock farms. Input/output relationships which are more repeatable can generally be more accurately established for crops. Also, on cropping farms annual production decisions have to be taken, whereas stock production is usually of a longer term nature. Significant changes usually occur during the life of a plan, especially if it includes livestock enterprises with long production cycles. Many people have found partial budgeting techniques more suitable than programming methods in such cases. Developments in the use of computers, however, are continuing.

(iii) Spreadsheets

The use of spreadsheet application by the agricultural industry, particularly for accounting and budgeting, is especially worthy of mention. A spreadsheet allows the computer screen to act like a large worksheet which has a built-in calculator. It is a grid of columns and rows, and each position on the grid can contain information in a cell. The cells are labelled using the column and row which cross at that point:

A range of items can be put into the cells. These are "text", "data" or "formulae".

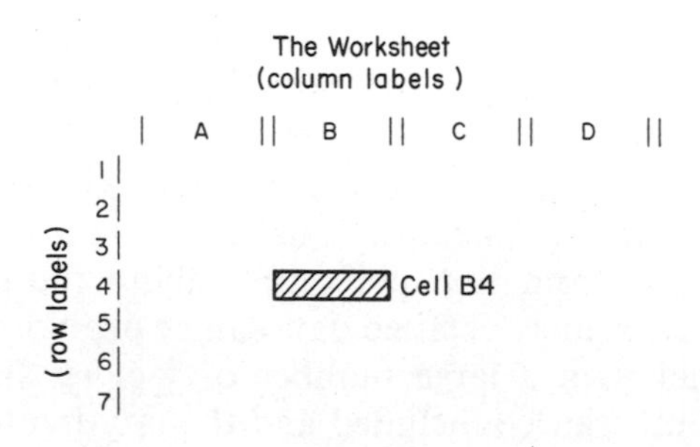

(a) A word or "text" item allows titles or headings to be given to results and in some cases formatting can take place using upper or lower case letters.

(b) "Data" can be entered into blank cells or written over existing data when a change is required (Note: contents of cells can be protected to prevent accidental alteration).

(c) Data manipulation is achieved through instructions applied to certain cells. For example:

(1) a "formula" can tell the computer to apply a constant, to add, subtract, divide, multiply, etc. so that there is interaction between data in one cell with that in one or more other cells. The result is then shown in the cell containing the formula;
(2) a "logical statement" forces the computer to check whether a particular situation exists and to act appropriately (e.g. if the number in cell B4 is less than that in D6 then enter zero in F5, but if it is greater then enter a one);
(3) a "look up" function requires the computer to establish an item of information from a table (this facilitates the use of tables e.g. a table showing the rate of substitution of dry matter between silage and concentrates when feeding cattle).

Each time an entry is made the spreadsheet recalculates. It is therefore ideal for budgeting to test the sensitivity of a plan to changes in items such as quantity and price.

It is preferable to use an existing spreadsheet application that has been developed by an expert. An inexperienced user would have to spend time becoming familiar with the package before developing any application. Several companies are now producing and selling a range of software applications for the agricultural industry.

CHAPTER 11

CAPITAL BUDGETING AND INVESTMENT APPRAISAL

CONTENTS

11.1. CASH FLOW BUDGETS

11.1.1. Cash Flows for the Whole Farm

The use of cash flows to help control the financial aspects of a business was referred to in Chapter 7. Cash flows can also be used to establish the potential viability of a plan, to assess the possible requirements for borrowed capital, and to help convince a bank manager, or other source of credit, that there is a good prospect of the business being able to repay any loan or overdraft which may be needed.

There is little point in proceeding with a plan if it is clear that the capital situation of the farm will not support it, and usually a new plan for an existing business must contribute positively to its financial viability before being adopted. Even where a farmer is proposing a change in policy which is less intensive and exacting on his efforts and management, a cash flow is essential to establish if he can continue to receive the same private drawings from the business.

Excessive optimism and pessimism must be avoided when projecting the financial data to reduce the chances of the plan failing through shortage of capital or of it not being adopted even though in reality it is viable. Every effort should be made to use figures which have the best chance of materialising. It may be advisable to do sensitivity checks to see what major differences from the projections could be withstood.

11.1.2. Cash Flow—Church Farm

Figure 100 shows a projected cash flow for the first year of Plan 4 for Church Farm (see section 10.3.4).

	Apr.-June £	July-Sept. £	Oct.-Dec. £	Jan.-Mar. £
Inflow				
Milk	19,365	17,386	25,142	26,287
Sheep and wool	-	3,626	12,586	8,157
Cattle	5,389	1,200	3,699	16,562
Grain, straw	20,000	2,500	4,600	13,515
Machinery sales	8,000	-	-	-
Total inflow (a)	52,754	24,712	46,027	64,521
Outflow				
Fertilisers and lime	4,000	-	-	21,369
Seeds	-	500	-	5,571
Other crop expenses	1,214	725	-	2,970
Feed	5,300	1,926	9,558	13,900
Vet. and medicines	479	725	962	1,000
Other stock expenses	1,200	975	1,700	1,270
Labour	6,000	6,000	5,500	5,500
Machinery – repairs	4,950	5,600	300	3,650
– fuel	460	399	100	331
– other	1,400	916	250	384
Electricity	275	275	500	450
Property + rent/rates	975	-	975	-
repairs	1,400	900	2,650	1,350
Telephone	120	120	120	140
Insurance	300	400	300	300
Mortgage	-	3,900	-	3,900
Loan interest	2,225	2,225	2,225	2,225
Miscellaneous	400	400	400	400
Livestock purchase: sheep	-	4,560	-	-
Capital exp. – machinery	12,000	-	-	-
Personal – private drawings	3,000	3,000	3,000	3,000
– taxation	-	2,600	-	3,100
OUTFLOW (B)	45,698	36,146	28,540	70,810
CASH FLOW (A - B)	7,056	−11,434	17,487	−6,289
Deduct interest	−927	−1,036	−966	−804
Net cash flow	6,129	−12,470	16,521	−7,093
OPENING BALANCE	−30,000	−23,871	−36,341	−19,820
CLOSING BALANCE	−23,871	−36,341	−19,820	−26,913

Fig. 100. Cash flow for Church Farm, Plan 4, 19.. to 19..

The following points from this example may help readers who have to construct cash flows:

(i) The bank overdraft recorded in last year's balance sheet gives the "opening balance" for the penultimate line of the cash flow.
(ii) Sundry creditors at £7,000 and sundry debtors at £8,000, which enter this cash flow can be seen from last year's balance sheet.
(iii) Some items purchased during the year may not be paid for within this financial year, but will enter next year's cash flow and not this one. Their cost will be shown as sundry creditors in this year's balance sheet. Equally the value of some items sold may not appear in this year's cash flow if the money has not been received.
(iv) Previous cash flows can help, providing a pattern of possible cash flows during the year for some items.
(v) Tax: the July tax is half of the payment assessed on agreed profit from the year before last, and the January tax is half that based on last year's profits.
(vi) The closing balance of one quarter becomes the opening balance of the next quarter, and the net cash flow is added or deducted, as appropriate, to obtain the new closing balance.
(vii) The interest used was 14%. The interest calculation for the first quarter, which illustrates the method for each quarter, was as follows:

$$\frac{(-30{,}000)+(-30{,}000+7{,}056)}{2}\times\frac{14}{100}\times\frac{1}{4}=\pounds 927$$

The cash flow suggests that the peak overdraft requirement occurs from July to September at −£36,341. Appropriate arrangements would have to be made with the bank manager. Subsequently the overdraft requirement falls, but rises again for the period Jan–March when fertiliser and seed are purchased. The closing overdraft is below that at the start of the year. Cash flows for subsequent years would have to be constructed to establish the longer term trend.

Cash flows can be shown graphically by plotting time against the closing balance. The positive closing balances will appear above the horizontal axis and the negative balances below.

11.2. APPRAISING NEW CAPITAL INVESTMENT

11.2.1. The Aim of Investment

The aim of investment is to not only retrieve the original capital employed, but also to obtain an additional return which, after deduction of all costs, is sufficiently high to justify the risks involved. Two important points must be established when appraising any investment. These are the life of the project and the return acceptable.

Several methods can be adopted to assess the merits of an investment and the following will be discussed here:

(i) Accountants rate of return method.
(ii) Pay back method.
(iii) Annuity charge method.
(iv) Sinking fund rate of return.
(v) Discounted cash flow methods.

11.2.2. Accountants Rate of Return Method (see also Sections 5.2.2 and 5.3.1)

The rate of return is the extra annual margin (profit) derived from the investment after deducting depreciation allowances, but not interest, expressed as a percentage of the capital employed.

The capital employed is sometimes taken as the initial capital invested, and sometimes, for items which depreciate, as half the initial capital, so producing an average capital investment. The reason for the use of average capital, i.e. assuming that an amount equivalent to the depreciation is made available each year to the business for investment, was discussed in Section 10.2.4. A further possibility is to consider trade-in values and to calculate the return on [(initial capital − trade in value) divided by two + the trade in value]. A further method which can be used to calculate the average capital of an asset which is depreciated by the reducing balance method is illustrated in the next section.

Taking the initial capital is the severest test. It also avoids the chance of influencing a decision in favour of a project which depreciates, such as a building, compared to breeding livestock, which does not.

The rate of return technique is extremely widely used, but anyone employing it must recognise its limitations. It attempts to get a general picture and ignores the timing of cash flows. Over their life two projects may have identical returns, but one might produce higher returns in initial years and the other higher returns later.

Example

A farmer decides to reduce his cereal area by 16 ha, to adapt an existing building to house sheep, and introduce a flock of 300 ewes.

Data he employs:	
Cost of adapting shed for sheep	£6,000
Extra machinery for making hay	£2,600
Initial cost of buying sheep	£21,800
Average working capital for sheep	£4,000
Average working capital saved from barley	£2,000
Cost of temporary lambing assistant	£300
Present gross margin from barley	£300/ha
Expected gross margin for sheep	£700/ha

He considers that he will still require all his existing cereal growing machinery for the reduced area of crop which remains.

The project is considered over a 10 year period with the buildings depreciated at 10% per annum on a straight line basis, and the machinery depreciated at 20% per annum on a reducing balance method.

Extra building and machinery: depreciated values (£)

Year	0	1	2	3	4	5	6	7	8	9	10
Building	6,000	5,400	4,800	4,200	3,600	3,000	2,400	1,800	1,200	600	0
Machinery	2,600	2,080	1,664	1,331	1,065	852	682	546	437	350	280

The average capital invested in the building, because it is depreciated on a straight line basis, is:

$$\left[\frac{\text{Initial capital}-\text{Written down value}}{2}\right]+\text{Written down value}=\left[\frac{6{,}000-0}{2}\right]+0=£3{,}000$$

£6000

0

WDV

Time

Initial capital

£

WDV

Time

These two diagrams explain the formula. The one on the left shows the example above where the Written down value (WDV) is £0. The one on the right shows a case where the WDV has not reached £0.

When calculating the average investment in machinery many people accept a straight line depreciation for convenience in calculation. Earlier in the book in Chapter 10 this procedure was adopted. However, for taxation purposes, and usually for management purposes, machinery is depreciated using the reducing balance method. The diagram for this is of the following general character:

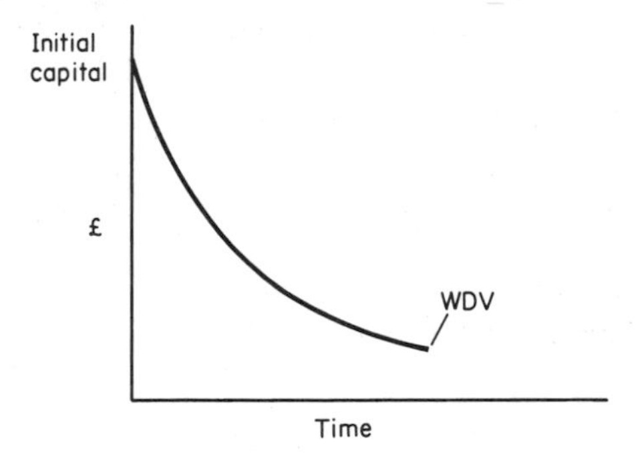

If extreme accuracy is required in calculating the average capital invested in an asset depreciated by the reducing balance method the formula is somewhat more complicated than that for the straight line basis. It is:

$$\left[A\times\frac{1-x^{n}}{1-x}\times\frac{1+x}{2}\right]\div n$$

Where A = initial cost of the machine

n = number of years machine depreciated

$x=1-\frac{d}{100}$ in which d is the depreciation rate

Thus for a 20% depreciation $x = 1 - \frac{20}{100} = 0.8$

The average investment in the machinery for the example given above is:

$$\left[2{,}600 \times \frac{1-0.8^{10}}{1-0.8} \times \frac{1+0.8}{2}\right] \div 10 = £1{,}044$$

If the average capital had been calculated using straight line depreciation then assuming that the machinery had a trade in value of £280 the average capital would have been £1,440.

The increased capital requirement would be:

	Initial capital		Average capital	
	£	£	£	£
Buildings	6,000		3,000	
Machinery	2,600		1,044	
Sheep	21,800		21,800	
Working capital	4,000	34,400	4,000	29,844
Less:				
Barley working capital		2,000		2,000
		32,400		27,844

The additional management and investment income is:

	£	£
Gross margin from sheep 16 ha × £700		11,200
Less:		
Loss of barley gross margin 16 ha × £300	4,800	
Additional labour	300	
Depreciation of building over 10 years	600	
Depreciation of machinery @ 20%	520	6,220
Additional MII		4,980

$$\text{Return on initial capital} = \frac{4{,}980}{32{,}400} \times 100 = 15.37\%$$

$$\text{Return on average capital} = \frac{4{,}980}{27{,}844} \times 100 = 17.88\%$$

11.2.3. Pay Back Method

This method calculates the number of years required to retrieve the initial capital investment. Essentially it is based on the cash flow.

The basic formula to use is:

$$\frac{\text{Initial capital}}{\text{Sum of net cash flows}} = \text{years for pay back}$$

The pay back technique can be employed to rank alternative investments. The investor may decide to invest in a project with the shortest pay back period from a list of possibilities or decide not to invest in a project which has a pay back period longer than say 5 years because of uncertainty about the future. Basically it is a simple method, but it tends to favour projects with a high return in early years and neglects cash flows after the pay back period. It is possible for one project to have a longer pay back period than another, but once it really gets going it might produce better net cash flows than the other.

11.2.4. Annuity Charge Technique

This technique involves an annual charge which incorporates both an interest charge and an additional element to recover the initial cost of the project over its life. The project can be justified if each year the net cash flow is greater than this annual charge.

The technique is employed by the Agricultural Mortgage Corporation, Scottish Agricultural Securities, and building societies for repayment of mortgages. The annual charges to the borrower are equal throughout the life of the loan. Special amortisation tables are used to calculate the annual payment (see Appendix A).

Thus £1,000 borrowed for 20 years at 15% costs

$$\text{Annual charge} = 1{,}000 \times \frac{16^*}{100} = £160$$

(*This can be found in Appendix A if the line 20 years and the column 15% is observed.)

The system means that in the initial years most of the annual payment is used to cover interest and very little to repay the principal. As the years go by the interest portion progressively gets less and the capital repayment grows. Example: £1,000 borrowed for 20 years at 15%; annual payment £160:

Year	Principal left (£)	Interest @ 15% (£)	Amount of principal paid off (£)
1	1000.00	150.00	10.00
2	990.00	148.50	11.50
3	978.50	146.77	13.23
4	965.65	144.79	15.21
		etc., to 20 years	

For taxation purposes only the interest is allowed as a deduction from profits and not the repayment of principal. Thus in the early years more tax can be saved.

With house mortgages this can be unfortunate because a young person may not have a high enough salary to pay high levels of tax and so cannot derive maximum benefit of tax saving from the mortgage interest. Later in life, when he has a higher salary, the interest, and hence the tax saving, have fallen.

To overcome this many people take out an insurance policy with the same length of life as the mortgage. This policy is one that gives bonuses each year. At the end of the loan period the value of the policy should equal or even exceed the cost of the mortgage. The policy remains with the building society who charge interest on the mortgage during its life and no repayment of principal at this stage. This produces a high tax relief throughout the life of the loan.

At the end the building society take the amount of the mortgage from the matured insurance policy and, if there is anything left, gives the rest to the borrower. The system does cost more in the initial years but generally is cheaper in the long run.

Repayment Period

The period within which a loan has to be repaid can be critical to a business. Reconsider the case of the farmer mentioned in Section 11.2.2 above.

His initial capital requirement was £32,400 and the change in cash flow, neglecting interest and taxation, from his present situation is £6,100 (i.e. MII £4,980 + depreciation on new machinery and buildings, which would not enter the cash flow £1,120).

Period 1: Assume that the farmer had to pay back the capital over 15 years at 14% interest (See Appendix A).

$$\text{The annual charge} = £32{,}400 \times \frac{16.3}{100} = £5{,}281$$

The net cash flow is £6,100 so that it could meet this annual charge of £5,281 and still leave £819.

Period 2: Now assume that the capital has to be paid back over 10 years at 14% interest.

$$\text{The annual charge is } £32{,}400 \times \frac{19.2}{100} = £6{,}221$$

In this case the annual charge exceeds the net cash flow. Subject to any taxation relief on the interest this means that the rest of the business would have to subsidise this project for 10 years. Afterwards the farm would benefit from the full net cash flow and would have the asset of a sheep flock plus the building adapted for housing sheep. In this case it is probable that the taxation relief attributable to the interest could have a major influence on the farmer's decision about whether to go ahead with the project.

A weakness of the annuity charge technique is that it does not allow for variations in the net cash flow which might take place over the years.

11.2.5. Sinking Fund Rate of Return

This technique assumes that once a project has started a sum of money is invested each year outwith the business at minimum risk. The sum chosen is such that the annual investments, together with interest, exactly equals the initial cost of the project once it comes to the end of its life.

It is one way of ensuring that there is sufficient capital to, for example, buy a new machine once the old one has worn out, but this only applies if there is no increase in the cost. This practice has comparatively little application in agriculture.

11.2.6. Discounted Cash Flow

The discounted cash flow technique has several advantages over other appraisal techniques presented here. Variations in net cash flow can be fully considered. It recognises the two dimensions of capital—quantity and time—and that cash in hand is worth more than cash received at any time in the future. Not only can returns from some period ahead have the uncertainty that they may not materialise, but cash in hand has the benefit that it can immediately be reinvested.

The technique is based on the fact that cash can accumulate interest on a compound basis, and could be described as compound interest in reverse. For example, £1,000 invested on a compound interest basis at 10% for 4 years would produce £1,464. It follows that £1,464 received in 4 years time discounted at 10% has a present value of £1,000. In other words if someone was to offer the choice of £1,000 today which must be invested for 4 years in a project with a return of 10% compound, or guarantee £1,464 in 4 years, there is no difference between the two financially.

Factors for converting future cash flows to present values are shown in Appendix C. In the case of the above example it would be established as follows:

$$\frac{1{,}000}{1{,}464} = 0.683.$$

In Appendix C 0.68 will be found on the line 4 years and in the column of discount rate 10%.

An important point to note is that the technique does not directly concern itself with inflation.

The discounted cash flow technique can be employed in two ways:

(i) To find the net present value of the sum of cash flows from an investment.
(ii) To find the discounted yield of a project.

(i) Net Present Value (NPV)

Consider the case of the farmer used in the earlier appraisal examples and assume that the project life is 10 years. To demonstrate that this technique will allow for variations in net cash flows assume that further injections of capital are required as follows: £1,000 in year 3; £700 in year 6 and £600 in year 7. The increase in net cash flow for the sheep compared to the barley is used to establish the "cash" produced each year.

The selection of 10 years as the life of the project is somewhat arbitrary. Factors like obsolescence and the need for change often help determine the period chosen.

The NPV can now be calculated for the net cash flow. Debate could arise over the discount factor to use. This reflects the interest rate assumed. One way is to relate it to

Summary

Year	Capital required £	Cash produced £	Net cash flow £
0	32,400	—	−32,400
1	—	6,100	+ 6,100
2	—	6,100	+ 6,100
3	1,000	6,100	+ 5,100
4	—	6,100	+ 6,100
5	—	6,100	+ 6,100
6	700	6,100	+ 5,400
7	600	6,100	+ 5,500
8	—	6,100	+ 6,100
9	—	6,100	+ 6,100
10	—	31,900*	+31,900
			+52,100

(* £6,100+£21,800 from sheep+£4,000 working capital: all assumed to be available)

the cost of borrowing money, or to use the opportunity cost of returns available from the next most profitable investment.

The NPV of the net cash flows discounted at 12% is as follows:

Year	Net cash flow (£)	Discount factor*	Present value (£)
0	−32,400	1	−32,400
1	+ 6,100	0.892	+ 5,441
2	+ 6,100	0.797	+ 4,862
3	+ 5,100	0.711	+ 3,626
4	+ 6,100	0.635	+ 3,874
5	+ 6,100	0.567	+ 3,459
6	+ 5,400	0.506	+ 2,732
7	+ 5,500	0.452	+ 2,486
8	+ 6,100	0.403	+ 2,458
9	+ 6,100	0.360	+ 2,196
10	+31,900	0.321	+10,240
		NPV	+£8,974

* See Appendix C

An investment showing a positive NPV is usually acceptable although this does depend upon the discount rate used and whether it really reflects interest rates that will be paid. Where two or more projects are competing for available funds the one showing the highest NPV is, as far as the results of this technique are concerned, the best. Such factors as the degree of risk associated with each possible investment would, however, be carefully considered.

If the net cash flow had been identical throughout the life of the project it would have been possible to use another set of tables shown in Appendix D which are called the "annuity factors". Thus if £8,360 had been the net cash flow for 10 years using the discount rate of 12% the NPV would have been £8,360×5.65=£47,234. The advantage of this is that the calculation is quicker. However, a simple computer programme can be employed to undertake the discounting for variable cash flows in a very limited amount of time.

(ii) Discounted Yield

One method of overcoming the problem of selecting the discount rate to be used in the NPV calculation is to calculate the discounted yield of the project. This is the rate which would make the NPV of the particular project equal to zero. Other names for it are "internal rate of return" and "discounted cash flow rate".

A project can be considered acceptable if its discounted yield exceeds the opportunity cost of the capital invested in it. Where more than one project is being considered, the one with the highest discounted yield is indicated by this technique to be the best.

Interpolation is used to establish the discounted yield. All this means is that a discount rate should be used which will make the NPV positive and another rate which will make the NPV negative. The rate which will bring the NPV to zero is found as shown below.

Consider the project illustrated in the NPV example above. It was shown that at a discount rate of 12% the NPV was +£8,974. A discount rate of 20% produces the following results:

Year	Net cash flow (£)	Discount factor	Present value (£)
0	−32,400	1	−32,400
1	+ 6,100	0.833	+ 5,081
2	+ 6,100	0.694	+ 4,233
3	+ 5,100	0.578	+ 2,948
4	+ 6,100	0.482	+ 2,940
5	+ 6,100	0.401	+ 2,446
6	+ 5,400	0.334	+ 1,804
7	+ 5,500	0.279	+ 1,535
8	+ 6,100	0.232	+ 1,415
9	+ 6,100	0.193	+ 1,177
10	+31,900	0.161	+ 5,136
			− 3,685

NPV at 12% rate	=+£8,974
NPV at 20% rate	=−£3,685
Difference	= £12,659

Discounted yield = lower of two rates used

$$+\left(\frac{\text{NPV of lower rate}}{\text{Difference in NPV}}\times\begin{matrix}\text{Difference between}\\ \text{discount rates used}\end{matrix}\right)=12+\left(\frac{8{,}974}{12{,}659}\times 8\right)=17.67\%$$

The acceptability of this figure has to be judged against returns from other investments, the relative risks involved, and interest rates.

There is the possibility that a project might produce substantial negative cash flows late in its life. This type of case illustrates a limitation of the discounted yield technique because discounting here would appear to be fatuous. This can be overcome by using the "extended yield" method. Any large negative cash flows are first discounted by one year using a realistic interest rate and then subtracted from the previous year's cash flow.

Example

Assume that the net cash flows for years 6, 7 and 8 of a project are as follows and that the interest rate is 12%:

Year:	6	7	8
Net cash flow	+£15,000	+1000	−£4000

"Bring back" year 8 to year 7:

$$£+1{,}000-\frac{4{,}000}{1.12}=1{,}000-3{,}571=-2{,}571$$

The answer is still negative so this must be "brought back" to year 6:

$$£+15{,}000-\frac{2{,}571}{1.12}=+15{,}000-2{,}296=+12{,}704$$

The cash flow would now be presented as:

Year:	6	7	8
	+£12,704	0	0

11.2.7. Critique of Investment Appraisal

In making a decision on whether an investment is worthwhile it is easy to concentrate on the arithmetical calculations, but in reality these form the easiest part. The more difficult task is to forecast accurately for several years ahead.

Forecasting is hazardous, and the further ahead the prediction the greater the chance of errors. Before a decision is reached attention has to be focused on some of the points mentioned at various stages elsewhere in the book, notably:

(i) Risk Factors

These cannot easily be built into the financial investment appraisal but should be carefully considered outwith the financial calculations. Questions that could be asked include: What are the chances of changes occurring in future cash flows because of political decisions, or from an inefficient worker, or from a disease outbreak, or from changes in marketing? Equally it is important to ask if this is a "once and for all" opportunity, e.g. the purchase of a neighbouring farm.

It is not easy to assess the importance of factors such as these, but they and many others should certainly be considered to be at least as important as the simple financial assessment. On many occasions the answers will depend upon the subjective judgements of the particular farmer.

(ii) Personal and Social Factors

Care has to be taken to establish what problems may be caused in the widest sense by an investment. For example, if the silage or the pig slurry pit, is to be sited to the windward side of houses, what will be the reaction? The attitude of the wife to an investment which means a "tightening of the belt" until the project has had time to get fully under way can be equally important in many households. Reducing the labour force may cause resentment locally and alter the farmer's standing in the community.

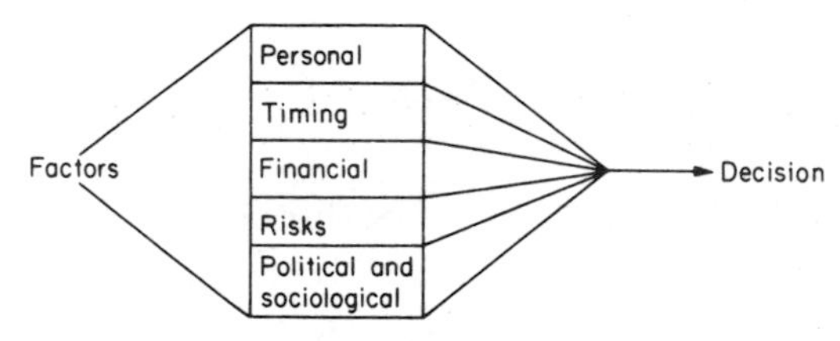

FIG. 101. Choice: to invest capital in the business or to refrain/delay.

(iii) Effect of Inflation

In times of inflation "good" money is repaid in money worth less, and as a result interest rates can be high in an inflationary period. If interest on borrowed capital is included in a cash flow used for investment appraisal then it should be remembered that repayment of the amount borrowed may become relatively easier in later years provided inflation continues.

It must be stressed that all the above "surround" factors (Fig. 101) need to be taken into account alongside the financial appraisal before a decision to go ahead with an investment is taken or declined.

CHAPTER 12

IMPLEMENTATION, MARKETING, CONTROL

CONTENTS

12.1. IMPLEMENTATION OF A PLAN

12.1.1. "Programme of Action"

When a plan is produced the basic strategy is formulated and a great deal of thought is given to tactics. There could be cases, usually where profit is not the main objective, of farmers who decide to farm in a certain way and start planning from the tactical rather than the strategic point. Generally, however, before a plan is finally adopted time should be spent developing a "programme of action" so that it can be put into operation efficiently. This will involve the acquisition of additional resources, goods and services, the organisation of production programmes, and the disposal of produce. Implementation involves putting this "programme" into operation, and gives rise to the need for control in the light of actual developments.

12.1.2. Obtaining Goods and Services

Most aspects under this heading have been referred to at points elsewhere in the book and they will only be mentioned briefly here. Negotiation and arrangements for loans or of an overdraft may be necessary at an early stage. The next step could be the employment of suitable staff. This will require the preparation of job and person descriptions, formulation of advertisements, interviews, and the preparation of contracts of employment.

An ordering and receiving procedure must be adopted which can involve the acquisition, interpretation, and selection of tenders. A major factor governing success is that goods and services should be available when required for production. Delays in delivery must be considered, and when they do arrive on the farm goods should, as far as is practicable, be stored to facilitate subsequent handling.

They should be obtained at least cost, but with due regard to quality, and also the best interests of the business. Thus, considerable savings can be made by purchasing fertiliser in the summer for winter delivery, since this helps the manufacturer's cash flow and production programme. However, the opportunity cost of this policy to the farmer may be higher than the benefit if greater returns could have been produced from investment elsewhere. Continuity of supplies is essential for certain items, and for others after sales service is a major factor.

Discounts for buying in bulk, the possibility of joining a buying group, buying out of season, and "opportunity purchasing" can all reduce costs, but each needs careful evaluation so that the full effects on the business are understood.

12.1.3. Organising Production

Although the strategy and basic tactics of the production programme will have already been formulated, planning of day-to-day operations is an on-going process requiring decisions in relation to all the variable factors which can effect farming, such as the weather, crop and animal performance, ill-health of staff, and market prices. Most farmers spend time at night planning the next day, perhaps having alternative plans in mind in case of bad weather or some other contingency. Priorities have to be evaluated and appropriate decisions taken.

Usually it will not just be the next day that is considered, but the pattern of events over a period so that production plans can be operated as effectively as possible.

Each day orders will have to be given to staff and some time devoted to communicating with them with reference to progress and other aspects. Emergencies may arise requiring quick decisions and possibly amendments to plans.

Part of the day will be spent monitoring the work in hand. For example, checking that soil conditions really are right for drilling, that enough seed and fertiliser are being delivered to the field, and so on. Frequently it will be necessary to spend some time looking at crop and animal growth and health, and other factors which will determine the work programme over the next period.

The major functions of management when the plan is operating therefore include the selection and implementation of production techniques within the framework of the plan; the need for attention to those factors that can be influenced by the farmer or his staff, and which effect profitability; the recording or monitoring of performance; the organisation and general management of the work force; and the acquisition and allocation of finance and resources so that efficient production can take place.

Marketing and the maintenance of an appropriate cash flow is integral to all this, and control is a logical development.

12.2. MARKETING

12.2.1. Introduction

Marketing involves both buying and selling, and much of Section 12.1.2 on obtaining goods and services could equally well be included here. The farmer must produce products of a quality and type which are in demand, and he has to recognise the need for close integration between production and marketing.

Marketing embraces grading, processing, packaging, transport, and advertising. Some of these functions are undertaken on the farm, depending upon the product.

The British farmer is frequently criticised in relation to marketing and it is therefore worthwhile to spend some time to see how standards could be improved. Many of the basic economic concepts such as supply and demand are relevant. It is essential to discuss when the farmer can operate on his own and when he should act in conjunction with others, but first it is necessary to look closer at what marketing involves.

12.2.2. Some Principles of Marketing

There is money to be made out of farm products until they reach the final consumers in the form in which they will be used. Apart from one or two exceptions the farm is only at the start of a long production chain. This can involve several "middle men", some of whom can make more money out of the products than the farmer, although most undertake a necessary function in getting products to those who will eventually use them.

Apart from cases where items are in extremely short supply relative to demand it is the final customer who largely dictates the type and quality of products which will sell best. Admittedly with television advertising and other marketing methods it is possible to influence housewives in their purchases. The poultry industry has made significant achievements in this context, but they have benefited from the fact that, because of improved production methods, they have been able to produce meat which is very competitive with pork, beef, and lamb.

Although the farmer does not usually trade directly with the housewife he must keep abreast of consumer trends because the middle men will only purchase what they can sell. The advent of the supermarket and new packaging methods, coupled with a change in society which has produced an increased requirement for easily prepared foods, attractive to look at if not always to eat, has changed demand. There are those who suggest that the farming industry has not fully realised this. They argue, for example, that there are too many beef breeds and systems of production, designed more to suit the farmer than the consumer, and point out that large beef producers, willing to supply the type of animal required direct to supermarkets, will provide increasing competition for smaller producers. Smaller producers retort that there will always be a place for quality beef.

12.2.3. Time, Place, Form

When farm plans are formulated the general policy of what to produce, how much to produce, and what quality to produce will be decided. When production is in operation it is essential to ensure that the products are produced at the right time, that they are presented for sale at the right place, and in the right form. Several references were made to these aspects in Chapter 9 when the importance of product prices to profits was stressed.

Deciding the right time to sell can greatly influence the success of marketing with many items. Market intelligence can be of great assistance, but such factors as the maintenance of a satisfactory cash flow, and the cost of keeping the product on the farm longer relative to price changes, must be kept in mind. With perishable goods, in particular, the timing of production to suit market demands is essential. However, this can also be important when finishing stock if the best prices are to be obtained and animals are not to be kept too long.

The right place can mean the right auction market for the particular product, or it might mean a contract which effectively means a sale direct from the farm.

The right form including grading, quality, and packaging, is particularly important for some products. Vegetable producers recognise this probably more than most. The farmer who sells potatoes must produce samples which are clean, healthy, and of the right size. Even the livestock producer must present his stock to market in such a way that they are attractive to buyers. This includes such aspects as even groups of store animals of similar breed.

12.2.4. Contracts

There is some evidence to suggest that an increased amount of produce is being sold on contract. Advantages claimed for producers include: more stable prices and guaranteed markets; the fact that timing of production can be related to the contract; the benefit that information is available on the quality of product demanded; that less time is spent on marketing; and the absence of marketing and haulage costs.

Disadvantages include lack of freedom in marketing and sometimes lower prices than elsewhere. There is also the high standard of quality demanded for some products, and the added pressure to maintain supplies necessary to comply with the contract.

Contracts vary in scale and formality. Gentleman's agreements between farmers, such as those between a weaner producer and a pig fattener, are the simplest types. There are merits in having more formal arrangements so that if the contract becomes unfavourable to one party he cannot withdraw. However, before agreeing the price in a formal contract the farmer has to have sound information about his production costs and ability to meet the contract.

The contract should detail: its duration; a clear statement of the product, including quality and/or size; price including bonuses for quality and delivery on time; the quantities to be delivered at specific times and who pays the transport; appropriate penalty clauses; and details of the parties to the contract.

With some contracts the farmer has little opportunity to discuss price, or other terms, and must accept what is being offered. In certain cases, such as with vining peas, the farmer has to agree to most of the production decisions being taken out of his hands. Other contracts go further; e.g. egg production contracts, where a feed company supplies the building, feed, and birds in return for a percentage of eggs. The latter is an example of vertical integration.

12.2.5. Selling Direct to the Consumer

Few farmers are big enough to have the scale, organisation, time, and all the other factors necessary to sell to the consumer and so eliminate the "middle men". There are exceptions including those with dairy rounds; farmers who grow soft fruit and advertise "pick your own"; farm gate sales of potatoes and eggs; even farmers who have their own farm shops or who develop freezer supply centres. Through such initiative some farmers make very satisfactory profits. However, total sales by such outlets account for a small proportion of national production, and as more stringent regulations with reference to the sale of certain goods are enforced there could be some decline.

12.2.6. Market Intelligence

The aim of market intelligence is to provide information to those who trade which will help them make decisions with reference to the best time and place to undertake their transactions, and to give some guide to the most appropriate form of goods or products to sell. Such information can help forecast supply, demand, and prices ranging from the immediate to long term future, although the accuracy in the latter case will not be high. Details of recent prices can also provide a guide to the success or failure of sales or purchases undertaken of late.

The sources of information were discussed under forecasting in Chapter 2. What has to be stressed here is that because of the long term nature of agricultural production many farmers are committed; they have the products and must sell them within a limited period, although their flexibility varies from product to product with the possibility of storage. A knowledge of the short term past and future can have different degrees of value according to this flexibility. Unfortunately, there will be few cases where the daily supply and prices for products are known before they leave the farm for market.

The value of information for the medium term future will also vary with the product and the length of its production cycle. With cases such as broilers, or certain crops, the farmer can consider reducing production where an over supply situation is forecast. Data on the area of crops sown in a year and subsequently estimated yields, or figures for stock numbers at various stages of production can give a guide to supplies coming on to the market at particular times. These can form a basis for evaluating forward contract prices, and also for the farmer to look at marketing strategy as well as tactics.

The futures market, mentioned under cereals, can be one guide to possible prices, and provide a means for the farmer to cushion himself against the possibility of price falls.

Although there are increased dangers of inaccuracies with long term forecasts there is greater opportunity for the farmer to change his production programme. However, the farmer must carefully consider the costs and benefits of making any change, and estimate how many other farmers might amend their production and so alter the supply/demand situation from that being predicted.

12.2.7. Co-operation

A co-operative can be described as a group working together towards a common aim for mutual benefit.

The Co-operative Development Board of Food From Britain aims to promote and assist the formation of co-operatives. Grants for feasibility studies, management training, and other items are available. In Less Favoured Areas grants are provided to co-operative forage groups.

A co-operative can be constituted as a company, in which case it comes under the Companies Acts, or more usually as a Society under the Industrial and Provident Societies Act. Production groups can operate as partnerships. Rigid regulations exist for both companies and societies. In both cases the members' liability is limited to their share holding, and any operating surplus must be paid to members in proportion to their trading in the year.

There are many different types of co-operative but they can be conveniently classified under the headings production, marketing, supply, and service co-operatives. To this could be added worker training groups.

(i) Production Co-operatives

This category includes groups for livestock improvement or production, vegetable and crop production, and in some cases they form part of integrated production-marketing groups. Machinery syndicates can also be included under this heading, the benefits of which were discussed in Chapter 8.

In many cases members can learn from each other. This is particularly the case where a new line of production is undertaken by a farmer. Some farmers even claim social benefits from the meetings of familes involved.

(ii) Marketing Co-operatives

Successful selling involves a considerable amount of skill and time. The individual farmer has to devote much of his energies to production and may not be in a position to undertake the detailed study essential to many commercial decisions. Co-operative marketing organisations can employ specialists who should have the necessary skill and information to recognise the right moment to sell and to establish what might sell best in the future. If mistakes are made the effects can be pooled in a co-operative venture so that they are not so severe to the individual producer.

Marketing groups may be extremely simple or very highly organised. For example, a grain-marketing group can be established without capital by farmers who retain the grain on their farms until sale, but who sell collectively through one outlet and so benefit from selling in large quantities. The other possibility is to have central co-operative dressing, drying, and storage facilities. These can be particularly attractive, in terms of cost per tonne stored, for farmers who have to replace their existing plants. Similar advantages can be claimed for potato co-operatives.

The marketing manager can market larger quantities of products of the same quality, and find suitable outlets for produce of different qualities. He has the advantage over the individual farmer in being able to ensure continuity of supply. In many cases some of the roles of the "middle men" in the marketing chain can be undertaken, and the profits which they would make can be obtained for the members of the co-operative.

Full commitment by farmers to a marketing group is essential if they are to receive the benefits expected. Members cannot decide to sell the best part of their produce themselves and the poorest part to the co-operative.

Some co-operative bodies do not actually sell, but act on behalf of members by promoting their produce, providing advice, and arranging meetings on production and marketing.

(iii) Supplying Co-operatives

Looked at in the broad sense this heading includes bodies ranging in size from those who each year sell many millions of pounds worth of goods, to those consisting of just a few farmers. The larger bodies operate in the interests of their members, who are farmers, and the profits are retained in the agricultural industry.

Small farmer groups can save by purchasing in bulk. Members are usually encouraged to discuss their requirements and may learn from each other. Their secretary, usually one of the farmers, approaches merchants asking for quotations for the supply of the goods required. The group should evaluate the tenders to see if there are any limitations in any of them, such as delivery problems, and take a decision according to the best interests of the majority.

(iv) Service Co-operatives

Farm secretarial services, milling and mixing groups, pest control, and many other service type groups come under this heading. In the future farmers may join forces to use computer facilities.

12.2.8. Evaluation of Co-operation

Many of the benefits of co-operation have been outlined above including the ability to negotiate from a position of strength. This extends to credit which might be more readily available to a group than to individuals. It is also claimed that co-operative organisations provide greater security and give their members increased chances of survival in difficult times.

A co-operative is, however, only as good as its members and staff. Many people do not like to lose their individuality and could not accept a decision which would benefit the majority, but which might not entirely suit them. A good co-operative calls for a high degree of loyalty and commitment. When prices are high some producers will want to sell elsewhere than through the marketing group, but they quickly return to the fold when produce is difficult to sell. Great problems can arise when members lose interest and fail to put their full energies into promoting the interests of the particular group. Some farmer members have to undertake new roles for which they do not have the experience or perhaps ability, although many are highly successful.

A large number of farmers are not prepared to take the risk of committing themselves to a co-operative venture until they have seen the benefits to others, but some co-operatives require large scale membership before they are successful, e.g. cull cow marketing.

Some of the regulations governing societies can be a handicap especially when in competition with the big non-co-operative companies with high risk capital. In a society members' capital remains at par value, and when the need arises it might be difficult to raise extra money from members. This can result in the necessity for high borrowing with attendant problems when interest rates are high. To help compensate the "supply co-operatives" may ask members for a higher commitment to their trade with them.

Clearly there are benefits and limitations of co-operation. However, at a stage when British farmers are facing increased competition from other countries, many of which have a high degree of co-operation, it would be unwise for any farmer to be so insular that he could not find time to evaluate the possibilities for his business.

12.2.9. Marketing Boards

Probably the best known of the British marketing boards are those associated with milk. They are concerned with procurement, processing, packaging, and promotion. When established they produced order out of chaos, and whilst there are those who wish to see an end to their control, so that prices of milk could be negotiated individually, the Boards protect the majority from marketing competition, which might result in a fall in the average price of milk. The Milk Boards also provide many services for farmers, notably the AI service which has helped to improve British cattle.

The Potato Marketing Board (PMB) and others which have been established from time to time, have all had their critics, in particular because of the loss of freedom of the individual. Although it is a dangerous generalisation, it is possible that the best and most aggressive people at marketing might have done better without the boards, but the majority, and certainly the weak, may have suffered without their controls. In some cases, such as that already mentioned in Chapter 9 for the PMB, they have attempted to equate supply with demand by regulatory controls.

12.3. BUDGETARY CONTROL

12.3.1. The Aim of Control Measures

One of the most important functions of management is to check on the performance of a plan once it has been implemented and then to take any necessary action. Predicting the future for the numerous factors relating to a business is extremely difficult and it is unlikely that budgets will work out exactly in every respect. If results are not monitored and appropriate measures adopted at an early stage it may be too late to exploit favourable trends or to minimise unsatisfactory aspects before the position becomes too serious or possibly irretrievable.

Many components play their part in influencing the performance of an enterprise, and even more influence a whole farm business. It is not surprising, therefore, that the operation or achievement of some factors can easily be masked by others. Budgetary control aims to uncover the true situation.

12.3.2. Budgetary Control in Management

Figure 102 shows the place of budgetary control in management.

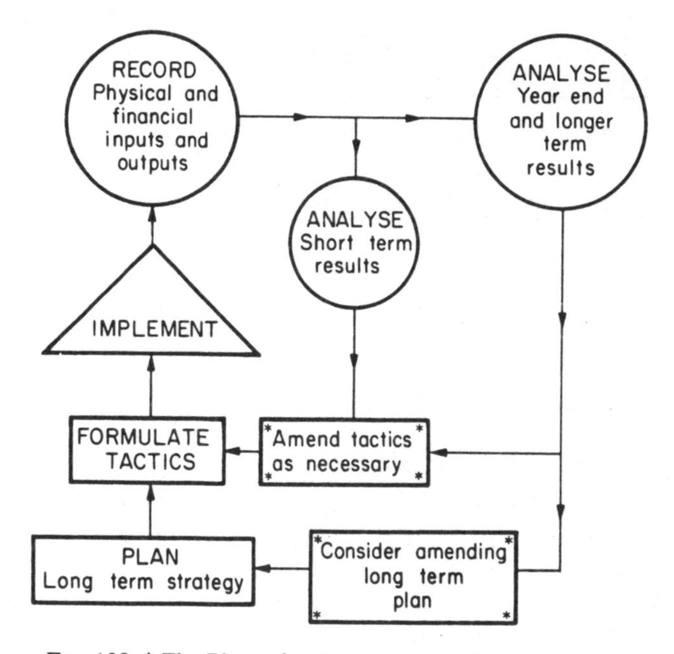

FIG. 102. * The Place of budgetary control in management.

The diagram is intended to indicate that the process of formulating strategy and tactics, implementation, recording, analysis of short and long term results, and amending strategy and tactics as necessary, are all parts of one dynamic process.

Budgetary control can be defined as "the establishment of budgets to comply with objectives for a business and the comparison of actual results with budgeted data with the aim of helping to achieve those objectives or to provide a basis for their modification". This latter point recognises the fact that sometimes the objectives are not always attainable.

Examination of performance can be at the enterprise or at the whole farm level. It must be realised that both physical and financial inputs and outputs can be considered.

Wheat gross margin per hectare						
	Budget			Actual		
	Quantity	Price	£	Quantity	Price	£
Grain sold	4.5 t	£100	450	5 t	£98	490
Seed	200 kg	18p	36	220 kg	20p	44
Fertiliser	100.40.40	—	44	100.40.40	—	46
Spray	—	—	18	—	—	24
Other	—	—	3	—	—	4
Variable costs			101			118
Gross margin			349			372

FIG. 103. Gross margin layout for control (straw burnt).

Projections for gross margins can be written in such a form that the actual results can be entered alongside Fig. 103.

12.3.3. Intervals for Control

A farmer has to be careful to balance the time he devotes to record keeping, production and marketing. Improved secretarial and computer facilities are helping some farmers with their office work but many still provide all or a large element of the manual labour as well as the management expertise on their farm. Control can be divided into three stages although in practice there is some fusion between the three.

(i) Planning Stage

Good planning prior to commencing production can minimise the need for control within the actual production process. This involves establishing those factors which contribute to performance and specifying the inputs necessary to achieve the desired results. If, for example, a particular liveweight gain is required from a group of stock or a given plant population is wanted in the spring for a winter cereal those factors under the farmer's control which contribute to the desired performance must be considered together with the levels and timing of inputs.

(ii) Production Stage

When production is under way the monitoring of those factors which the farmer can influence should be considered. Care must be taken not to select more factors than the farmer's time and management ability will allow him to control and the potential economic reward for his efforts must be examined.

With most cropping enterprises it might be argued that comparatively little can be done until the marketing stage, once the crop has been established. However, there is a need to maintain good husbandry, and to assess the cost/benefit of additional fertiliser and spray applications.

Many livestock enterprises, lend themselves to control techniques, particularly in relation to nutrition and performance. Those activities with a rapid turnover such as broilers, finishing pigs, and veal calves, in addition to requiring control of physical factors are more likely to require short term financial control than those whose turnover is long. The main physical records might be limited to those factors which have most influence on profitability, but where profit margins are small some of the less significant factors should be watched with equal care.

Cash flow budgets for the whole farm can be undertaken monthly or quarterly, but if the turnover is high, and especially where cash is "tight" or the business is comparatively large, there is increased pressure to select short term intervals for the budgets and appraisal of results.

In the case of enterprises with rapid turnover the data from physical and financial performance, coupled with knowledge of the cash flow situation and estimates of future input and output prices, can be employed to decide the scale of future production. With finishing pigs, for example, it might on some occasions be desirable from the financial standpoint to, at least temporarily, cut back on weaner purchases. It must be remembered, however, that most of the fixed costs will still be on-going.

(iii) Historical Analysis Stage

When a production cycle or financial year has been completed there is merit in reviewing the performance achieved in order to establish what can be done to repeat or improve the results attained when the next cycle takes place or year begins.

It can be seen that good control involves an integration of husbandry and management expertise. However, the difficulty of predicting the future in agriculture, even with the advantage of previous results on the same farm, must not be underestimated. This is because of its biological nature and the way in which market forces can change.

Budgets can be revised on the basis of past experience, but there is usually merit in subjecting the amended budgets to sensitivity tests which reflect the uncertainty of future performance.

12.3.4. Examples of Short Term Control

If budgetary control is taken in its widest sense the majority of farmers practice control far more regularly than might be thought, particularly at the enterprise level. For example, a farmer may plan a rationing system for the winter feeding of his dairy cows, but then find that the silage quality, as indicated by the cows' performance, is not as expected. He will therefore amend the rations.

A typical example of a monitoring system is shown in Fig. 104. This consists of projected daily yields for a dairy herd prepared by computer analysis of data for the herd in question, and information from a large number of cows in other herds.

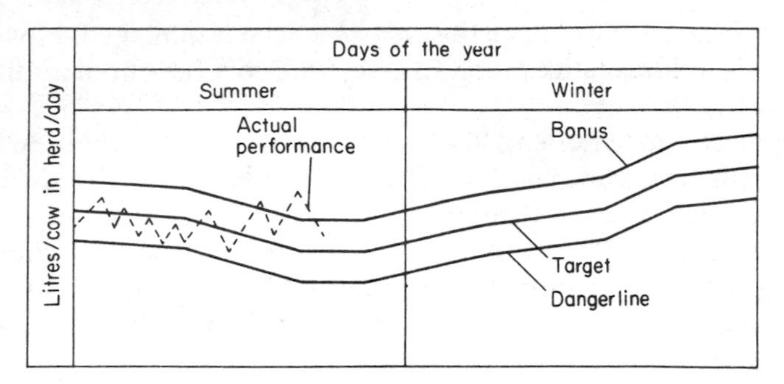

FIG. 104. Short term control for milk production.

Three lines are drawn on a "control board", at the start of the year, in different colours. One of these shows the daily herd production necessary to obtain the target yield for the forthcoming year. One of the others is the "bonus line" and shows daily herd production which would result in an annual average yield per cow of 500 l above target. The other is the "danger line" and shows production which would result in annual yields 500 l below the target.

The farmer has to supply details of calvings in the previous and forthcoming year, the number of cows and the number of lactations they have produced, together with expected numbers of heifers to be introduced. This data, and information based on statistics obtained from a large number of cows showing such things as yield in relation to lactation number and month of calving, are used to obtain projections. A "fertility factor" relating to the calving interval is also used.

The actual results are plotted on the chart as the fourth line. This shows when things are going wrong and the farmer may quickly be able to do something about them. It may be that he is just expecting too much from his grass or the silage, or it might be something outwith his control, such as a period of adverse weather.

The reader will probably be able to think of a considerable number of other examples at this level of control. These could include such things as the amount of feed required per kg liveweight gain in meat producing animals or feed costs per £100 of output.

There are many recording systems that the farmer might employ such as those issued by the commercial feed and fertiliser companies, Milk Marketing Boards and the Meat and Livestock Commission.

Some farmers convey the results of such monitoring to their staff and will try to foster an understanding of the reasons for changes in plans from time to time.

12.3.5. Cash Flow—Budgetary Control

The aim of cash flow budgetary control is to help "manage" the cash inflows and cash outflows of a business so that the cumulative cash position can be monitored. This is important to such things as overdraft limits, the optimum use of funds available, and the prospective demands for cash, especially in the immediate future.

In many cases appropriate management of a cash flow can influence the harmony of the relationship between a farmer and his bank manager. This may not only be important to the funding of the business at present but also to the possibility of securing additional finance in the future. The farmer will also wish to ensure, as far as he can, that interest payments are at a level commensurate with good business practice. It is therefore important to use cash flows to help control the credit and debt situation. In this context deciding when to pay bills and ensuring prompt receipt of money owed to the business is highly relevant.

The benefit of delaying the payment of outstanding accounts must be compared with any interest which the supplier may add, the risk that if not paid promptly he may not continue to supply, and the opportunity cost, or alternative use, of the funds used to pay the bills.

If the farmer does not receive money on time which is owed to him he is effectively helping to finance someone else's business. Unfortunately where debts are known to be "bad" then they have to be "written off" and charged as an expense in the profit and loss account. No entry is made for these in the cash flow and at the same time they must be removed from the debtors entry in the closing balance sheet.

Within the general context of a discussion on credit and debt it is worth mentioning that many farmers fail to manage their VAT returns to best advantage. If they keep their VAT records on a purchases and sales basis, instead of recording them when payment and receipts actually occur, the VAT refunds to which they are entitled may be received at an earlier date.

The degree of liquidity to maintain or, if borrowing, the margin to leave before overdraft limits are exceeded, must be decided by each farmer for his business. Some farmers feel very strongly that they must pay accounts on time but this desire is not universal. Others may wish to have funds available which will allow them to take advantage of bargains, or of purchasing supplies "out of season" when prices can be low. Equally some will want to ensure that the funds of the business will allow marketing of goods at the most optimum time rather than be faced with the possibility of being pressured into early sale by cash demands. They may maintain a degree of liquidity to cover such possibilities.

Each farmer's individual attitude can influence the way in which a cash flow is managed, but the difficult financial situation of many businesses increases the need to put opportunity costs ahead of any personal views such as "not owing money".

As mentioned above quarterly cash flows are appropriate to some businesses but a monthly basis is more appropriate to others. A layout which has three headings, budget, actual and deviation, facilitates control, Fig. 105.

Some care is necessary when using projected cash flow sheets to establish the deviations from predictions. It is possible that the particular debtor or creditor situation at the time of examination may be responsible. In other words, a cheque may not have been received or a bill may not have been paid when predicted.

Examination of items on the farm might also alter the significance of any deviation. For example, bullocks might still be unsold at a time when the projections indicate that they should have been marketed. Fertiliser may have been bought earlier than anticipated and be in store.

A farmer who uses a cash flow for very "rigid" cash control (e.g. to keep within his borrowing limits or to keep borrowing to a reasonable level to minimise interest payments) might sell produce earlier than planned or delay purchase of some items.

	April - June			July
	Budget	Actual	Deviation	Budget
	£	£	£	
Inflow				
Cattle	17,000	14,000	−3,000	
Milk	24,000	25,000	+1,000	
Sheep	8,000	9,000	+1,000	
Wheat	7,000	8,000	+1,000	
Total inflow	56,000	56,000	-	

FIG. 105. Layout of cash flow to facilitate control.

Care is necessary here. If all the items have been sold early there will be nothing else to sell and trouble could arise later. A delay in purchase may mean that when items are bought prices may have risen. In other words any change in policy must be viewed against longer term implications.

Generally the further ahead of actual events that cash flow budgets are prepared the greater the chance of them being wrong. This fact must be recognised when using such budgets in control. With careful use, however, cash flows can be used as one factor in maximising return on capital by ensuring that funds available to the business are always "working". The cereal farmer, for example, who stores his grain, hopefully to increase the market price, may market it so that money is received just in time to pay bills such as wages, rather than leave money in the bank ready to pay future bills.

12.3.6. Paying Accounts on Time

(i) Discount Available

It was established above that timing in relation to the payment of accounts is one of the integral parts of cash flow control. Before leaving the subject it is therefore worthwhile looking at discounts. Merchants who supply goods are just as anxious as farmers to manage their cash flows to best advantage. Sometimes they offer a discount of from 2 to 5% for prompt payment of accounts. The farmer may wish to calculate whether it is to his advantage to pay on time or to delay payment but lose the discount. If the farmer has to borrow money from the bank through his overdraft then the bank interest charge must be matched against the cost of losing the discount.

The following formula can be used to evaluate the situation:

$$y = \frac{A \times \frac{d}{100}}{n} \times \frac{365}{t} \times 100$$

where:

y = the annual percentage interest rate which the farmer will effectively pay if he does not receive the discount
d = the percentage discount available
A = gross amount of account without discount
n = net amount of invoice if paid promptly
t = the time in days between when the farmer normally pays his accounts and the time he must pay to receive the discount

Example

A farmer buys something for £1,000. He normally pays his accounts 31 days after receiving them but is told that if he pays within 10 days he can receive a 3% discount and pay £970.

$$y=\frac{1{,}000\times\frac{3}{100}}{970}\times\frac{365}{21}\times 100=53.75\%$$

The annual percentage rate which the farmer is charged if he does not pay in 10 days is clearly higher than interest rates payable on a bank overdraft. It would pay him to borrow the money to receive the discount.

(ii) Credit Charges Stated

Many merchants now issue invoices which do not give a discount but which include a credit charge that does not have to be paid if the account is settled within a given .period. The following example issued by a company which supplied a farm with sheep vaccine is typical:

	£
Total goods value	400
Total VAT value	60
Net invoice value	460
Credit charge (5%)	23
Gross invoice	483

The invoice, dated on the first of the month, stated that the credit charge could be deducted if the account was settled by the end of the month.

Effectively this means that the business has the use of £460 interest free for, say, 30 days which is extremely attractive. If the account was settled promptly on day 1 using money borrowed on bank overdraft with interest rates of 14% there would be a cost of

$$\left(460\times\frac{14}{100}\right)\times\frac{30}{365}=£5.29.$$

Alternatively if the farmer had a positive bank balance and he used his own money to pay the account there would be an opportunity cost.

Failure to pay the account by day 30 is, however, very expensive. Consider the case where the farmer pays the account 4 days late on day 34. He has now had the use of £460 for an extra 4 days, but this will have cost him £23. The interest charge for these 4 days can be calculated using the following formula:

$$y=\frac{G-L}{L}\times\frac{365}{t}\times 100$$

where:

y =APR
G=Gross invoice
L =Net invoice
t =number of days "late" in paying

$$y=\frac{483-460}{460}\times\frac{365}{4}\times 100=456\%$$

The longer he delayed payment, however, the lower the percentage would be. For example if he did not pay until 70 days after receiving the account, i.e. 40 days late, the figure would be 45.6%, and if he did not settle it until it was 140 days late it would be 13%. Clearly the company would press for payment before this and the farmer would prejudice further supplies from that organisation.

The importance of considering the credit terms offered by companies is obviously an area for careful consideration by farmers or managers.

12.3.7. Variance Analysis

The technique used to check in detail where results actually vary from the budget is known as variance analysis. This falls under three main headings:

(i) Plan or volume differences
(ii) Quantity differences—(yield or use/inputs)
(iii) Price differences

The aim is to establish how each of these three factors has influenced the results and to see if one is masking the results of another. The objective is then to see if anything can be done in future, using the outcome of this analysis, to improve profits.

(i) Plan Differences

If the plan has not been implemented correctly results will differ from budgets. Thus if a farmer plans to have 90 suckler cows and he only keeps 70, or if he plans to grow 20 ha of potatoes and he grows 22 ha, there will be plan differences.

(ii) Quantity Differences

Examples: a farmer budgets for a yield of milk per cow per annum of 5,750 l but the actual yield is 6,000 l; or he budgets to use 0.30 kg of concentrate per litre of milk produced but uses 0.38 kg.

(iii) Price Differences

Prices of many inputs and outputs will change from the time budgets are made and so influence results.

The difference between the budget and the actual result is known as the "variance" or sometimes the "deviation".

Example

Assume that a farmer budgets to grow 60 ha of winter wheat with a yield of 6 t/ha to be sold at £96 per t. In practice he grows 63 ha with a yield of 5.5 t sold at £100 per t. Undertake variance analysis on the output. The answer is shown in Fig. 106.

This simple example shows that although the actual output was fractionally higher

	Budget	Actual	Variance
	£	£	£
Output	34,560	34,650	+90

Analysis:	+£	−£	Final variance
Plan variance [+3 ha] +3 (6t × £96)	+1,728		
Quantity (yield) variance [−0.5t] −0.5 (63 ha × £96)		−3,024	
Price variance [+ £4 per tonne] +4 (63 × 5.5t)	+1,386		
	+3,114	−3,024	+90

FIG. 106. Variance analysis

than the budget, the increased area and price together masked the significant difference in yield.

Note from this example that:

Variance due to plan = Difference in area or stock numbers × Budget yield (or input) per ha or animal × Budget price

Variance due to quantity = Difference in yield (or input) per ha or animal × Actual area or stock numbers × budget price

Variance due to price = Difference in price × Actual area or stock numbers × Actual yield (or input) per ha or animal.

The reader must be careful to recognise that a positive variance is favourable for outputs but unfavourable for costs. Conversely a negative variance for outputs is unfavourable but favourable for costs. If the above example had been extended to include the variable costs of growing the wheat a positive variance in the fertiliser costs would have been unfavourable because more would have been used than was allowed for in the budget. However, if there was a negative variance for fertiliser, less would have been used. To overcome this problem some industries use the terms favourable and unfavourable rather than plus and negative signs.

Students are sometimes confused by the formulae for calculating the effect of plan, quantity and price. The following diagrams can help alleviate this confusion. Consider

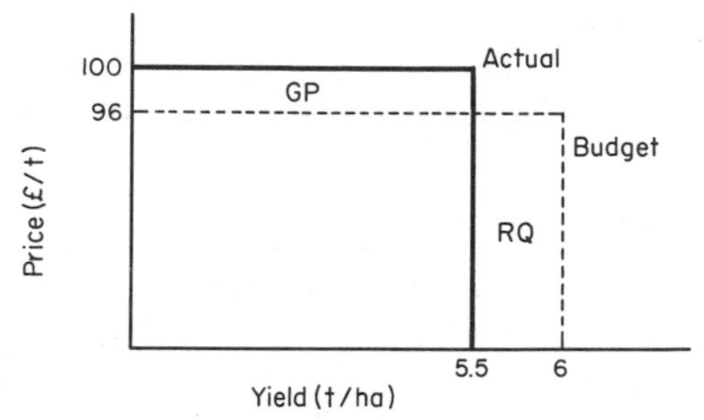

FIG. 107. Influence on output of variations in price and quantity—(Price increase, Quantity decrease).

just two factors, price and quantity, at first to simplify the issue, (Fig. 107), and base the diagram on one hectare.

The area of the rectangle RQ represents the reduction per hectare due to the reduction in quantity. Thus it is the reduction in yield × the original price per tonne.

The rectangle GP represents the gain per hectare due to price. This is the actual yield × the increase in price.

This example, where the price is above the budget and the quantity below predictions, is straightforward. Unfortunately, if the price as well as the quantity are above budget then the picture shown in Fig. 108 is presented. In this case the budgeted price was £96 per tonne and the yield 6 t per hectare, whereas the actual results were £100 and 6.5 t respectively.

It can be seen that a so-called "grey area" is created. Area P is definitely attributable to the price increase and area Q is derived from the quantity increase, but the grey area is produced by the combination of the price and the quantity increase in spite of

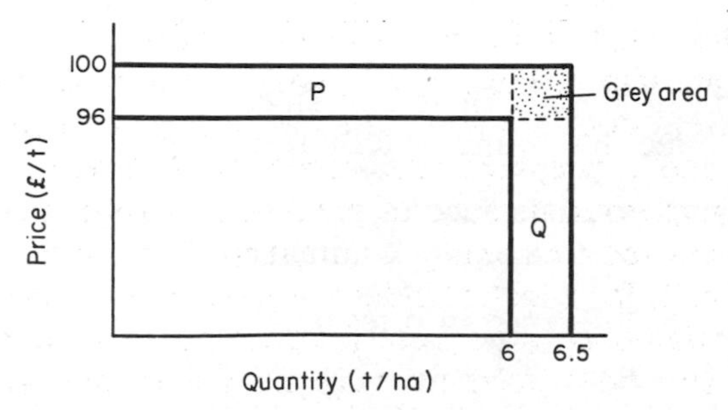

FIG. 108. Influence on output of variances in price and quantity (Price and quantity both above budget).

the fact that the formulae shown earlier indicate that all the benefit represented by this grey area is due to the price increase.

In almost every instance in farming the importance of allocating this grey area is so low because of the small amounts of money involved that the formulae given above are retained and the area is attributed to the price differential.

Variance analysis is not a technique which should simply be dismissed because of its time consuming nature and its grey areas. The cost/benefit of undertaking such an exercise must obviously be considered but it does provide a means of identifying, with a reasonable degree of accuracy, how each component factor of "plan, quantity and price" has contributed to a result.

[A more detailed study of variance analysis can be found in the following book: HORNGREN, C. T., *Cost Accounting: A Managerial Emphasis* (Prentice-Hall, fifth edition 1982)].

12.3.8. Action

The farmer must clearly establish the reasons for failure or success in relation to the budget, and must check whether these reasons are because of long term trends or of a short term nature. Very great care must be taken to consider the full impact before making changes. Partial budgets can be employed to assess the effects of amendments being considered and to produce new target figures.

12.3.9. Evaluation of Budgetary Control

Budgetary control is potentially a valuable technique for helping to attain the objectives for a business provided sufficient observations and records have been completed. To be fully effective it must be considered at the planning and production stages as well as when the year end or "historical review" takes place. It is advisable to restrict data to that which will be valuable but the introduction of suitable computer "software" has greatly assisted the recording and use of information which is necessary.

At the production stage deviations from budgets can be observed quickly after they occur and appropriate remedial action taken at an early point. The technique can help reduce the failures resulting from over optimism or pessimism in a long term plan.

Budgetary control is especially valuable in situations where there is rapid turnover. It can foster better investment and marketing together with better planning for taxation, as well as helping to minimise interest charges.

Properly used it can help farmers to maintain better control of debtors, but care must be taken because it does not automatically take note of stocks in hand.

When budgets have to be prepared to replace those already in existence great care is necessary in selecting appropriate data. There is some danger that farmers may "chop and change" their plans too frequently, so negating some of the benefits of control.

Budgetary control calls for a high standard of judgement in decision making, and it is this which differentiates the good from the poor business manager.

Whilst there will still be a need for husbandry and management expertise it is probable that in the future the developments in the use of computers will go past the stage of data recording and the level of analysis currently practiced. Simulation models requiring information about the interrelationship of performance factors will play their part in the improved control systems by assisting with "short term" and "long term" decisions.

12.4. THE FUTURE

It is extremely difficult to predict the future. In the last edition of this book it was stated that the only sure thing was that there would be change, and change there has been. However, in spite of the economic problems facing large sectors of the industry there will still be a satisfactory future for many farmers. They must be prepared to adapt to change where necessary and every attempt must be made to make optimal use of resources available.

Fixed cost levels have already attained great significance and will continue to dominate decisions taken for a high proportion of farm businesses. For those with a high level of borrowed capital interest rates will remain very important especially if they are high in relation to inflation. Farmers who bought their farms with the aid of fixed interest mortgages when land prices were high face special problems.

Machinery replacement policies must be reviewed on many farms and greater attention paid in some cases to maintenance and repair. Consideration should be given to the formation of syndicates or the operation of "machinery rings". The latter not only enable those with excess machine capacity to contract out their equipment to farmers who cannot afford to purchase machinery but frequently help make more efficient use of labour resources.

The level of product prices in comparison to input costs will very much determine the degree of challenge faced by farmers. Government and EEC policies will, however, be the most significant factor. The question of whether they will adopt "Protectionist" policies, with some form of support to keep people on the land, especially in difficult farming areas, or whether they will adopt "Efficiency" policies, will be of major concern to many.

Farmers will have to farm in an era during which conservation of the countryside will be given high priority. It is probable that new regulations relating to a range of production techniques will be introduced. These could include greater control over farm effluent and the use of chemicals, together with changes relating to animal welfare.

Research and development centres and leading farmers will introduce new methods of production. All farmers should take an objective view of such methods and evaluate them carefully to establish the cost/benefit to their particular farm and to see if they would enhance business viability.

Many farmers will have to change their production techniques and even the products they produce. The resourceful will adapt and remain in business in some cases by employing "novel" ideas and in other cases by adapting old methods. A number will improve their marketing and others will receive the benefit of "added value" for their products by processing them on the farm or co-operatively. Others will take advantage of the increased leisure time available to the public and cater for this need in a variety of ways. The majority, however, will continue with a more traditional approach to the selection of their farm activities, albeit that some of these may have to change.

With the reduction in the cost of "hardware" more farmers will employ computers to assist in business management. Computers will also play an increasing role in assisting production through such things as automatic control of environments, grain drying, and by improving the performance of farm machinery.

Political factors will have a major impact on the number of farmers who remain in business. At the same time they will influence the levels of profit which can be attained, but the quality of management will, in the final analysis, determine profits achieved within the economic environment created by the politicians.

APPENDIX A

ANNUAL COST OF A £100 LOAN*

Life of loan years	Rate of interest (%)									
	7	9	10	11	12	13	14	15	16	20
1	107	109	110	111	112	113	114	115	116	120
2	55.3	56.9	57.7	58.4	59.3	59.9	60.7	61.5	62.3	65.5
3	38.1	39.5	40.2	40.9	41.6	42.4	43.1	43.7	44.5	47.5
4	29.5	30.9	31.5	32.2	33.0	33.6	34.3	35.1	35.7	38.6
5	24.4	25.8	26.4	27.1	27.8	28.4	29.1	29.9	30.5	33.4
8	16.8	18.1	18.8	19.4	20.2	20.8	21.6	22.3	23.0	26.1
10	14.2	15.6	16.3	17.0	17.7	18.4	19.2	20.0	20.7	23.9
15	11.0	12.4	13.2	13.9	14.7	15.5	16.3	17.1	17.9	21.4
20	9.4	11.0	11.7	12.6	13.4	14.2	15.1	16.0	16.9	20.5
25	8.6	10.2	11.0	11.9	12.8	13.6	14.5	15.5	16.4	20.2
30	8.1	9.7	10.6	11.5	12.4	13.3	14.3	15.2	16.2	20.1

* This assumes equal annual payments. The interest is paid on the outstanding balance, and the loan is just repaid by the end of the specified loan period. Method: multiply the amount of the loan by the relevant factor and divide by 100. Thus the annual charge on a loan of £6000 at 10% for 20 years is £6000 $\times \frac{11.7}{100}$ = £702.

APPENDIX B

SINKING FUND TABLE

Capital accruing to a regular annual investment of £100.

	Percentage rate of interest									
Years	5	7	8	9	10	11	12	14	16	20
1	100	100	100	100	100	100	100	100	100	100
2	205	207	208	209	210	211	212	214	216	220
3	315	321	325	328	331	334	337	344	351	364
4	431	444	451	457	464	471	478	492	507	537
5	553	575	587	598	611	623	635	661	688	744
6	680	715	734	752	772	791	812	854	898	993
7	814	865	892	920	949	978	1,009	1,073	1,141	1,292
8	955	1,026	1,064	1,103	1,144	1,186	1,230	1,323	1,424	1,650
9	1,103	1,198	1,249	1,302	1,358	1,416	1,478	1,609	1,752	2,080
10	1,258	1,382	1,449	1,519	1,594	1,672	1,755	1,934	2,132	2,596
11	1,421	1,578	1,665	1,756	1,853	1,956	2,066	2,305	2,573	3,215
12	1,592	1,789	1,898	2,014	2,138	2,271	2,413	2,727	3,085	3,958
13	1,771	2,014	2,150	2,295	2,452	2,621	2,803	3,209	3,679	4,850
14	1,960	2,255	2,422	2,602	2,798	3,010	3,239	3,758	4,367	5,920
15	2,158	2,513	2,715	2,936	3,177	3,441	3,728	4,384	5,166	7,204
20	3,307	4,100	4,576	5,116	5,728	6,420	7,205	9,103	11,538	18,669
25	4,773	6,325	7,311	8,470	9,835	11,441	13,333	18,187	24,921	47,198
30	6,644	9,446	11,328	13,631	16,449	19,902	24,133	35,679	53,031	118,190
40	12,080	19,964	25,906	33,788	44,259	58,183	76,709	134,200	236,080	734,390

Example: The capital accruing after 10 years to the regular annual investment of £600 at 11% would be £1672 × 6 = £10,032.

APPENDIX C

DISCOUNT TABLES

Discount factors for calculating the present value of future cash flow where cash flows are *irregular.*

	Percentage										
Years	1	4	5	6	7	8	9	10	11	12	13
1	0.990	0.961	0.952	0.943	0.934	0.925	0.917	0.909	0.900	0.892	0.884
2	0.980	0.924	0.907	0.889	0.873	0.857	0.841	0.826	0.811	0.797	0.783
3	0.970	0.888	0.863	0.839	0.816	0.793	0.772	0.751	0.731	0.711	0.693
4	0.960	0.854	0.822	0.792	0.762	0.735	0.708	0.683	0.658	0.635	0.613
5	0.951	0.821	0.783	0.747	0.712	0.680	0.649	0.620	0.593	0.567	0.542
6	0.942	0.790	0.746	0.704	0.666	0.630	0.596	0.564	0.534	0.506	0.480
7	0.932	0.759	0.710	0.665	0.622	0.583	0.547	0.513	0.481	0.452	0.425
8	0.923	0.730	0.676	0.627	0.582	0.540	0.501	0.466	0.433	0.403	0.376
9	0.914	0.702	0.644	0.591	0.543	0.500	0.460	0.424	0.390	0.360	0.332
10	0.905	0.683	0.613	0.558	0.508	0.463	0.422	0.385	0.352	0.321	0.294
11	0.896	0.649	0.584	0.526	0.475	0.428	0.387	0.350	0.317	0.287	0.260
12	0.887	0.624	0.556	0.496	0.444	0.397	0.355	0.318	0.285	0.256	0.230
13	0.873	0.600	0.530	0.468	0.414	0.367	0.326	0.289	0.257	0.229	0.204
14	0.869	0.577	0.505	0.442	0.387	0.340	0.299	0.263	0.231	0.204	0.180
15	0.861	0.555	0.481	0.417	0.362	0.315	0.274	0.239	0.209	0.182	0.159

	Percentage										
Years	14	15	16	17	18	19	20	25	30	35	40
1	0.887	0.869	0.862	0.854	0.847	0.840	0.833	0.800	0.769	0.740	0.714
2	0.769	0.756	0.743	0.730	0.718	0.706	0.694	0.640	0.591	0.548	0.510
3	0.674	0.657	0.640	0.624	0.608	0.593	0.578	0.512	0.455	0.406	0.354
4	0.592	0.571	0.552	0.533	0.515	0.498	0.482	0.409	0.350	0.301	0.260
5	0.519	0.497	0.476	0.456	0.437	0.419	0.401	0.327	0.269	0.223	0.185
6	0.455	0.432	0.410	0.389	0.370	0.352	0.334	0.262	0.207	0.165	0.132
7	0.399	0.375	0.353	0.333	0.313	0.295	0.279	0.209	0.159	0.122	0.094
8	0.350	0.326	0.305	0.284	0.266	0.248	0.232	0.167	0.122	0.090	0.067
9	0.307	0.284	0.262	0.243	0.225	0.208	0.193	0.134	0.094	0.067	0.048
10	0.269	0.247	0.226	0.208	0.191	0.175	0.161	0.107	0.072	0.049	0.034
11	0.236	0.214	0.195	0.177	0.161	0.147	0.134	0.085	0.055	0.036	0.024
12	0.207	0.186	0.168	0.151	0.137	0.124	0.112	0.068	0.042	0.027	0.017
13	0.182	0.162	0.145	0.129	0.116	0.104	0.093	0.054	0.033	0.020	0.012
14	0.159	0.141	0.125	0.111	0.098	0.087	0.077	0.043	0.025	0.014	0.008
15	0.140	0.122	0.107	0.094	0.083	0.073	0.064	0.035	0.019	0.011	0.006

APPENDIX D

DISCOUNT TABLES

Discount factors for calculating the Present Value of future cash flows where cash flows are *regular*.

Years	Percentage											
	1	2	3	4	5	6	7	8	9	10	11	12
1	0.99	0.98	0.97	0.96	0.95	0.94	0.93	0.92	0.91	0.90	0.90	0.89
2	1.97	1.94	1.91	1.88	1.85	1.83	1.80	1.78	1.75	1.73	1.71	1.69
3	2.94	2.88	2.82	2.77	2.72	2.67	2.62	2.57	2.53	2.48	2.44	2.40
4	3.90	3.80	3.71	3.62	3.54	3.46	3.38	3.31	3.23	3.16	3.10	3.03
5	4.85	4.71	4.57	4.45	4.32	4.21	4.10	3.99	3.88	3.79	3.69	3.60
6	5.79	5.60	5.41	5.24	5.07	4.91	4.76	4.62	4.48	4.35	4.23	4.11
7	6.72	6.47	6.23	6.00	5.78	5.58	5.38	5.20	5.03	4.86	4.71	4.56
8	7.65	7.32	7.01	6.73	6.46	6.20	5.97	5.74	5.53	5.33	5.16	4.96
9	8.56	8.16	7.78	7.43	7.10	6.80	6.51	6.24	5.99	5.75	5.53	5.32
10	9.47	8.98	8.53	8.11	7.72	7.36	7.02	6.71	6.41	6.14	5.88	5.65
11	10.36	9.78	9.25	8.76	8.30	7.88	7.49	7.13	6.80	6.49	6.20	5.93
12	11.25	10.57	9.95	9.38	8.86	8.38	7.94	7.53	7.16	6.81	6.49	6.19
13	12.13	11.34	10.63	9.98	9.39	8.85	8.35	7.90	7.48	7.10	6.74	6.42
14	13.00	12.10	11.29	10.56	9.89	9.29	8.74	8.24	7.78	7.36	6.93	6.62
15	13.86	12.84	11.93	11.11	10.37	9.71	9.10	8.55	8.06	7.60	7.19	6.81

Years	Percentage											
	13	14	15	16	17	18	19	20	25	30	35	40
1	0.88	0.87	0.86	0.86	0.85	0.84	0.84	0.83	0.80	0.76	0.74	0.71
2	1.66	1.64	1.62	1.60	1.58	1.56	1.54	1.52	1.44	1.36	1.28	1.22
3	2.36	2.32	2.28	2.24	2.20	2.17	2.13	2.10	1.95	1.81	1.69	1.58
4	2.97	2.91	2.85	2.79	2.74	2.69	2.63	2.58	2.36	2.16	1.99	1.84
5	3.51	3.43	3.35	3.27	3.19	3.12	3.05	2.99	2.68	2.43	2.21	2.03
6	3.99	3.88	3.78	3.68	3.58	3.49	3.40	3.32	2.95	2.64	2.38	2.16
7	4.42	4.28	4.16	4.03	3.82	3.81	3.70	3.60	3.16	2.80	2.50	2.26
8	4.79	4.63	4.48	4.34	4.20	4.07	3.95	3.83	3.32	2.92	2.59	2.33
9	5.13	4.94	4.77	4.60	4.45	4.30	4.16	4.03	3.46	3.01	2.66	2.37
10	5.42	5.21	5.01	4.83	4.65	4.49	4.33	4.19	3.57	3.09	2.71	2.41
11	5.68	5.45	5.23	5.02	4.83	4.65	4.48	4.32	3.65	3.14	2.75	2.43
12	5.91	5.66	5.42	5.19	4.93	4.79	4.61	4.43	3.72	3.19	2.77	2.45
13	6.12	5.84	5.58	5.34	5.11	4.90	4.71	4.53	3.78	3.22	2.79	2.46
14	6.30	6.00	5.72	5.46	5.22	5.00	4.80	4.61	3.82	3.24	2.81	2.47
15	6.46	6.14	5.84	5.57	5.32	5.09	4.87	4.67	3.85	3.26	2.82	2.48

Example: The present value of £350 received for 6 years at 11% discount rate is £350 × 4.23 = £1,481.

INDEX